PERGAMON INTERNATIONAL LIBRARY
of Science, Technology, Engineering and Social Studies
The 1000-volume original paperback library in aid of education, industrial training and the enjoyment of leisure
Publisher: Robert Maxwell, M.C.

TECHNOLOGY, THE UNIVERSITY AND THE COMMUNITY

Other Pergamon Titles

Burrington, *How to Find Out About the Social Sciences*
Durey, *Staff Management in University and College Libraries*
The European Universities 1975-1985 (In English and French editions)
Long, *Participation and the Community*
Rabinowitch, *Views of Science, Technology and Development*
Steg, *Should We Limit Science and Technology*
Tropman et al, *Strategic Perspectives in Social Policy*
Lewin & Shakun, *Policy Sciences: Methodologies and Cases*

TECHNOLOGY, THE UNIVERSITY AND THE COMMUNITY

A Study of the Regional Role of Engineering Colleges

By
George Bugliarello
Polytechnic Institute of New York
and
Harold A. Simon
University of Illinois at Chicago Circle
with Vivian G. Cardwell, Clarke H. P. Schneider
and Numerous Collaborators

Pergamon Press

New York/Toronto/Oxford/Sydney/Frankfurt/Paris

Pergamon Press Offices:

U. S. A.	Pergamon Press Inc., Maxwell House, Fairview Park, Elmsford, New York 10523, U.S.A.
U. K.	Pergamon Press Ltd., Headington Hill Hall, Oxford OX3 0BW, England
CANADA	Pergamon of Canada, Ltd., 207 Queen's Quay West, Toronto 1, Canada
AUSTRALIA	Pergamon Press (Aust.) Pty. Ltd., 19a Boundary Street, Rushcutters Bay, N.S.W. 2011, Australia
FRANCE	Pergamon Press SARL, 24 rue des Ecoles, 75240 Paris, Cedex 05, France
WEST GERMANY	Pergamon Press GMbH, Postfach 6242 Kronberg/TS, West Germany

Library of Congress Cataloging in Publication Data

Bugliarello, George.
Technology, the university, and the community.

Includes bibliographical references.
1. Engineering schools—United States. 2. Community and college—United States. I. Simon, Harold A. joint author. II. Title.
T73.B89 1974 607'.1173 74-8299
ISBN 0-08-017872-3

Printed in the United States of America

CONTENTS

THE AUTHORS

George Bugliarello, currently President of the Polytechnic Institute of New York, was Dean of the College of Engineering at the University of Illinois at Chicago Circle. He holds a Dr. Ing. Degree from the Univeristy of Padua, an M.S. from the University of Minnesota and an Sc.D. from M.I.T.

Harold A. Simon is currently Professor of Heat Transfer and Fluid Mechanics in the Energy Engineering Department of the University of Illinois at Chicago Circle. He received his Ph.D. at the University of Minnesota and his earlier education in England and South Africa.

Vivian G. Cardwell is Assistant to the Dean of the College of Engineering at the University of Illinois at Chicago Circle. She holds an M.B.A. from the University of Chicago, has held positions with the Federal Government and is coauthor of the book, "Women in Engineering."

Clarke H. P. Schneider is Assistant Professor in the College of Urban Sciences and Assistant Director of the Urban Systems Laboratory in the College of Engineering at the University of Illinois at Chicago Circle. He received his M.S. and Ph.D. at Northwestern University and an A.B. at Dartmouth College.

PREFACE

This study was performed under a grant from the National Science Foundation to the College of Engineering of the University of Illinois at Chicago Circle. The report is the result of the collaborative effort of many institutions and individuals. It was written by George Bugliarello, Project Director, and Harold A. Simon, with the assistance of Vivian G. Cardwell and Clarke H. P. Schneider, who also coordinated the efforts of all the contributors, organized the Conference which constituted a phase of the study and edited the Conference proceedings.

Major contributions to the direction and implementation of the study were made by Thomas W. Knowles and Joseph L. Schofer. Other major contributors were the members of the project steering committee and the technical working committee, consultants, the Conference participants, and the Director and Deputy Director of the Office of Intergovernmental Science Programs of the National Science Foundation, M. Frank Hersman and Robert C. Crawford.

The study was carried out in three phases. Phase I was devoted to the problems of developing models and associated studies of the interaction between the university (the school of engineering) and its environment. The individual studies carried out by the technical working committee toward this end are reported in Appendix I, Project Reports.

Phase II centered on a Conference (February 14 and 15, 1972), which gathered participants from throughout the country who were concerned with areas pertinent to the present study. The purpose of the Conference was to provide a critique of the project's progress at that time and to give direction for the conclusion of the project. The proceedings of the Conference consist of the project papers marked by asterisk in the Appendix I contents and all the papers contained in Appendix II.

Phase III of the project focused on the development of mechanisms for increasing the interaction between the universities, or their colleges of engineering, and the local region. The mechanisms proposed reflect the experience gained through the course of the study, the ideas put forth at the Conference, suggestions from consultants and information gathered from interviews with legislators, local government officials and civic leaders.

All the engineering schools in Illinois and three government units actively participated in carrying out the project. They are listed below with the abbreviated designation given to each in the body of the report.

Engineering Schools	*Abbreviated Designation*
College of Engineering and Technology Bradley University	BRADLEY
College of Engineering and Physical Sciences Illinois Institute of Technology	IIT
The Technological Institute Northwestern University	NU
School of Engineering and Technology Southern Illinois University at Carbondale	SIUC
College of Engineering University of Illinois at Chicago Circle	UICC
College of Engineering University of Illinois at Urbana-Champaign	URBANA

Government Units
Department of Public Works
City of Chicago

Illinois Legislative Council

State of Illinois Board of Higher Education

(The Illinois Legislative Council is the research arm of the Illinois Legislature; the State of Illinois Board of Higher Education is the coordinating board for all public institutions of higher education in Illinois.)

A Steering Committee* with representatives from the participating institutions was established to plan the endeavor:

George Bugliarello,** Chairman and Project Director
Dean, College of Engineering
University of Illinois at Chicago Circle

Peter Chiarulli, Dean
College of Engineering and Physical Sciences
Illinois Institute of Technology

Representative John H. Conolly
Chairman, Illinois Legislative Council

William L. Day
Director of Research
Illinois Legislative Council

John J. Desmond, Associate Director
Engineering Experiment Station
University of Illinois at Urbana-Champaign

Daniel C. Drucker, Dean
College of Engineering
University of Illinois at Urbana-Champaign

*Titles and affiliations are as of the beginning of the study.
**Currently President, Polytechnic Institute of New York

Thomas B. Jefferson, Dean
School of Engineering and Technology
Southern Illinois University at Carbondale

Elizabeth J. McLean
Assistant Commissioner of Public Works
City of Chicago

David Mintzer, Acting Dean
The Technological Institute
Northwestern University

Bichara B. Muvdi, Acting Dean
College of Engineering and Technology
Bradley University

Milton Pikarsky
Commissioner of Public Works
City of Chicago

Steven B. Sample
Deputy Director for Programs
State of Illinois Board of Higher Education

A Technical Working Committee provided the working link among the cooperating institutions and carried out the tasks of the project:

M. Kenneth Blanchard, Coordinator
University Industry Relations
College of Engineering
University of Illinois at Urbana-Champaign

Vivivan G. Cardwell, Assistant to the Dean
Engineering Administration
University of Illinois at Chicago Circle

Paul Chan, Research Assistant
Information Engineering Department
University of Illinois at Chicago Circle

Hugh Folk, Professor
Department of Economics and Institute of Labor
& Industrial Relations
University of Illinois at Urbana-Champaign

William J. Geekie
Department of Children & Family Services
State of Illinois

Christopher Gibbons, Research Assistant
Systems Engineering Department
University of Illinois at Chicago Circle

Robert G. Granda
Illinois Legislative Council

Fred J. Gunther, Research Assistant
Information Engineering Department
University of Illinois at Chicago Circle

William J. Haga, Research Assistant
Department of Sociology
University of Illinois at Urbana-Champaign

Thomas W. Knowles, Assistant Professor
Industrial & Systems Engineering Department
Illinois Institute of Technology

Judy Millaney, Research Assistant
Department of Public Works
City of Chicago

Nancy L. Nihan, Assistant Professor
Systems Engineering Department
University of Illinois at Chicago Circle

William D. O'Neill, Associate Professor
Systems Engineering Department
University of Illinois at Chicago Circle

Bradley Parlin, Research Assistant
Institute of Labor & Industrial Relations
University of Illinois at Urbana-Champaign

Eugene Ryan, Research Assistant
Information Engineering Department
University of Illinois at Chicago Circle

Amado San Mateo, Research Assistant
Civil Engineering Department
Northwestern University

Clarke H. P. Schneider, Assistant Professor
Center for Urban Studies
University of Illinois at Chicago Circle

Joseph L. Schofer, Associate Professor
Civil Engineering Department
Northwestern University

Harold A. Simon, Professor
Energy Engineering Department
University of Illinois at Chicago Circle

Spencer B. Smith, Professor and Chairman
Industrial & Systems Engineering Department
Illinois Institute of Technology

Thomas M. Smith, Assistant Professor
Information Engineering Department
University of Illinois at Chicago Circle

Nelson Wax, Research Professor
Coordinated Science Lab
University of Illinois at Urbana-Champaign

Advice was also sought from five consultants:

Vinton Bacon, Professor
Civil Engineering
University of Wisconsin at Milwaukee

Samuel S. Baxter, Former Commissioner
Department of Water
City of Philadelphia

Lynton K. Caldwell, Professor
Political Science Department
Indiana University

Robert S. Green, Dean
College of Engineering
The Ohio State University

Merritt A. Williamson, Director
Engineering Management Program
Vanderbilt University

THE HIGHLIGHTS OF THIS STUDY

The engineering college or, more generally, the university represents today a complex and unique resource, whose full potential in service of the local community—from the state and region to the city and the neighborhood—is neither well understood nor effectively utilized.

Although this study focuses on the regional role of engineering colleges, it is also conceived as a first step toward better understanding the regional role of universities in general. The study analyzes the problem and suggests some mechanisms for its solution.

As a community resource, the engineering college performs two paramount roles: it supplies the community with an important human resource and it represents a unique information resource. Several additional roles can be recognized, of which at times neither the community nor the college is fully aware, but which are inherent in the particular nature of the college as an institution.

The role of the college as a supplier of human resources is well known and a traditional one. However, when it comes to many new and complex community problems, such as urban decay or the delivery of social services, the impact of the human output from the college will be very limited until the college has undergone a period of self-education, during which the fundamentals and the methodologies pertinent to such problems are defined, and responsive academic programs developed.

As an information resource, the services which the college can perform for the community are determined by three distinct problem levels which require, in order of increasing complexity, information extraction, manipulation and generation. The corresponding college activities can be typified as search for information, consulting and research.

Currently the interaction of the college with the local community at each of these three levels is very limited and skewed. The problem lies both in the lack of a determined effort by the college to become a major and active information resource for the community, and in the lack of appropriate interfacial mechanisms between the college, the community, and local and state governments.

We suggest the necessity for three specific kinds of mechanisms.

1. *Any* university or college can, and should, establish an information center capable of providing the community with information and easy access to other resources—human and physical—within the university. Such centers can provide a basic but very inexpensive link between the university and the community. Centers for housing data banks and other well codified information on specific subjects of interest to the community, such as building codes or energy inventories, are also recommended.

2. Universities or colleges *with appropriate capabilities* should be encouraged, if they are not already doing so, to take a second more sophisticated and aggressive step in linking to the community. This can occur through centers (we suggest Information, Monitoring and Innovation Centers) that actively identify community needs and opportunities for innovation and mobilize the resources of the university or the college for the appropriate response.

3. As a third step, existing university groups active in specified problem areas, such as housing, health care delivery and urban transportation, need to be identified as parts of state and interstate networks. These groups can perform a crucial function of relating Federal programs to local needs and should, where appropriate, be identified as State-Federal Centers. Where lacking, such centers can be created with the assistance of ad hoc "Interface Boards."

The success of these centers will depend critically on their resources, adaptability, acceptability and governance. The most appropriate institutional location should be determined by the problem level and the resources required. An information service may be controlled totally by the university, while complex research activities may involve control shared between the university and the source of the requisite funds.

THE PROBLEM

Recently, the Council of State Governments[1] commented that "...the state universities have a major potential to bring applied science to the service of state governments...this potential is not being tapped effectively."

Also recently, the Committee on Intergovernmental Science Relations[2] included among its recommendations the statement that "New institutional arrangements be created between academic institutions and State and local governments...to deal with the application of science and technology to State and local problems."

A recent study of the problem of developing science and technology capabilities in state legislatures indicates[3] that "...the limited data base...does not permit a conclusion to be drawn about the relationship between the presence of a university in a legislative district and/or a legislator's reliance upon university based expertise. It is also not clear whether a legislator is likely to draw upon the expertise of universities or other research organizations if they are situated in or near the capitol city."

It has been pointed out, furthermore, that current state mechanisms for science and technology advice generally suffer from a low degree of permanence and stability* which is prejudicial to the states. Finally, the cities also have immense and urgent problems which could

*George Bugliarello, "State Science and Technology Interaction: A Cybernetic View," Appendix III-1.

greatly benefit from advice and assistance from the universities—but which the universities thus far have been largely unable to provide. The problem extends to the inability of the university—as perceived by the city—to provide technical personnel with training responsive to the challenges of the city.*

These views reflect the growing concern that the talents available in the universities are not having the kind of impact on policy-making bodies in both the private and public sectors that they could have.

The problem then is to identify mechanisms whereby these interactions could be enhanced. We are essentially concerned about the linking of entities that have very different characteristics.** Clearly, a necessary prerequisite is a thorough understanding of what a university is and how it functions and also of what kinds of problems arise in the outside world to which the university can respond. This study is an attempt at understanding the nature of the *interaction* between the university and its environment and at proposing mechanisms for achieving a more effective interaction. The focus is on one of the units of the university which perhaps can contribute the most to such an interaction—the engineering college. However, many of the considerations and approaches in this study should be valid for the university as a whole.

*Judy Millaney, "Engineering Colleges and Their Impact on the City of Chicago," Appendix I-13.

**Merritt A. Williamson, "Summary Report of National Evaluation Committee on Summer Faculty-Student Program in 'Application of Interdisciplinary University Resources to Local and State Government Problems,'" Appendix III-2.

CHAPTER 1

The Engineering College and its Environments

DEFINITIONS

The Engineering College

A first look at engineering colleges impresses one more with the differences among colleges than with their similarities. Each college has its own character, distinguished by such features as source of finance, geographical location, enrollment size and traditionally accepted involvements. Equally important are the type and nature of the university of which the college is a part. The concern here is with the college viewed in its local context as a resource to be exploited to the full. As such, as one accentuates the resource aspect of the college activities, there is merit in looking for a conceptualization that is sufficiently broad to be valid for any engineering college.

The essential character of an engineering college stems from two main resources: human beings and information. All the desirable characteristics and potential that an engineering college holds for its environment arise from the multifaceted aspects that the interaction of these two resources can assume. The college itself is made up of people; true they are located in a physical facility and usually provided with engineering equipment, but the functioning of the college and its interactions with other institutions primarily depends on its people. The store of human resource, depicted in Fig. 1 and labelled as the engineering college, is distinguished from other groups of people in a very special sense. The faculty of an engineering college comprises a human resource that has been developed and matured by their education and experience, enabling them to maintain an active link with the store of technological information also shown in Fig. 1. They are adept at extracting from, manipulating and adding to, the information resource, and herein lies their most direct and important value to society. The coupling of these abilities with a number of distinctive institutional characteristics of the college as part of the university can also make the college an institution that assists in policy making processes at many levels, and can exert an enlightened and basically impartial role among contrasting advocacies in our technological society.

A closer examination of engineering colleges will reveal several differences among them. Apart from the matter of size, the effectiveness of the links with different parts of the information pool vary depending upon the interests of the faculty involved, who in turn depend, at least in part, on the goals of the college itself.

The main function of an engineering college is that of education. Fig. 1 shows this function as the acceptance into the college of undeveloped human resource material which is eventually released, in a partially developed form, into the region served. Of course the admission is not merely limited to new students, since various levels of continuing education are also encompassed.

This simple model ignores the fine structure of a college's interaction with its environment. It merely groups the surrounding institutions as constituting the public and private sectors. Some of these institutions, such as a large industry, may have their own effective links with the information resource. Others may only connect in a very tenuous way. A large amount of technological information has been suitably codified so as to be readily available in handbooks. However, much of the technical literature is not in such a convenient form and so will be unavailable to potential users in industry or government, unless skilled personnel are at hand to act as information finders and translators. It is conceivable then that if effective links are established with the engineering college, information resources not available in the ordinary way can be released to a potential user.

The interaction of the engineering college faculty with the information pool extends beyond that of mere translation. By their own efforts to discover and codify new knowledge, the engineering faculty constantly replenish and increase the effectiveness of their link with the information pool, and they add to its content. This heightens and

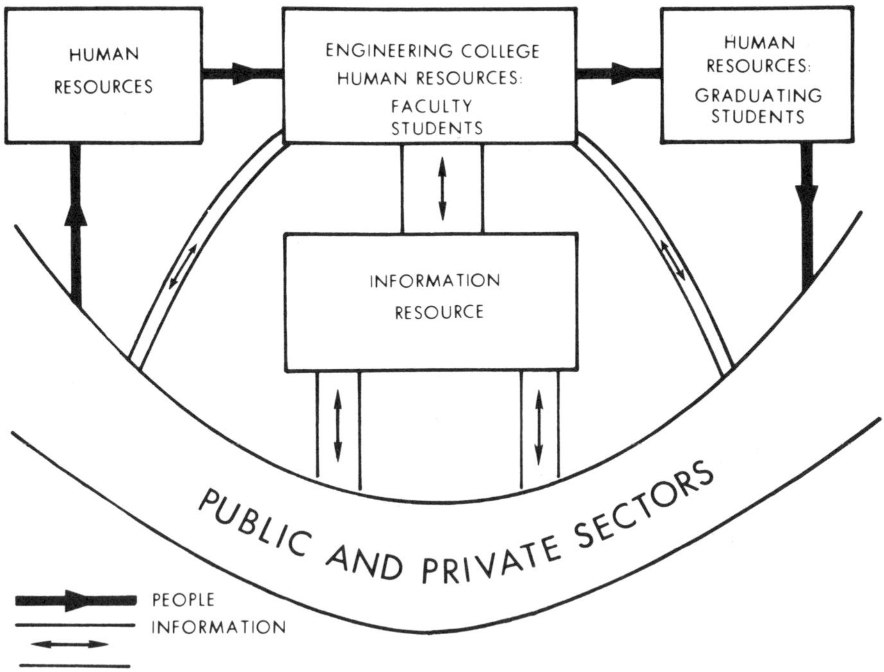

Fig. 1 People and information linkages between a College of Engineering and its environment.

improves their role not only as educators but as skilled manipulators and suppliers of information.

To enable the links shown in Fig. 1 to convey either human resource material or information, several supportive flows and activities are necessary. The college is housed, equipped and supported by fiscal flows emanating from individuals and private or public institutions at the local or national level. Not only are these flows necessary to make the operation functional, but the source of funds provides a strong determinant in the establishment of the goals of a college.* A model of these fiscal flows for the university as a whole, both within the university and between the university and the community, has been formulated by Caffrey and Isaacs.[4] Specific information as to the specific fiscal flows between an engineering college and the community is today lacking.

*Warren B. Cheston, "Higher Education in the United States," Appendix II-2.

In addition to these fiscal flows, which are necessary to make a college operational, it must be realized that a number of other economic mechanisms arise from the operation of an institution of higher learning and link such an institution to its local region. For instance, B. Hudson* identifies five separate economic mechanisms of this kind (Fig. 2).

Finally, what are the significant differences between an engineering college and the other units of the university? In the next chapter we briefly attempt to answer some of these questions.

Regions

In discussing the regional role of an academic institution one must recognize from the outset that the term "region" may assume several meanings. For instance, a narrow definition could refer to a political jurisdiction, as Caffrey and Isaacs do[4]—the city, county, state or nation. However, in the case of a regional institution such as Southern Illinois University at Carbondale, located near the meeting point of several states, political subdivisions are inadequate to describe the region from which the recruiters seek the bulk of the SIUC engineering and technology students (Fig. 3).

The region involved in an impact study of the type considered in this report is essentially an open system, which becomes more open as the region studied becomes smaller. **A number of factors contribute to making the system an open one. Graduates from the university may be expected to seek employment in a large area, as exemplified in the case of graduates from the University of California at Los Angeles in Table 1. Also, some interaction with a local industry may influence branches of the company located throughout the nation. Finally, research conducted on a local urban problem may well generate ripples that will eventually affect other cities in this country or reach even further afield.

It is thus apparent that the regions of interaction are variable both in space and time, and this variability is reflected in the current study. The question of regionalization has been studied at length by Hartshorne[5], among others.

*Barclay Hudson, "Regional Economic Effects of Higher Education Institutions," Appendix 11-5.

**Ibid.

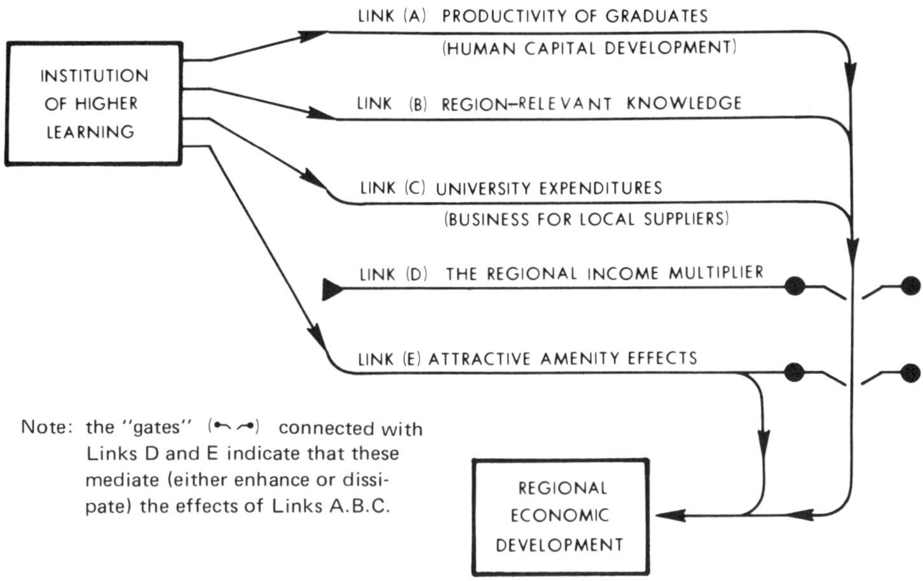

Fig. 2 The five link model of the economic role of a regional institution of higher learning.*

*Hudson, Appendix II-5

TRADITIONAL CONCEPTION OF ROLES

Teaching

Without a need for higher education the universities would not exist. A large part of the acquisition of an education by the students and, for that matter, by the teacher is derived from the teaching function. Traditionally this has been viewed as one facet of the pursuit of truth that earmarks the activity of a university. However, the form has undergone considerable change, and nowhere more than in the engineering colleges. Many of these colleges have their origins in schools that were established to train apprentices in specific areas of application. Full membership in the university community came as the colleges began to appreciate and cater to the breadth of the educational

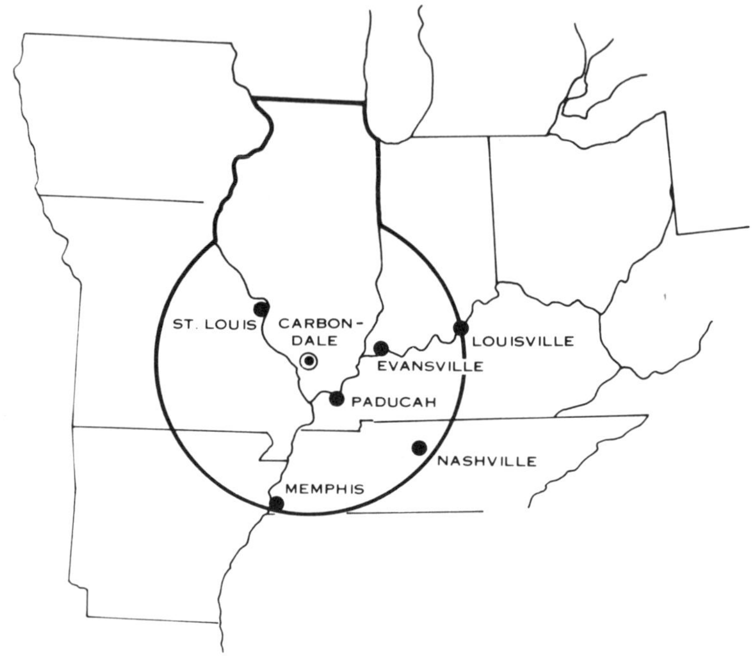

Fig. 3 Southern Illinois recruiting area for 90% of the engineering and technology graduates from Southern Illinois University at Carbondale.*

*Thomas B. Jefferson, "Statement of the Unique Historical and Geographic Factors which Impinge on the Regional Role of the Engineering Programs at Southern Illinois University at Carbondale," Appendix I-16.

needs of such programs. In the last fifty years programs have changed from courses with a strong flavor of specialized application, to others that concentrate on fundamental principles and on the techniques of their use. The direction of change has been towards sound scholarship as perceived by the academic community. The claim has always been that a good grasp of broad fundamentals is of the greatest value to an engineer. This is directly at the core of the criticism that engineers have retreated to their "ivory towers" while neglecting the real world. However, good engineers do have a vivid sense of reality and appreciate fully that, as expressed by Hardy Cross[6], "Scholarship for engineers means first, that they know accurately what they are talking about." Engineering faculties have refined their pro-

Table 1 Percentage Distribution of UCLA Graduates by Concentric Regions as of 1972.*

	REGION					
	(1) Westwood	(2) West L.A.	(3) L.A. City	(4) L.A. County	(5) California	(6) Total USA
Population of region (1970) (Cumulative)**	3.6×10^4	5.1×10^5	2.8×10^6	7×10^6	2×10^7	2×10^8
1970-71 *Alumni Distribution* −cumulative −marginal*** (n=6848) †	6.4 6.4	35.2 28.8	47.4 12.2	78.1 30.7	94.4 16.3	100.0 5.6
1960-64 *Alumni Distribution* −cumulative −marginal (n=11,980)	2.5 2.5	19.5 17.0	31.1 11.6	64.2 33.1	81.4 17.2	100.0 18.6
1950-54 *Alumni Distribution* −cumulative −marginal (n=9,423)	1.8 1.8	15.6 13.8	24.4 8.8	59.1 34.7	81.8 22.7	100 18.2

*Hudson, Appendix II-5.
**"Cumulative" denotes the entire alumni population in the region.
***"Marginal" refers only to the increment in the number of alumni added by moving from one region to the next larger region.
†The sample (n) does not include persons with unreliable addresses. This may bias the sample slightly in the direction of exaggerating the proportion of graduates living in the local area, but the size of error is unknown.

grams as the educational needs have become better defined. For example, today's engineering student is likely to spend from one eighth to one quarter of his time on humanities and social science subjects, a response encouraged in the last 30 years or so, which reflects an increasing consciousness of the role that technology plays in society.

In the report of a workshop carried out as part of this study,* the "need for conscious policy decisions to build flexibility into the engineering educational system" is emphasized. Uncertainties of the future dictate that undergraduate education must stress fundamentals and that "premature specialization should be avoided." However, this should not imply that early specialization in engineering methodology, such as the design process, should be entirely neglected.

Another view is expressed by Johnson[7], who considers that teaching can become a public service only if it offers formal or informal courses off campus through continuing education and extension divisions. He considers that the claim that teaching *is* service is a "definitional escape hatch." We do not agree that teaching is not service, but what is it? We acknowledge that it cannot be the only service that the university provides to satisfy the needs of the community. In spite of these semantic difficulties, Johnson's concerns are obvious and valid. He is looking for a stronger problem orientation than has been traditionally the case, with an eye on today's problems in society.

University faculties have responded in the past to changing demands, but teaching programs, by the very weight of their organizational structure, have tended to change rather slowly. Many consider this to be a desirable trend, as each shift is only made after due consideration and trial. Thus only persistent trends are reflected in curriculum changes. On the other hand, excessive prudence or slowness in responding to a specific need can lead to establishment of courses and curricula out of phase with the need. Two examples are the development in universities of a host of courses in numerical analysis and the development of aerospace engineering programs, long after the need had peaked in both areas.

In general, curricula have evolved slowly over the years in response to several different pressures. Some of these, though not ranked according to their relative importance, are associated with:

1. the faculty's own view of the knowledge that will equip a student in the best possible way for his future. This has tended to foster programs that reflect a balance between sound presentations of several fundamental concepts, with opportunities for specialization coming in later years.

*Hugh Folk and Betty M. Vetter, "Report on Workshop IV—Technological Manpower Retraining: The Role of the Engineering College in Retraining Engineers as Technological Needs and Employment Patterns Shift," Appendix II-15.

2. the minimum requirements, which are greatly influenced by the standards set for entrance into advanced graduate programs. This has led to the common criticism that educators are more concerned with compressing additional new knowledge into the undergraduate programs, than with developing a sensitivity to societal problems.
3. the individual student's developing view of his own interests and commitment. This has tended to generate programs with inherent flexibility permitting many options to the student.
4. the requirements of potential employers. Here the tendency has been to preserve traditionally accepted areas of study, such as Mechanical or Civil Engineering, even though this leads to duplication and awkwardness in the design of programs, as the boundaries between many of these areas become blurred.
5. the guidelines set down by the various accrediting bodies.
6. the standards set for entry into the profession as typified by the Professional Engineering Licensing Examinations.

It is largely through the undergraduate and graduate programs that students are developed as a human resource, able by their acquired knowledge and skill to use and generate information for the solution of problems in the real world. The current of criticism reflected by Johnson[7] points to what is seen as insufficient sensitivity to societal problems and the criteria which are realistic in that realm. It does appear that engineering students are best equipped to deal with the technological world, while the interface with societal issues is treated in a more haphazard fashion.

Research

Certainly over the last 20 years or so the universities have strived for excellence and prestige by intensifying their searches for scholars for their staffs. The academic community judges scholarship, to a large extent, by an individual's quality and number of publications. Also, by virtue of its visibility, this criterion is used to judge performance for promotion purposes. Hence research plays an important role on the university scene.

This accent is not surprising. The quest for new knowledge or information has always been recognized as a key university function, and outstanding scholars have mostly achieved their status by conducting research with a spirit of free inquiry. The lack of vested interest is frequently cited as the main reason for encouraging university research. Unfortunately, this latter statement is no longer as true as it might once have been. There is no doubt that academic research has followed the demand, particularly where this has been backed up by dollar support. In engineering this is exemplified by the post-Sputnik era.

Johnson[7] is concerned with research as a back-up resource for the applied end of the knowledge-to-action spectrum and looks for a suitable "problem focus" able to concentrate the resource on immediate public service. Some institutions, such as the University of Wisconsin, have a strong public-service orientation, while others seem to have less. New or "emerging" colleges—those that have not established themselves as yet and are still seeking a direction for a principal thrust—probably have the greatest opportunity for developing such a thrust.*

As the report of a workshop in the course of this study points out,** very well established schools with strong faculties and well established research programs, as well as possibly an experiment station, can encounter a great deal of inertia when they consider changing directions to respond to regional needs through research. Rather, there is a tendency to talk about small or large programs which may be additions to what is being done already. On the other hand, the "emerging" colleges can approach the problem of response to local and national needs in different ways.

The workshop report also indicates that, regardless of the type of institution, there is acceptance of a number of significant premises:

1. Research, teaching and public service are inter-related.
2. Engineering needs to be problem oriented and based on situations within a region.
3. Engineering cannot operate in public service in isolation from the rest of the university.
4. Each institution must design its own program or relations with the region, depending on its traditions, abilities and the nature of the region it serves.

*Lawrence N. Canjar, "Report on Workshop I—Methodology for Stimulating the Regional Role of an Engineering College: Industrial-Economic-Academic Mechanisms," Appendix II-12.

**Ibid.

5. Effective public service should be the responsibility of the faculty and be recognized and rewarded.
6. The development of the college must be planned with public service to the region in mind. (Those colleges which are already developed will have to rethink their goals.)
7. Engineering curricula can influence and be influenced by effective public service.

The last item is most important. It is not enough for the faculty to continue to teach their traditional courses in their traditional way and try to function as public servants. If they are truly involved in the region's or community's problems, somehow, this will be reflected in changes in their courses and curricula on campus.

Any discussion of research must recognize that it is largely this activity that keeps academic personnel effective as a human resource. Committee V[8] in its study of graduate education for the State of Illinois Board of Higher Education states that, "Research is an integral part of a good engineering program," and "It provides a means for the faculty to stay up-to-date and to maintain updated courses. . ."

It is clear that research, whether sponsored or not, is a vital component of the every-day activity of the faculty of an engineering college. The productivity of such research depends upon faculty commitment and the extent of the financial support. In many respects faculty research is very efficient. On the other hand, since in most cases it is only a part time activity for teaching faculty and since graduate student aid is also part time and highly variable, faculty research tends to be more protracted than might be anticipated by the sponsors of the research projects.

It is indeed fully consistent with the nature of the university as an institution that this should be so. But then, clearly, different approaches and additional mechanisms are necessary if the university is also to provide research that responds rapidly to current problems. (One difficulty arises because of the lack of well established and academically accepted journals that cater to the publishing of public service activities.) Traditionally the degree of "current problem focus" has been dictated either by faculty motivation or origin of financial support and usually a mixture of both. For research of this kind to be fostered in the university, it must find a ready market, and not encounter unusual obstacles in its reporting through publications.

Service

Service, which, as we have indicated, is strongly interrelated to the other activities of the university, has been interpreted in two distinct ways: service directed to the public and service directed internally to the operating needs of the university. The present study is concerned primarily with the first—the arena of public service as identified by Johnson.[7] The image is one of university-associated human resources aiding the solution of societal problems. Johnson quotes Professor J. T. Bonner of Michigan State University, who visualizes university functions as having three capacities, that of teacher, researcher and socially responsible organization. Public service occurs where these three capacities overlap. The conceptualization offered in Fig. 1 permits a definition of service that includes any positive regional impact, either through the delivery of human resource or information.

The internal service component within the university often entails a large amount of faculty time spent on the organizational activity required to run departments, generate programs, plan for the future and generally contribute to institutional maintenance and development. Faculty members always complain that their time is absorbed in committee meetings; though often wasteful, such involvements are certainly necessary. Tables 2, 3 and 4 show the distribution of faculty time at a typical comprehensive state university and reveal the extensive involvement of the senior faculty in administrative duties. This comprises the executive function which falls within the faculty's domain and usually impinges in some sense on educational policy. This is direct service to the university.

Returning to the first view of service, Johnson reflects, "Just as research enriches teaching, so service adds its dimension to both. . .the test of relevance, the antidote for overspecialization, and the freshness of immediacy." He stresses the unity of the knowledge spectrum involving the discovery, delivery and application of knowledge.

Bugliarello* comments on the inevitability of universities becoming involved in operating services for the community but stresses the need for caution lest other university functions become weakened. He also observes that the involvement of university personnel in the formulation of policies at the national, state and local levels has been weak and haphazard, and that their vast potential in this respect remains untapped.

*George Bugliarello, "The Professional School and Chicago Circle," Appendix III-3.

Table 2 Academic Staff Effort by Function, Source of Funds, and Rank at a State University.*

	CAMPUS TEACHING	ADMINIS-TRATION	DEPARTMENT RESEARCH	ORGANIZED RESEARCH	LEAVE	OTHER
			(from general funds)	(separately budgeted)		
PROF	55.0%	15.1%	7.0%	11.3%	4.8%	6.8%
ASSOC PROF	58.8%	6.9%	4.9%	9.3%	5.2%	14.9%
ASST PROF	60.6%	5.4%	6.2%	12.7%	1.3%	13.8%
INSTR	58.2%	1.1%	2.6%	4.7%	—	33.4%
ASST	49.0%	2.7%	2.5%	38.1%	—	7.7%
OTHER	6.2%	13.4%	1.3%	28.6%	—	50.5%
ALL RANKS	46.0%	7.4%	3.9%	23.2%	1.6%	17.9%

*Extracted from Statistical Records of the Institution.

Table 3 Academic Staff Effort by Function, Source of Funds, and Rank in an Engineering College at a State University.*

	CAMPUS TEACHING	ADMINIS-TRATION	DEPARTMENT RESEARCH	ORGANIZED RESEARCH	LEAVE	OTHER
			(from general funds)	(separately budgeted)		
PROF	60.2%	10.4%	16.1%	10.0%	1.4%	1.9%
ASSOC PROF	71.2%	2.9%	11.8%	10.1%	3.4%	.6%
ASST PROF	68.1%	1.3%	12.2%	18.3%	—	0.1%
INSTR	77.5%	1.7%	5.0%	15.8%	—	—
ASST	29.8%	0.4%	1.4%	68.1%	—	0.3%
OTHER	5.6%	11.0%	2.9%	76.0%	—	4.5%
ALL RANKS	43.1%	4.4%	7.3%	43.2%	0.7%	1.2%

*Extracted from Statistical Records of the Institution.

Table 4 Academic Staff Effort by Function, Source of Funds, and Rank in an Engineering Department at a State University.*

	CAMPUS TEACHING	ADMINIS-TRATION	DEPARTMENT RESEARCH	ORGANIZED RESEARCH	LEAVE	OTHER
			(from general funds)	(separately budgeted)	5.1%	3.5%
PROF	41.7%	13.8%	30.5%	5.4%	—	1.2%
ASSOC PROF	72.6%	—	16.4%	9.8%	—	—
ASST PROF	58.7%	—	26.9%	14.4%	—	—
INSTR	100%	—	—	—	—	—
ASST	34.0%	—	0.8%	65.2%	—	—
OTHER	7.3%	3.3%	6.0%	83.4%		
ALL RANKS	40.6%	3.8%	12.4%	40.8%	1.3%	1.1%

*Extracted from Statistical Records of the Institution.

Much of the core of the present study concerns this aspect of application and tries to ascertain whether or not this end of the spectrum can be better exploited with the human resources available in engineering colleges.

OTHER ROLES

Teaching, research and service to the university are expected of every faculty member in the course of executing his or her obligation to the university. However, many faculty members also participate in other activities, which deserve to be singled out because of the options they provide for additional links between the engineering college and the public.

Additional Teaching

In addition to carrying out their normal teaching responsibilities, faculty members may be engaged in several other kinds of teaching functions. These may range from invited lectures, which prestigious faculty are called upon to give away from the home campus, to short courses, seminars and debates, to full courses offered off campus or outside of normal working hours. Off-campus courses may be given utilizing telecommunication technology so that they can be received simultaneously at more than one location; they are usually organized through special divisions of the university, such as an Extension Division or a Division of Industrial Liaison.

Faculty members may also be engaged in teaching practicing engineers in continuing education programs, or high school students in orientation or college preparatory courses. The objective, as in all teaching, is the development of human resources.

Through cooperative education programs, whereby the student alternates periods of study in college with periods of work in industry, industry assists the faculty in educating students and, in a sense, becomes an extension of the university by its offer of cooperation. The students stand to achieve motivation and maturity, and so become further developed as human resources. Such programs are difficult to manage effectively, and to be successful demand a great amount of time from both the faculty and the industry involved.

Applied Research

When they are able to, engineering faculty members provide professional services to industry and other segments of the community outside the university. Enlightened universities, and particularly professional colleges, encourage this, though they usually set limits to the time that may be spent (the general norm is equivalent to 1 day per week). Involvements of this kind are considered desirable not only to keep the faculty up-to-date with progress in the practice of the profession, but also because useful links are established and public relations are improved. University facilities such as computers or testing equipment can be used in such work.

Applied research involves the transfer and sometimes the generation of information, and may take several forms. The enormous body of technical literature is often not applied for lack of interpreters. Academics, conversant with the literature and familiar with the language of analysis, play an important role by making this body of information available, through their consulting work, in forms directly applicable to their customers' problems. Other problems may require testing and analysis.

Special Services

The professionally oriented or professionally concerned faculty may also consult or volunteer assistance to the public or private sector where their opinion or advice, based on broad experience in an area, can be valuable. Examples of this type of special services include advice given to policy-making bodies or expert testimony given in court.

Many engineering faculty are active in professional societies, which often provide a convenient vehicle for the communication to the public of their members' concerns and opinions on professional and public issues. When such communication occurs, it too must be regarded as a significant public service. However, at times some societies reflect the particular concerns of practicing engineers, making it difficult for academic voices to be heard.

IMPACTS ON REGIONAL DEVELOPMENT

The outputs from the engineering college, as either human resources or information, depicted in Fig. 1, are absorbed by the external region. If the role of the college is to be evaluated fully, its impacts on the development of the region need to be examined.

These impacts range from local economic effects that arise simply because of the concentration of potential customers for local business, to changing land values in the environs, to the encouragement for the development of a particular industry, to an increasing demand for education (a feedback process).* However, a regional economy is an open system, and the engineering college has the potential for attracting resources in the form of funds, personnel and information, from outside the immediate region. Ultimately, these resources flow back to the region, enhanced by having been organized and focused by the college. Potentials for fostering regional growth can depend significantly on the extent to which the college—or more generally the university—can attract these external resources.

There are several examples of engineering departments within a college and local industry evolving together in a mutually beneficial way. A case in point is the Chemical Engineering Department at the University of Delaware and the Dupont Company. In other cases, government research and a university have interacted strongly, as exemplified by the Department of Aeronautics and Astronautics at Stanford and the NASA Ames Laboratory, or by the Department of Civil Engineering at Urbana-Champaign and the Corps of Engineers Construction Research Laboratory. Other examples are the Engineering Experiment Stations, such as the one at the University of Illinois at Urbana-Champaign, which particularly during its earlier history had a predominantly regional orientation. Battelle Northwest, which benefits from proximity to the University of Washington in Seattle, is still another example of a situation in which an engineering school has exerted a beneficial regional impact.

The pattern of interaction in any one of these cases merits detailed study, to determine the crucial actions and mechanisms which nurture the beneficial interactions. Above all there is a strong need for new directions in endeavoring to cultivate the engineering schools as instruments of public policy at the local level, as effectively as has occurred at the national level in the areas of defense and the space program.

*Hudson, Appendix II-5.

CHAPTER 2

Dimensions and Models for the Multiple Roles of the Engineering College

INTRODUCTION

The first section in the previous chapter described the functional areas that have a bearing on the regional role problem and provided a framework for this study. In this section the important dimensions are identified and the areas of interaction are represented in a more formal sense, by the collection of data and the construction of models. The problem is an extremely intricate one when all its various aspects are taken into account; hence little has been achieved thus far in the sense of formal mathematical models which can be used for predictive purposes. However, for the purpose of gaining an understanding of the situation, the preliminaries of information collection and the construction of models of various complexities, either qualitative or quantitative, are essential.

A first phase in a modeling process is the development of descriptive or conceptual models, in order to structure the problem and the study process, to identify and understand impacts and relationships, and to organize measurement and data collection processes.* Fig. 4 provides an example of a conceptual model of the region-engineering college interaction, a model which has the simplest possible configura-

*Joseph L. Schofer and Amado San Mateo, "Toward the Development of Models for Studying the Regional Role of Engineering Colleges," Appendix I-2.

tion for the interaction. It encompasses all that is involved but provides little more information than is contained in the title of this book. A more detailed conceptualization, identifying a number of individual modules, such as regional interface, student flow, career and faculty time allocation, is summarized in Figs. 5 and 6. The model is described in Appendix I-2.

It is convenient at this point to propose a structure or set of pigeonholes*, which will encompass the range of regional impacts of interest. To this purpose a matrix which identifies the actors and activities can be developed, as shown in Fig. 7. Each cell of the matrix (which may involve more cells than shown) represents a specific type of impact of the engineering college. For example, if one wishes to determine impacts of the college on the student body, the matrix indicates that they are influenced by other students, faculty, administration and staff, through the mediums of knowledge diffusion, community activities, life style and so on. A similar matrix may be drawn to examine the impact of the college on any other group influenced, be it external or internal to the college: faculty, shopkeepers, industrial employees, etc. Figure 7 reveals the three dimensional character of the matrix, the same set of actors and activities being pertinent for any group influenced by the college.

With this as a basis, each cell may be considered with regard to the importance of the particular impact which it represents. For instance, an important cell in this study is that which represents the influence of the faculty on the state, through public service.

The fundamental activity of a college is the interaction of students and faculty within the framework provided by the college. Groups which support this framework within the college provide the fiscal and managerial controls necessary to satisfactorily operate the college. The faculty's main task is clearly the instruction of students or, more generally, the diffusion of skills and ideas. This in fact occurs through publication, as well as teaching, and the subsequent influence of graduates in the community. The students may reside at home, in college-provided units, or in apartments within some distance of the college. They attend classes, laboratories, and recreational activities on campus while residing in the locality of the college.

The major activities of the college take place usually within its physical facilities and are supported by a series of fiscal transfers. Funds

*Clarke H. P. Schneider and Joseph L. Schofer, "Impact Identification Model," Appendix I-1.

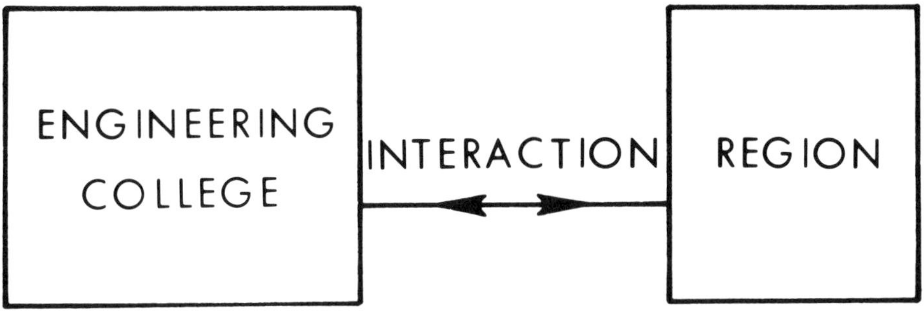

Fig. 4 Simplest model of region–College of Engineering interaction.

come into the college from students, appropriations, subsidies, grants, contracts, donations and so forth. A portion is paid in wages and salaries to faculty and staff. Other payments go to purchases of supplies and services.

Research activities are pursued under the aegis of the college and, as such, are more or less directly linked to it; research through grant, contract, internal funding, or individual prerogative indicates an investigation on a relatively long term basis. On the other hand, public service activities may be carried out either under the aegis of the college, or directly by the individual faculty member.

Consulting and community activities are in most cases individual undertakings of members of the faculty. Consulting is a paid or unpaid activity where the technical expertise of a faculty member is applied to solve a problem or fill a need of an outside group or agency; a particular skill is applied to provide an answer to a technical question. This activity may be distinguished from research in three ways. First, it generally does not constitute an expansion of present technical knowledge, though there are cases where faculty members, through consulting, gain experience and knowledge that can be generalized for teaching purposes; second, it is an activity of the faculty rather than the college; and third, it is usually short term. The types of skill and knowledge which are provided by faculty members acting in this capacity make consulting a significant element of regional impact.

Community activities of the faculty are those for which technical expertise is not generally needed or utilized; serving on a church board or being the leader of a boy scout troop are examples. Since

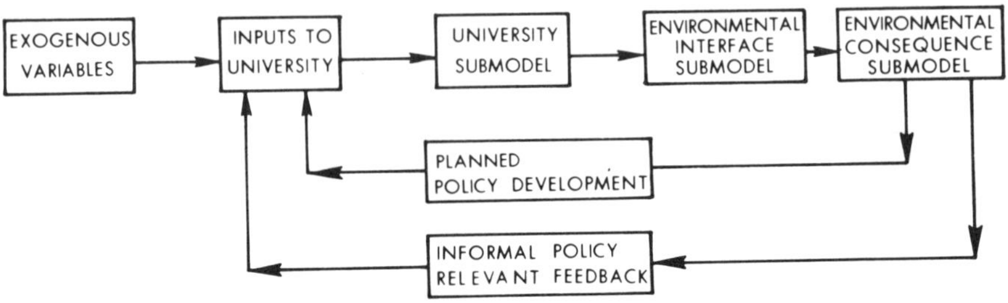

Fig. 5 An overall model of inputs, outputs, and feedback for a university.*

*Schofer and San Mateo, Appendix I-2.

these activities do not utilize the engineering knowledge of the faculty members, they do not make the impact of an engineering college, or a university, any different from that of any other fairly large employer within the region.

In a very real sense, public and most private colleges and universities are themselves "public service" developments of bodies like the state government, a particular religious order or alumni of the institution. In this light, teaching is itself a public service and so are many

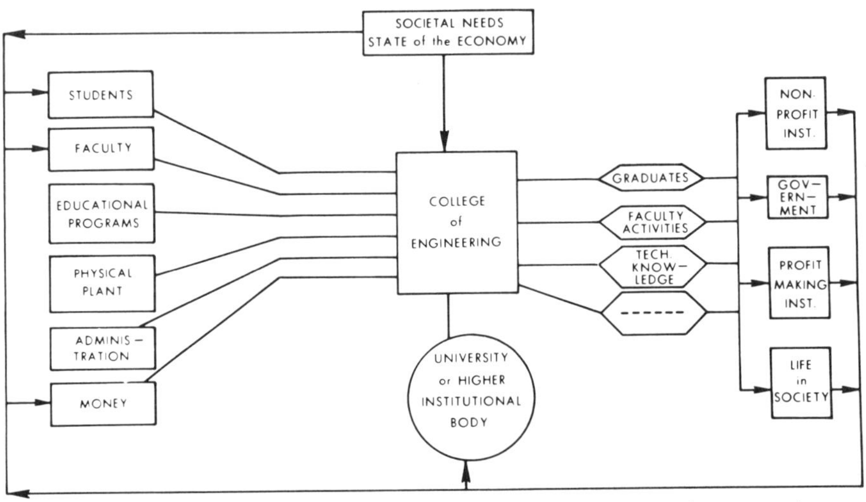

Fig. 6 More detailed conceptual model identifying inputs and outputs.*

*Schofer and San Mateo, Appendix I-2.

ancillary activities. In the case of a professional college, public service may imply also the delivery of technical expertise without charge to elements in the community. Frequently the groups served in this way are non-profit organizations. Thus, the definition of exactly what constitutes public service is by necessity somewhat vague. Furthermore, the demand for college involvement in community service is not universal. Some private laboratories and consulting companies may in fact feel threatened by the public service capabilities of engineering colleges.

External agencies or groups which are affected by the engineering college may fruitfully be broken down both geographically and by function. Thus at local, regional and national levels, government, industry and business, and other institutions (including the general public) are influenced by the activities of the college. In this context, these groups, which are the recipients of the impacts of the engineering colleges, represent the region.

In brief, we may then say that *the activities of the people of the college constitute the main areas of impact upon the region.* Of course, the presence of the physical facilities of the campus also affects the region, but not in ways substantially different from any other occupant of that amount of land, visited daily by a similar number of people, and with the relatively low levels of pollution typical of the operation of an academic establishment. Some of these facilities, such as libraries, gymnasia and laboratories, are often available for utilization by other inhabitants of the region and so represent impacts of the college.

The main feature which distinguishes an engineering college from other institutions is its accumulation of human resources, as described in the previous chapter. It follows that those cells in the matrix in Fig. 7 which have to do with faculty and student involvements should receive the greatest attention. It is their impacts through the seven columns in the matrix, in the region and within subregions, that are of interest. The groups influenced by the actions of the college's human resources include governments, governmental agencies and planning groups, and regional industry.

The graduation of students influences not only the manpower needs of the state, but also the graduates' own attitudes and socioeconomic status. The impact of college in general on students' attitudes is analyzed by a recent survey by the American Council on Education[9], based on a large sample of students who entered college in 1967. The survey shows the changes in attitudes and goals occur-

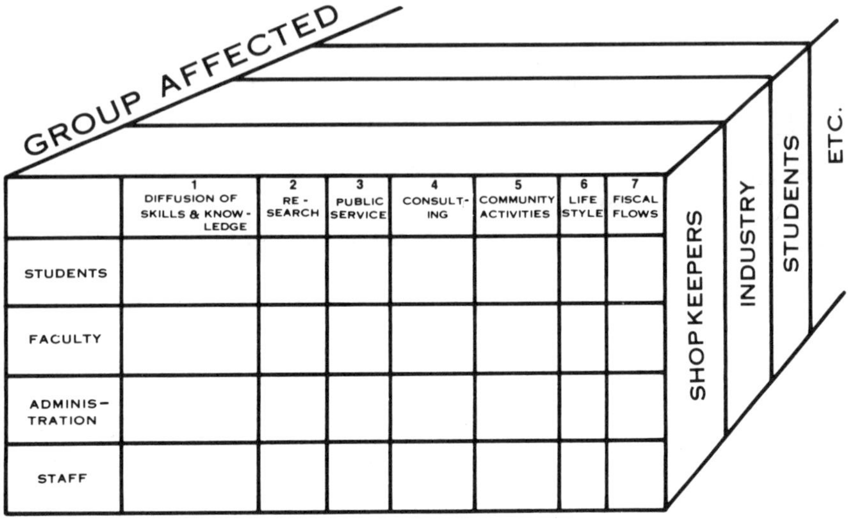

Fig. 7 Engineering college impact matrix, identifying actors and activities.*

*Schneider and Schofer, Appendix I-1.

ring after four years in college (Tables 5 and 6). (Incidentally, the high decrease in the interest in engineering as a career choice is noteworthy.)

Impacts are not easily described, so a plurality of models is needed.* Some possibilities are discussed by Caffrey and Isaacs[4]; others have been investigated in the present study, as discussed in subsequent sections. The outcome of a modeling study could be an extremely simple and direct statement. For example, Caffrey** mentions the hypothesis which proposes that all the capital spent on an institution since it was founded, excluding land, gives a good estimate of its annual budget. Simple conclusions of this kind may contain something of value but they must be treated with great caution. Extreme difficulties are also encountered in modeling political inputs.

The impact studies by Caffrey and Isaacs[4] have limited themselves to modeling the economic interaction between a university or college

*Morton J. Marcus, "The Economic Impact of Educational Institutions: Part II," Appendix II-11.

**John Caffrey, "The Economic Impact of Educational Institutions: Part I," Appendix II-10.

Table 5 1967 Freshmen 4 Years Later — Choice of Discipline and Career.*

MAJOR FIELD	PLANNED *1967*	ACTUAL *1971*	*%CHANGE*	*CAREER CHOICE*	PLANNED *1967*	ACTUAL *1971*	*%CHANGE*
Agriculture (incl. forestry)	2.4%	2.3%	− 4	Artist (incl. performer)	5.8%	5.7%	− 2
Biological sciences	3.7	4.5	+22	Businessman	11.2	13.6	+21
Business	16.2	16.3	+ 1	Clergyman	1.2	0.8	−33
Education	10.5	13.0	+24	College teacher	1.2	3.3	+170
Engineering	9.5	7.5	−21	Doctor (M.D. or D.D.S.)	4.2	2.4	−43
English	4.0	6.0	+50	Secondary school educator	14.1	12.8	− 9
Health professions (non-M.D.)	5.2	4.5	−13	Elementary school teacher	8.3	9.8	+18
History, political science	6.7	8.3	+24	Engineer	8.4	5.1	−39
Other humanities	4.6	3.9	−15	Farmer or forester	1.9	2.2	+16
Fine arts	8.6	7.8	− 9	Health professional (non-M.D.)	4.3	3.3	−23
Mathematics, statistics	4.2	3.0	−19	Lawyer	3.5	3.7	+ 6
Physical sciences	3.0	2.7	−10	Nurse	2.5	2.4	− 4
Pre-professional	6.7	1.2	−82	Research scientist	2.9	1.9	−35
Psychology, sociology, anthropology	7.8	12.3	+58	Other	20.5	26.7	+30
Other technical fields	2.6	3.9	+50	Undecided	10.1	7.2	−29
Other non-technical fields	2.5	2.6	+ 4				
Undecided	1.8	0.6	−67				

Data from American Council on Education
*Robert L. Jacobson. "Data Show the Effects of College," The Chronicle of Higher Education, 2 April 1973.

Table 6 1967 Freshmen and 4 Years Later: Attitudes and Objectives*

ATTITUDES: AGREE STRONGLY OR SOMEWHAT THAT:	*1967*	*1971*
Faculty members are more competent than students to set curriculum	85.3%	61.1%
Married women best confined to home	56.6%	23.5%
Scientists should publish findings, regardless of consequences	43.7%	52.2%
Chief benefit of college is monetary	56.0%	36.7%
Faculty promotions should be based in part on student evaluations	62.2%	81.5%
Disadvantaged students should have preferential admissions treatment	43.3%	36.5%
OBJECTIVES CONSIDERED ESSENTIAL OR VERY IMPORTANT:		
Achieving in performing arts	11.3%	8.5%
Becoming an authority in subject area	67.8%	51.0%
Being recognized by peers in field	41.2%	32.0%
Becoming an expert in finance	11.5%	10.6%
Having administrative responsibility	24.9%	26.3%
Being very well-off financially	43.5%	29.8%
Helping others in difficulty	61.8%	68.9%
Becoming an outstanding athlete	13.8%	6.5%
Becoming a community leader	23.8%	17.9%
Contributing to scientific theory	11.5%	7.5%
Writing original works	13.8%	13.7%
Creating works of art	15.5%	20.2%
Keeping up with political affairs	50.6%	52.9%
Succeeding in own business	46.4%	29.4%
Developing a meaningful philosophy of life	82.9%	81.9%

Data from American Council on Education

*Jacobson, Ibid.

and its surrounding community. The purpose was the provision of "a commonsense guide for persons interested in finding out whether it costs a community more or less than it gains economically by having a college or university in its midst." The studies do not aim at answering the question, but rather at providing a paradigm, a systematic approach for identifying costs and benefits in what is acknowledged to be an extremely complex situation. A wide range of variables is included, leading to a flexibility that permits application to institutions other than a university. Still, several effects such as the long-range influences of a source of skilled manpower or the socio-economic impacts on the graduates themselves are omitted.

A continuing research program, in which many aspects of the overall operation have been modeled,[10,11,12,] has been conducted by the University of California. However, the main thrust of the study is not concerned with community interactions; in assessing the allocation of resources, the study deals with aggregates, examining the entire university structure rather than a specific type of college.

DIMENSIONS AND MODELS

Education

The Academic Progress of Students. A good starting point for identifying and modeling the college is the central activity of the colleges—the education of the student. The main questions to be answered are: Who enters an engineering college? What happens to them in an engineering college? Who graduates?*

An Analysis of a Cohort of Students. As part of the current study, the career of an identifiable group (a "cohort") of engineering students at URBANA was analyzed as the group progressed through the institution**, in order to obtain specific information concerning engineering students. The "cohort" approach enables one to include the influence of time, as opposed to the "cross section" approach which views the total situation at one given time. The first question concerns the origin of the entering freshmen in engineering. An early study by

*William J. Haga, "Who Goes to Engineering School? Who Graduates? A Cohort Analysis," Appendix I-3.
**Ibid.

Trow[13], conducted in the fifties, concluded that engineering has a special appeal for bright boys from the lower middle class. Trow speculated that these boys have limited aspirations and view engineering as being attainable for them.

The analysis focuses on the 909 freshmen that entered engineering at URBANA in 1966 and examines their profile in some detail. The educational path of the freshmen is shown in Fig. 8. Of the cohort who were freshmen in fall 1966, 42% had dropped out by the spring of 1971 (a rate typical of the university as a whole), 30% had graduated with a B.S. degree in engineering, 15% had switched fields and graduated from other areas, 4% had switched fields and were still in school, an the remaining 9% were still in engineering school. Table 7 shows an increasing efficiency of operation from 1965 through 1968 for both URBANA and UICC. Of course, students who drop out need not necessarily be viewed as waste, without a complete study of their subsequent efforts.

The most significant conclusions are:

• The modal choice of the engineering freshmen was that of "researcher," though among the entire entering class at the university the most popular choice was "producer of services" (Table 8).

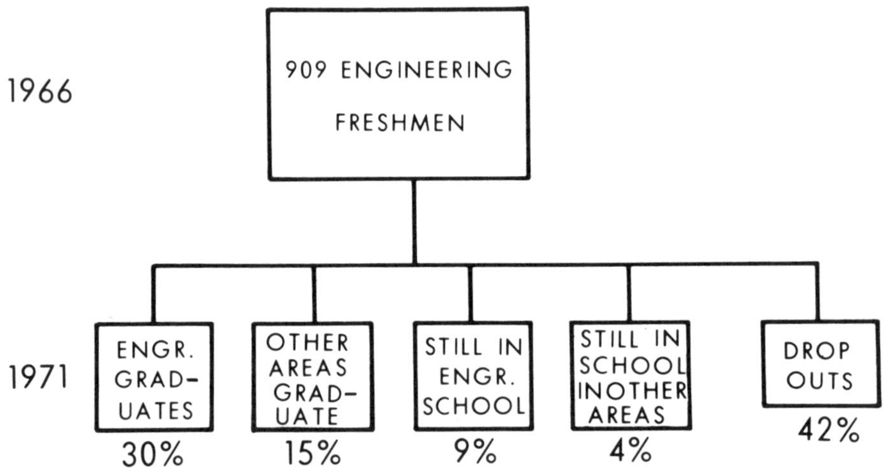

Fig. 8 Educational path of the Fall, 1966 engineering freshmen at the University of Illinois at Urbana-Champaign.*

*Haga, Appendix I-3.

Table 7 Comparisons of Cumulative Drop Rates after Two Years of College in the Colleges of Engineering at the University of Illinois at Urbana-Champaign and at the University of Illinois at Chicago Circle (UICC). The ACT and HSPR medians are for beginning freshmen and do not reflect the existence of transfer students from other colleges.*

	URBANA			UICC		
Year	ACT:C**	HSPR	Drop Rate	ACT: C	HSPR	Drop Rate
1965	27	88	24.1	24	79	37.5
1966	27	86	26.8	25	78	40.3
1967	26	87	22.6	24	77	30.7
1968	28	90	14.9	24	77	27.5

*Jane Loeb, "Recent Trends in College Retention, Drop, and Graduation Rates," Office of School and College Relations, Research Memoranda 72-3 and 72-9, University of Illinois, February and April, 1972.
**ACT: C denotes American College Test: Composite. HSPR denotes High School Percentile Rank.

Table 8 Vocational Role Preferences of Freshmen, University of Illinois-Urbana, Fall, 1966.*

	All 1966 Freshmen	Engineering Grads 1971	Other** Grads 1971	Still in Engrg. School 1971	Others** Still in School 1971	College** Drop Outs
Researcher	17.6%	41.0%	33.9%	33.9%	29.4%	35.9%
Teacher	17.5%	0.5%	—	1.6%	—	0.7%
Administrator	6.3%	6.9%	10.1%	4.8%	8.8%	6.3%
Salesman	2.0%	4.1%	1.8%	3.2%	2.9%	2.3%
Producer of Services	24.8%	16.1%	19.3%	12.9%	23.5%	19.6%
None of These	9.3%	6.9%	9.2%	14.5%	2.9%	10.0%
Combinations	4.7%	4.6%	4.6%	6.5%	5.9%	4.7%
Undecided	17.9%	19.8%	21.1%	22.6%	26.5%	20.6%

*Haga, Appendix I-3.
**Started as engineering frosh, Fall 1966.

• Engineering students have higher motivation for achievement in their chosen profession than their peers in other disciplines.

• Engineering students give their first goal as job training and mind development, with increased income listed as their second choice (Table 9). However, the second choice probably elicited a more honest response than the first, according to Haga.

• The more successful engineering students (*at URBANA*) come from suburban areas and almost not at all from large central cities.

• Engineering students tend to come from the lower economic strata, as concluded also by Trow.[13] Those who switch to other fields tend to come from more affluent families than those who persist in engineering, but those who remain in engineering beyond 5 years come from the most affluent category of the engineering cohort.

• There are only minor differences—possibly of no statistical significance—between the marital status and dating patterns of engineers and other students. The differences seem to indicate, however, less social sophistication of the engineers (Table 10). One would have been led to expect greater disparity between the engineers and the total freshmen percentages than the figures in the table show. Within the engineering cohort, the percentage differences are more telling. Those who ultimately earned their engineering degrees had the smallest percentage of the cohort involved in serious relationships with girls and the correspondingly largest percentage of those who did not date at all.

It is also significant to note that:

• The high school educational preparation of engineering students at URBANA has been improving with time; furthermore, students who enter engineering are educationally superior. In 1966, 68% were in the 80-100 percentiles of their secondary classes. The trends show a marked increase in high school standing for engineering students in the late sixties.

In the absence of similar analyses, the general validity of the results of this case study for other institutions remains to be assessed.

Admissions Policies. It is obvious that those who enter a university are to a large extent influenced by the policies of the institution with regard to admissions. The ivy league institutions and the private engineering colleges such as M.I.T., for instance, draw from quite a different sample than state universities (indeed, it has been said that they now tend to draw more the affluent and the poor, rather than the middle class).

Table 9 Second Most Important College Goal of Freshmen, University of Illinois-Urbana, Fall, 1966.*

	All 1966 Freshmen	Engineering Grads 1971	Other** Grads 1971	Still in Engrg. School 1971	Others** Still in School 1971	College** Drop Outs
Enjoy Life	2.7%	4.6%	3.7%	4.8%	—	2.0%
Develop Mind	28.5%	23.5%	23.9%	29.0%	29.4%	27.2%
Job Training	29.5%	23.0%	24.8%	22.6%	35.3%	24.3%
Desirable Marriage	1.2%	0.5%	—	—	—	0.3%
Raise Income	17.5%	32.7%	31.2%	21.0%	29.4%	31.9%
Develop Ethics	0.5%	—	—	—	—	2.0%
Develop Culture	6.6%	3.7%	5.5%	6.5%	2.9%	3.7%
Develop Personality	6.6%	5.5%	5.5%	9.7%	2.9%	4.7%
Develop Philosophy	4.5%	4.1%	2.8%	3.2%	—	2.3%
Some Other Goal	2.4%	2.3%	2.8%	3.2%	—	1.7%

**Haga, Appendix I-3.

**Started as engineering frosh, Fall 1966.

Table 10 Dating Status in 1971 of Those Students Who Entered as Freshmen, University of Illinois-Urbana, Fall, 1966.*

	All Those Who in 1966 were Freshmen	Graduating in Engineering in 1971
Date Same Person	18.8%	22.6%
Date Several People	53.1%	47.0%
Don't Date	14.9%	18.4%

*Haga, Appendix I-3.

A particular problem confronts those public institutions that tend to draw from the city, such as the City University of New York (CUNY) and the University of Illinois at Chicago Circle (UICC).

The most significant aspect in this regard at CUNY is the result of the policy of open admissions, inaugurated in the Fall of 1970. UICC, like CUNY, also tends to offer a comprehensive university education to those from an urban area who could not otherwise afford it. Its admission policies are in effect different from those of URBANA. It is instructive to compare the trends in cumulative drop rates in engineering after two years of college at both campuses of the University of Illinois URBANA and UICC, as a function of ratings in college admission tests (Table 7).[14]

Cohen[15], in an analysis of the admissions to a unit of CUNY, the Borough of Manhattan Community College (BMCC), reports that the open admissions policy led to an increase of students admitted from officially designated poverty areas from 47% in 1968 to 57% in 1970. Total withdrawals from courses in the college ran 75% above the 1968 rate. Poverty area residents had lower drop out rates from college (16.2%) than average but withdrew from more courses (6.9% withdrawal from two or more courses). In Fall 1968 over 90% of all entrants came directly from high school—the rest being either working, or unemployed, or in the services at the time of admission. In Fall of 1970 this had fallen to 70%. The high school averages in 1970 were 5.5 percentage points below the 1968 level and 21% entered with high school averages below 70%. The admission policy at CUNY represents an extreme case.

The differences in scores between CUNY and UICC reflect the different admission policies. It would appear, from the UICC results, that drop rates are being influenced by factors other than academic aptitude of the incoming freshmen, though the higher drop rate compared with URBANA probably reflects the lower selectivity of admissions at UICC.[14] It should also be noted that the large decrease in drop out rates over a four year period, observed on both campuses, also occurred in the other colleges of URBANA and UICC.

Growth Trends in Degree Production. For the purpose of constructing simple models, it is useful to review growth trends for engineering degrees granted at the B.S., M.S. and Ph.D. levels on a national basis. The trends show exponential growth functions for the B.S. and M.S. degrees, with oscillations that follow the national population growth trends but with roughly a 17 year lag as expected. The oscillations are not reflected in Ph.D. trends, which have been highly filtered by the stringent requirements for that degree.*

Some changes in the B.S. and M.S. curves are due to the varying demand for graduates. For example, a fairly clear upswing is apparent around 1965-66, probably stimulated by the space program. The curves for B.S. degrees for the individual engineering colleges almost all show a peaking in the 1957 to 1959 period, probably again reflecting the influence of the space program, compounded by the returning veterans from the Korean war. However, the pattern varies according to local conditions, such as limited capacity in the undergraduate program (Princeton) or lack of a graduate program (Northern Illinois University). For instance, at URBANA, B.S. degrees granted show peaking around 1960, reflecting the national trends; on the other hand, M.S. degrees peaked in about 1966, while the national trend continues to rise. Ph.D. degrees resemble the national trend. UICC and SIUC, being in an expansion stage, show only rapid growth.

The trends in Illinois compared with the rest of the nation (Table 11) show that at all levels below the Ph.D. Illinois produces fewer degrees than might be expected given the size of the population. Both on a per capita basis and on the basis of tax dollars contributed to the Federal government, the State of Illinois is under-represented in terms of engineering faculty and engineering degrees granted. The State is also under-represented in terms of Federal support of research and development per Ph.D. awarded in science and engineering. This finding is in agreement with the results in Reference 16.

The relative importance of graduate programs in the six schools which form the basis for the present study is shown for 1970 in Table 12. It should be noted that the engineering programs at SIUC and UICC are still relatively new, and in particular the numbers for UICC do not reflect as yet the involvement of what is planned to be a comprehensive campus. The most significant conclusion evident from Table 12 is that in Illinois the distribution of engineering programs does not follow the distribution of population in Illinois. Only 34.2% of the total engineering degrees are granted in the Chicago Area with 62.8% of the State's population residing there. The question relating

*Nancy L. Nihan and William D. O'Neill, "Regional Role of Engineering Education: Data and Models," Appendix I-5.

Table 11 Relative Standing of Illinois with Regard to Three Levels of Graduate Engineering Degrees Granted for Male and Female Recipients.*

Earned Engineering Degrees Conferred 1969-70

Degree		Number Granted in Illinois by the Six Graduate Engr. Degree Granting Institutions	Total Such Degrees Granted in the U.S.	Degrees Granted in Illinois as % of Total U.S.
Bachelor's	Total	1675	44,772	3.7
	Male	1653	44,434	3.7
	Female	22	338	6.5
Master's	Total	635	15,597	4.1
	Male	626	15,425	4.1
	Female	9	172	5.2
Doctor's	Total	257	3,681	7.0
	Male	255	3,657	7.0
	Female	2	24	8.3
Overall Total		2567	64,050	4.0

*Schneider, Appendix I-6.

to where the students go upon graduating will be discussed in a later section.

The pronounced sexual imbalance in engineering enrollment at all levels is indicated by the data in Table 11. For instance, in 1970 the ratio of female to male Ph.D.'s was .0066 in engineering, as compared to .119 for all fields. This factor exerts a profound influence on the attitude of society toward technology[17], which is perceived as an activity of interest only to the male segment of the population.

Other Student Activities. Apart from their participation in classroom undergraduate instruction, engineering students are often given the opportunity to elect a cooperative education program. Several variants are possible, but all such programs are characterized by the involvement and cooperation of industry in the educational process, with students working in industry for specified periods as part of their undergraduate program.* This is one area where the students have an impact on the community outside of the usual in-house activities.

A study at Virginia Polytechnic Institute** tested and confirmed the hypothesis that the engineering students engaged in the program (1) perform academically better than the non co-op students and (2) experience a lower rate of attrition. However, there are not sufficiently complete studies of this kind to assess with certainty whether or not such programs are always successful from the academic point of view. There is a major need for***:

- in depth case studies,
- and/or institution of experimental changes,
- monitoring and evaluating of ongoing programs,
- formal surveys of affected groups, including students and industry representatives.

In any case, however, cooperative programs clearly provide an important avenue for interaction between an engineering college and industry. Usually, as shown by the experience at NU, the impact on industry is largely localized geographically. An institution can regulate the flow of students into cooperative programs in a number of ways, including financial mechanisms such as tuition levels and financial aid policies.*

In addition to cooperative education programs with industry, there is also a need to provide experimental programs of internships for engineering students in local, state and Federal government agencies.

*Gary Erenrich, "Cooperative Engineering Education: A Case Study," Appendix I-4.
**Ibid.
***Ibid.

Table 12 The Relative Importance of Graduate, Engineering and Graduate Engineering Programs in Graduate Engineering Degree Granting Institutions in Illinois in 1970.*

	Total Degrees Granted	Graduate Degrees Granted	Graduate Degrees as % of Total Degrees	Total Engr. Degrees Granted	Engr. Degrees as % of Total Degrees Granted	Graduate Degrees in Engr. Granted	Graduate Engr. Degrees as % of Total Engr. Degrees Granted	Graduate Engr. Degrees as % of Graduate Degrees Granted
Bradley	1306	187	14.3	156	11.9	39	25	20.8
IIT	1080	443	41.0	464	42.9	174	37.5	39.3
NU	3528	1628	46.1	310	8.9	176	56.8	10.8
SIUC	4434	855	19.3	144	3.2	17	11.8	1.9
UICC	2617	130	5.0	309	11.8	19	6.1	14.6
URBANA	8614	3469	40.3	1184	13.7	467	39.4	13.5
	21579	6712	31.1	2567	11.9	892	34.7	13.3

At this moment internships with these agencies are customary for students in political or social science or in law but are rather alien to an engineering school and an engineering curriculum. Finally, thesis work and undergraduate research, design, and independent study activities can provide a considerable amount of field work, with the possibility of meaningful interaction with the community.

Extension Activities. The education of students on a regular full-time basis is not the only avenue employed by engineering colleges in the development of human resources. Graduates of several years ago may feel the need for updating their knowledge. Other graduates may wish to obtain an advanced degree. Persons in industry and in other sectors may seek knowledge in a special area or wish to be informed about some new development. In many cases these objectives must be compatible with the ability to pursue further training while retaining a normal full-time job. The response to such a need can occur in several different and not mutually exclusive ways:

- Offering courses on campus outside of working hours.
- Offering courses off campus.
- Transmission of courses (either regular or ad hoc, either live or "canned," either during working hours or outside of working hours) using a variety of telecommunication systems.
- Correspondence courses, using written teaching materials, tapes, etc.

Programs of this kind may be organized under a special group such as the Division of University Extension at URBANA and UICC or the Evening Division at IIT. The annual report of the University of Illinois Extension Division[18] states, ". . .available statistics indicate that credit-hour production in Extra-mural Classes and in Correspondence Courses is considerably less costly to the University than it is in on-campus instruction. The differential increases as the level of instruction rises. The preceding remarks can be verified by cost-study reports made by the University Bureau of Institutional Research and by this Division." At URBANA classes are carefully scheduled, and in several cases faculty members are flown to and from the site using university aircraft.

Offering courses on campus outside of working hours is the most common type of extension teaching. Practically every university engages in a program of this kind; in Illinois, among engineering schools, IIT has by far the largest program, where 3500 students enrolled in the Fall of 1972. A typical student profile at IIT is given in Table 13 for new students entering the program for the first time.

Extra-Mural Classes. The dimension of the off-campus effort can also be very substantial. Thus, in 1970-71 at URBANA, 642 courses (381 for credit and 261 non-credit) were offered to a total of 12,497 students, 5,308 of whom took classes for graduate credit. In all, the classes were offered in 91 cities and in 45 counties and covered most fields of study. Sixty-two of the 642 courses were in engineering, with 24 for credit. A study by Casper[19] of the distances non-engineering students travel to attend extra-mural courses indicates that each off-campus class tends to draw its students from an area of roughly equal radius, regardless of where the class is held; the radius seldom exceeds 14 miles and is about 8 to 10 miles on the average. This must also depend to a large extent on the foresight used in selecting the proper site for a class.

In a survey of people attending the extra-mural classes offered by URBANA, Casper[20] received roughly 50% replies. In the responding sample of 766, 583 were teachers and 95 were educational administrators. Most sought either an M.S. degree or professional improvement.

Telecommunication Systems offer an approach to extra-mural teaching which minimizes travel, both for the students and for the faculty. There is inevitably some loss of rapport, but for many telecommunications make the difference between some courses and none. At URBANA the University Extension Network (UNIVEX-Net) uses a pair of telephone lines to put an instructor in contact with several widely separated classrooms. Dialogue is possible between the instructor and all classrooms, as well as between classrooms. Also written information and pictures may be transmitted in either direction using Victor Electrowriter equipment. In the Spring of 1972, nine credit courses in engineering or mathematics were offered at educational and industrial receiving stations throughout the State. Receiving centers are quite expensive to establish, but the possibilities are intriguing. A cooperative inter-university program in engineering education between URBANA and the University of Wisconsin was expanded by using the UNIVEX-Net. This provided a wider range of continuing education courses to the engineers in both states. During 1970-71, 13 Wisconsin

Table 13 Profile of Students Registering in the Undergraduate Evening Division of IIT for the First Time.*

Semester 1, 1971-72

Number of new undergraduate applications received and permits issued	639
Number of new students of the above group that actually registered	461 (72.1%)
The following statistics concern only those who actually registered:	
% of Women	9.1%
Average Age (Range from 16 to 57 years)	25.5%
% Residing in Chicago	59.0%
% Initially intending to pursue a degree program	59.2%
% of Foreign Students (Foreign citizens)	13.4%
% with part or all of Tuition reimbursed by company	52.5%
% Attending under Veterans Bill	6.7%
% Taking a course (or courses) concurrently at another school	4.9%
Educational Background	
High School Only	10.6%
Some Junior College Work Only	9.1%
Associate of Arts Degree	3.7%
Some College Work but no Degree	48.8%
Bachelor's Degree or above	36.9%

142 Bachelors — 30.8%
24 Masters — 5.2%
4 Doctors — .86%
(1 M.D.)

Comparison with Semester 1, 1970-71

Number of new undergraduate applications received and permits issued 599
Number of new students of the above group that actually registered 434 (72.4%)

*Kenneth V. Henninger, "Profile of Students Registering in the Undergraduate Evening Division for the First Time," *Evening Division Newsletter*, Illinois Institute of Technology, 30 October 1972.

courses and 21 URBANA courses were scheduled. Eighty-nine Illinois residents registered for the Wisconsin courses.

The transmission of instruction (voice and graphs) by phone suffers serious limitations because of the narrow band of the signal. To overcome this deficiency, a number of universities exemplified by Stanford and Southern Methodist University (SMU), have developed, with considerable success, dedicated TV networks, often with audio feedback, to diffuse instruction off-campus to the surrounding community. In the case of Stanford, the major beneficiary has been the industry in the Palo Alto area, and in the case of SMU the industry in Dallas.

At this moment no university in Illinois has developed a dedicated TV system for off-campus instruction. The Bell Telephone Picturephone, now under rapid diffusion in Chicago, appears increasingly attractive (in spite of technical shortcomings due to its being designed with no consideration of possible use for the delivery of education). Another major development in Illinois is URBANA's PLATO[21], a computer assisted instruction system with the potential for several thousand terminals, which could be used effectively for extension activities also.

Correspondence Courses. In 1971 the URBANA program of correspondence courses had an enrollment of 5378 students or 1633 full-time equivalents. The enrollment has been decreasing, probably because of the increased number of extra-mural offerings of both junior and senior institutions in the state.[18] Studies indicate that students prefer personal contact with their instructor.

Short Courses and Special Offerings. Finally, an educational mechanism which plays an important regional role is that of specialist courses and conferences, usually offered at the university, often making use of invited experts, and catering to a very special clientele. For instance, in 1971-72, IIT offered five special programs in response to the expressed needs of business and industry in the Greater Chicago Area. The attendance was between 300 and 400.

At URBANA, special offerings have ranged from a two-week County Law Enforcement Training course, to a one-week course in Airport Runway Paving Design, a two-week course to train railroad inspectors, and to courses on computer applications. In 1971, 175 short courses were offered and attended by 18,211 persons. They covered business, ecology, engineering, education, labor-management, liberal arts and sciences. In engineering, 13 courses were given and attended by 2275 people. Depending on its nature, a conference can be supported by groups or agencies from local, national or even foreign areas.

At UICC examples of special courses are given in the Digest of Service Projects[22] prepared by Dekker and Wilson.

Agricultural extension advisors constitute a mechanism that has exerted a major influence on agriculture in Illinois. The advisors operate out of the Cooperative Extension Service at URBANA and provide education through field work throughout the State. This activity is a possible model for achieving greater regional involvement of engineering colleges. However, the parallel probably only exists for very small business operations and even then only to a limited extent as the techniques for assistance are quite different for industry and agriculture.

Faculty Activities. The 1969 National Science Board study on graduate education[23], based on the result of a major survey *in all fields*, concluded that the time distribution of the average graduate faculty member is as follows:

45% instruction
23% research
14% administrative
6% consulting
4% public service
7% other

(figures do not add to 100 because of rounding)

This distribution varies considerably with time, depending on the timetable of instruction. The instructional load, for example, may increase to 50% at the beginning and end of a semester and fall to less than 40% during the semester.[24]

A survey of 65 engineering faculty members at IIT* shows that the average number of days spent consulting per year is 28.4 for professors, 29.5 for associate professors and 15.4 for assistant professors. About 91% of the consulting (23 days per year) was for private clients, and 9% (2 days per year) for government. Most of the governmental consulting is done at the national level through connections at the National Science Foundation, the National Institutes of

*Thomas Knowles, "A Faculty Activities Model," Appendix I-7.

Table 14 Average Days Consulting/Year/Faculty Member — 1969-1971.*

SCHOOL	TYPE	CHICAGO	ILLINOIS	NATIONAL	TOTAL
	Private	0.9	4.1	0.2	5.2
BRADLEY	Govt.	0.0	0.4	0.8	1.2
	Total	0.9	4.5	1.1	6.4
	Private	14.0	0.8	8.7	23.4
IIT	Govt.	0.1	0.9	1.3	2.3
	Total	14.1	1.7	10.0	25.8
	Private	5.8	1.7	1.0	8.6
NU	Govt.	1.7	2.7	2.0	6.4
	Total	7.6	4.4	3.0	15.0
	Private	0.0	0.9	0.4	1.3
SIUC	Govt.	0.2	0.2	2.7	3.0
	Total	0.2	1.1	3.1	4.3
	Private	3.4	0.4	0.8	4.6
UICC	Govt.	2.2	0.2	2.5	5.0
	Total	5.6	0.6	3.3	9.6
	Private	5.7	1.5	2.1	9.3
COMPOSITE	Govt.	1.3.	1.4	1.9	4.6
	Total	7.0	2.8	4.1	13.9
	Govt.				3.6
URBANA	Other	—	—	—	9.2**
	Total				12.8

*Thomas W. Knowles, "Extra-University Activities of Engineering Faculties," Appendix I-8.
**Industrial 7.2 Professional 0.6 Other I.4

Health and the Department of Health, Education and Welfare, even though Chicago is so close. About 54% of all the consulting is done for clients in Chicago, 7% in Illinois and 39% at the national level. On the average, professors gave 3.8, associate professors 2.2 and assistant professors 0.9 lectures to groups outside of IIT. Most of these are given to other universities and professional societies, with almost none to local firms.

The summary results for the other universities are given in Table 14.* The table reveals two interesting results:

1. Though there is considerable variability from one institution to another, most faculty consulting is done for private business.
2. Roughly twice as much consulting is done at the local level as at the national level.

At IIT the distribution of faculty by rank is roughly in accord with the national average as shown in Table 15. However, its faculty does consulting in excess of the national average (roughly 10% as against 6% nationally), with almost 40% at the national level.

Service and Research

Supply of Human Resources. As indicated by the budget, the use of facilities and by the allocation of faculty time, the major service of an engineering college to the community is the supply of human resources. To understand the impact on the community that arises from this activity, one must study the origins and destinations of the human resource material; this is best done against a backdrop of the national employment picture.

Focusing on the situation of the last few years, the demand for engineers and scientists changed from 1968, when it was partially unfulfilled, to 1972 when the concentration on urban and social problems and the highlighting of the threats inherent in pollution engendered a falling popularity for engineers and scientists. This, coupled with a shift in the national economy resulting from a retreat from a period of war and the winding down of the space program, has exerted a profound effect on engineering employment. In 1967, 235,000 scientists and engineers were employed in the aerospace industry; a large decrease of some 35% was estimated by 1972.** Of the 90,000

*(For methodological reasons the data for URBANA are arranged in a different manner from those for the other institutions.)

**Betty M. Vetter, "The Changing Demand for Scientists and Engineers," Appendix II-7.

Table 15 Comparison cf Distribution of IIT Faculty by Rank (Average over 1965-1971) with National Average.

	PROFESSORS	ASSOCIATE PROFESSORS	ASSISTANT PROFESSORS
IIT*	42.4	30.5	27.1
NATIONAL**	44.0	26.0	22.0

*Knowles, Appendix, I-8.
***Graduate Education,* National Science Board, Washington, D. C.: National Science Foundation, 1969.

graduates produced in 1969, 40% were employed by aerospace or defense contractors. Since then the Federal budget has been directed increasingly to the areas of welfare and education. An analysis of the unemployment picture in 1971 showed that about 3% of all engineers were unemployed in June 1971, with an additional 1.7% unsuccessfully seeking engineering employment.* By mid 1973, however, the unemployment rates were again much lower and by mid 1974 it had been completely reversed. It is significant to note that even the high 1971 overall rate of unemployment, albeit traumatic in terms of past levels, was lower than for many other careers.[25] Also the older engineers suffer more than the younger ones in unemployment situations.

This highlights the role that engineering colleges can play in updating practicing engineers to avoid obsolescence. Vetter predicted that in the latter half of the seventies the surplus of scientists and engineers will have disappeared, but that the shortages will not be serious.* On the other hand, Engineering Manpower Commission reports predict that the shortage may be more serious than was earlier thought. These different predictions serve only to highlight the very imperfect state of the art in manpower forecasting, particularly in so variable a sector as technology.

Turning to Illinois, a state with a broad base industry, the reflection of the national trends on the local situation is evident. In 1960 almost two-thirds of engineering graduates from URBANA left the state,

*Vetter, Appendix II-7.

many to take jobs with defense and aerospace industries; by 1971 two-thirds were remaining within the State.* SIUC estimates that about 90% of their engineering graduates will be recruited in a multi-state area within a 200 mile circle centered on Carbondale, as well as in the remainder of Illinois.**

The loss of graduates has long been of concern to Illinois. The State Science Advisory Council was empowered to seek ways to encourage an increased scientific climate within Illinois. A survey conducted in 1970 when about 50% of the graduates were leaving the State, determined that the main reason given was the lack of employment opportunities in the State. (This was based on a 30% response.***)

A detailed investigation in conjunction with the development of a State manpower and educational forecasting system for Illinois† shows that while the growth of employment in Illinois is slower than the national trend, a shortage of engineers is likely to occur in the State in the 1975-79 period.

It should be emphasized at this point that in looking at employment opportunities in a state, it is important to analyze carefully the factors that play a role in the problem of shortage. For instance, will the need for engineers in the state be greater than the supply from the schools in the state? If so, that is a "local shortage" which might be made up by "imports" from elsewhere. However, if there is a general U.S. shortage of engineers, then the state shortage could be due to a local shortage, compounded by the deficiency of attracting imports—or simply due to the normal "local surplus" being lured away.

An analysis of employment opportunities for engineers displaced from the defense industry in Illinois, of the current demand for new graduates, and of the demand specifically for environmental engineers** shows that the rapid growth of government employment of environmental engineers in 1971/72 will level off and that private employers do not expect currently to hire many environmental engineers. Babcock et al† believe that short-term forecasting can be developed from a system of interviews similar to that described. Further issues include the acceptability of retraining and substitution and the

*Haga, Appendix I-3.
**Jefferson, Appendix I-16.
***Eric S. Whitman, "Environmental Factors Affecting Technical Employment," Appendix II-6.
†Michael Babcock et al, "Supply and Demand for Engineers in Illinois: 1975 and 1980," Appendix I-10.

feasibility of short-term forecasting based on a system of interviews.

An interesting view of human resources given by Hudson* considers the "human capital"—the capital embodied in a person and related to his or her economic value to the region. The graduate's influence in attracting resources, both human and financial, from outside the local region, provided of course he himself does not emigrate, is also a significant regional asset. In this respect the university in educating engineers acts as a triggering mechanism for the flow of resources.

Supply of Information.

Information Levels. In examining the role of a university or of one of its units, such as an engineering college, with regard to the supply of information, it is necessary at the outset to clarify the nature of the information pool depicted in Fig. 1 of Chapter 1 and of the interactions that occur with it. The information of use to professionals and decision makers can be categorized for our purpose in three levels:

1. The simplest one comprises that which is readily available, for example, in handbooks or textbooks. The quantity of information in this form is vast and often occurs in multiple-volume form. Included in this group are mathematical tables, demographic data, material properties, physical constants, and so on. A professional faculty, and students to a variable extent, know where this kind of information can be found and in most cases have several volumes pertinent to their area of interest available in their offices.

2. Another level of information arises from the application of relationships, laws and methodologies which have been discovered and successfully used over the years; for instance the fundamental concepts of science are given in this form. This type of information is found in professional publications, from monographs to the literature published in specialized journals, and is also embedded in the accumulated experience of years of practice and investigation. The faculty know where this information can be found and possess the skills needed to manipulate it to achieve various ends. Furthermore, the familiarity that a faculty member has with their own specialty often enables them to provide "on the spot" assessment of a problem, including order of magnitude calculation in the case of quantitative problems.

*Hudson, Appendix II-5.

Practicing graduates also have the requisite abilities at this level, but frequently they do not have the time to apply their skills fully to an ad hoc problem, as they are involved in other activities. However, their level of skill could be a critical factor in achieving communication between the university and industry.

3. The third level comprises information that is still *terra incognita*—not codified and requiring extraction before it can be used. Professional faculty are by definition equipped to extract such information. In the case of an engineering faculty, for example, two primary methods are available. First, equipment and instrumentation have been developed in many areas, often making it a simple matter to obtain experimentally specific information. Secondly, an engineering faculty, by virtue of its analytical and experimental skills, can plan and execute research to discover new information which may then be added to the existing pool.

Apart from the information that professional faculty impart to students as part of their educational function, the faculty themselves may be considered to be a source of information for the solution of problems.

Problem Levels. Problem solving is an information extraction or manipulation process, which can be categorized as occurring at three levels.

1. Problem Level I (information extraction). The problem merely requires the extraction of information that exists in a well codified form. The source material is mostly in handbooks or other publications, and a competent faculty member has little trouble with this type of problem, provided a reasonably well equipped library is available to him. In this grouping one may include, in the case of engineering, the use of the more common analytical expressions which are easily extracted from textbooks or the literature, aided by the engineer's experience with their application. The process is essentially one of extraction with the minimum of manipulation involved; basically the faculty member performs an information service. Expenditure of resources is small as little time, certainly not more than a day or two, is required. If the amount of data to be tapped is vast, entailing search procedures, there could be more substantial costs.

2. Problem Level II (information manipulation). The problem solver is required not only to extract well codified information or analytical expressions, but also to manipulate and synthetize information.

Source material embraces the professional literature as well as experimental or simulation results that can be obtained, according to well accepted practices. The faculty member transforms the information that he extracts into a form that can be applied to the problem at hand. The time required for such endeavors may be weeks or months and the resources required are usually of the order of thousands of dollars. Most faculty activities in problem solving occur at this level, the level of consulting, though some extend into the next problem level.

3. Problem Level III (information generation). Not only must information be extracted and manipulated but a program of analysis and/or testing and evaluation must be executed for the purpose of adding new knowledge to the information pool. The faculty are therefore treading on new ground. In science and engineering it is this process that is referred to as research and the time required could be months or years. The resources required vary considerably and range from the tens of thousands to the millions of dollars.

The characteristics of these three problem levels are summarized in Table 16, with the understanding that each higher level encompasses the previous levels.

Consulting. Most faculty consulting occurs at problem level II with some at level III. The form and organization vary considerably, from

Table 16 Problem Levels and Associated Resources.

PROBLEM LEVEL	INFORMATION PROCESS	TYPICAL ACTIVITY	APPROXIMATE SOLUTION TIME	APPROXIMATE RESOURCES $
I	Extraction	Information Service	Hours or Days	$0 - 10^3$
II	Manipulation	Consulting	Days, Weeks or Months	$10^3 - 10^4$
III	Generation	Research	Months or Years	$10^4 - 10^7$

the faculty member who makes individual contact with an industry or institution and deals with them on a personal basis, to the activity conducted under the auspices of a permanent organization such as an experiment station.

In engineering, as we have seen earlier, faculty spend on the average, nationally, about 6% of their time doing consulting work. In the main this involves the supply of information at problem level II to local industry, though, as one might expect, eminent faculty are in demand at the national and even international level. We recall that very little private consulting is done for government, particularly for state and local government.

Research. In addition to the consulting activities of individual faculty members, in a university there are several formal mechanisms that have been developed for fostering research, which in the university is usually a highly individual effort and rarely a team project. These mechanisms embrace a wide spectrum of possibilities ranging from administrative entities that fall completely within the university's control to entities that incorporate the flimsiest of links. Some of the most common ones include:

Grants. In general, the university prefers grants that are made in a completely unfettered fashion. The members of the university community decide how the money shall be used and so individual motivations are encouraged. This obviously creates problems in terms of an outside group that wants a response to a specific question, or the solution of a specific problem, and that needs the results by a specified and frequently very short time.

Contracts. Contracts are more problem-directed and are a mechanism for obviating the shortcomings of indirection that grants hold for many outside groups. However, on occasion difficulties arise in complying with certain of the provisions contained in contracts. For example, the proprietary nature of the results obtained can create problems with respect to publication in the literature.

Centers, Institutes and Programs in the University. As with the Center for Urban Studies at UICC, these are often created to cater to a particular problem area. Their effectiveness often depends on the amount of freedom and support granted to the director, chiefly through his or her budget. At Columbia the Institute for the Study of

Science in Human Affairs is far more institutionalized than the Science Policy Program at Cornell, where a more gradual start is being made in this area. The creation of a program is one way of getting started in an area where the future development is uncertain. Such centers or institutes must have whole-hearted cooperation for success. For instance, the Center for the Study of Science in the Future of Human Affairs at the State University of New York (SUNY) at Albany is struggling from lack of support from the legislature.[26]

Experimental Station. This mechanism entails an accumulation of resources—both people and equipment. The orientation is almost completely dependent on the source of funds. Two illustrative examples are furnished by the Station at URBANA which is primarily Federally oriented (although sizeable research activity is State funded) and the Station at Iowa State, which is more locally oriented.

Research Institutes. A number of universities have established separate research institutes, having exclusively a research and development function. The thrust of these activities can be a generalized one, or it can be focused on a particular set of problems. Examples of the former are the Stanford Research Institute (SRI) and the IIT Research Institute (IITRI), and of the latter the Lincoln Lab at MIT. The ties of these institutes with the university vary, but they tend in the main to be rather loose. The operation of an institute can be largely independent of the university with the employment of a large proportion of non-faculty personnel. This makes it easier to engage in classified or proprietary research. A major drawback from the viewpoint of the university is that the institutes tend to be autonomous and, in the measure they are so, fail to make a major contribution to the intellectual or teaching function of a university. Conversely, the expertise of the university staff may not be generally available to the institute.

Cooperation with Government Facilities. This is usually a very loose association, the main connection coming through the part-time employment of faculty and graduate students. For example, at URBANA the U.S. Corps of Engineers established recently a Construction Engineering Research Laboratory; the State of Illinois also established several of its Surveys—the Illinois Water Survey and the Illinois Geological Survey. On occasion, special efforts are made for widening the purpose of the cooperation. An example is the Argonne Laboratory Universities Association, which was set up to promote increasingly close and mutually beneficial relationships between Argonne and the university community.

University Consortia. At times a number of universities combine to share some of their resources in education, research, public service and planning. An example is the Committee on Institutional Cooperation (CIC), a far flung organization which brings together eleven major midwestern universities. Generally, the impact of these consortia on the research activities of their constituents has not been a major one.

RELEVANT EXPERIENCES IN ILLINOIS

Today no effective, formal linkages exist between engineering colleges and state or local governments that can aid the latter in their decision making process. Neither is there at this moment much demand for such linkages from the side of those who could most benefit from them. The situation in Illinois is revealing and not atypical*:

- Dr. Milton Thompson, Director of the Illinois State Museum, who acts as a formal liaison between the Illinois General Assembly and the Illinois Academy of Sciences, reports that only 4 or 5 legislators have requested assistance during the two years he has offered the service of transmitting research requests from legislators to members of the Academy.
- William Rice, Assistant to the President of the University of Illinois, offers assistance to legislators without limiting that assistance to engineering questions. In a year he received a total of only 12-15 requests and he cannot recall whether any of them dealt with engineering.
- Only isolated examples of technical advisory activities, such as those provided by the Library of Congress at the national level, exist at the state level.
- The profitable experience of the Institute for Environmental Quality, which is the only agency that has developed a network of voluntary and paid groups and individuals from engineering faculties to assist in research, consulting and legislative testimony, has demonstrated the importance of utilizing the resources that the colleges possess.

Likewise, City of Chicago Departments and Committees, such as Avi-

*Robert C. Granda, "Influence of Engineering Colleges on State Policy," Appendix I-11.

ation, Building and Zoning, Local Transportation, and Public Works, which are concerned with engineering problems, have had practically no professional contact with engineering colleges, although it was found that areas exist where such liaison could be fruitful.*

For the State of Illinois alone, currently the annual budget amounts to some six billion dollars. During its 1971-72 session the General Assembly of 235 members considered approximately 6000 bills, of which 250 were clearly science-engineering-related; roughly double that number would have benefited from some technological expertise. Thus it becomes sharply evident that there is potentially a high need for information and expertise. Only two members of the 1971 General Assembly, out of a total 113, were engineers. Several groups, such as the Illinois Commission on Atomic Energy and the Data Information Systems Commission, were set up to assist the General Assembly with technical questions. A successful example of interaction is the voluntary cooperation between engineering faculty and the State's Institute for Environmental Quality.**

An examination of the findings of Underwood and Fritschler[27] in their study of cities with populations greater than 25,000 shows that universities are the third most heavily used resource for science and technology advice, ranking after city departmental personnel and private consulting firms. However, more penetrating questions in the same survey seem to moderate this finding.

Meaningful research usually requires considerable skill, time and money. Individual faculty seek the requisite funding from any agencies that they feel are in the market. At IIT, in our survey, 95% of the funded research came from Federal agencies such as NSF, NIH, HEW, with a negligible amount of locally or industry sponsored research.***

In trying to cope with the lack of funds for locally oriented research, the Office of University-Industry Relations in the Engineering Experiment Station at URBANA endeavors to identify areas of common interest to the University and industry.[28] To this end several active programs that have been established, including a Physical Electronics Affiliates Program involving twelve companies, an Electromagnetics Program with four companies and a Fracture Control Program with eight companies. The companies contribute funds to

*Millaney, Appendix I-13.
**Granda, Appendix I-11.
***Knowles, Appendix I-7.

the university in an organized manner in exchange for a relatively continuous scrutiny of on-going, pertinent research. In two years the first program generated $470,000, the second $100,000 and the third $150,000. Other programs are developing. The same office, at URBANA, has been instrumental in setting up 15 research projects with companies such as: Sundstrand Aviation, the Detroit Edison Company, McDonnell Douglas Corporation, Warner Electric Brake and Clutch, Illinois Central Railroad, Decatur Electronics, Magnavox, American Gas Association, Caterpillar Tractor Company, and several others.

Research occurs at a more organized level with the creation of experimental stations. The arrangement at the University of Wisconsin is described in the present study.* There the Engineering Experimental Station is an integral part of the College of Engineering. Hence, the research conducted is intimately tied to on-going graduate education. The Station is viewed as an organized access to the University; it has its own funds, some of which are directed toward research specifically aimed at the needs of the State of Wisconsin.

The original idea of the Experiment Station at URBANA, from its inception in 1903, was to serve the needs and interests of the State in similar fashion to the Agricultural Experiment Station.** The Station changed from a State-oriented resource, prior to the Second World War, with a large percentage of its support immediately before the war coming from private industry, to a national resource after the war, with mostly Federal funding. The amount of private support remained about the same but became 5-10% of the total as the research efforts expanded. One aspect of healthy growth is attributed to the management of indirect cost funds during almost the entire history of the Station. Departments contributed to and received indirect costs enabling them to initiate new programs. This source of funding has reached more than one million dollars per year.In the early sixties there began a substantial inquiry into the role of engineering colleges in regional economic development. In 1964 the Station created the Office of University-Industry Relations mentioned above. The Industrial Affiliates Program is a mechanism, under development, to meet the regional need.

*Camden A. Coberly et al, "Public Service at the University of Wisconsin," Appendix II-8.

**John J. Desmond, "The Engineering Experiment Station at the University of Illinois," Appendix I-14.

The IIT Research Institute* was founded in 1936 to provide research, development and design services to industry and government. It is affiliated to IIT but exists as a separate corporation. During the early years almost all the research was contracted by industrial firms and trade associations. During World War II the emphasis began to shift to government contracts with the military. As a non-profit organization the Research Institute is able to approach each problem without bias, and the results are held in strict confidence.

In the early days the staff of the IIT Research Institute were part-time researchers, with teaching their main activity. Today IIT faculty participate at a much lower rate, and the Research Institute has a large full-time staff of researchers primarily concerned with applied research.

*Thomas Knowles, "Illinois Institute of Technology Research Institute," Appendix I-15.

CHAPTER 3

Toward a Greater Regional Involvement: Obstacles and Opportunities

The previous chapters have indicated on the one hand the dimensions of the university as a resource and the nature of the outside problems to which it is called to respond, and on the other hand the current lack of effective linkages between the university and its local and regional envrionment. In this chapter we address ourselves to the nature of the obstacles that need to be overcome if such linkages are to be developed and to the unfilled opportunities for a greater regional involvement of the engineering colleges. We shall examine the problem by continuing to think of the university in terms of human and information resources.

HUMAN RESOURCE DEVELOPMENT

A college's educational function occurs today chiefly at the undergraduate and graduate levels. What may be called peripheral levels occur in continuing education, cooperative education and field studies, as well as in any orientation or remedial activity that occurs prior to entering the normal undergraduate program.

Undergraduate Education

Several top level personnel in the City of Chicago have given their views as to what engineering curricula should look like to satisfy the needs of the "city."* Suggestions include a strong accent on management and economics, as well as an understanding of how a city functions and how industry works, all areas mostly neglected in the usual engineering programs. In general, the view comes across, and the same view is often expressed by industry, that graduates from a university should be of more immediate use to the employer. The discussion given in Workshop IV** recognizes this in stating that "market realities require that the undergraduate engineer have a specialized background and training to get a job." However, it is also concluded that premature specialization should be avoided, reiterating the importance of stressing fundamentals.

Considering the lack of success of retraining programs, with particular reference to the transfer of aerospace engineers into new areas, Vetter***, the Executive Director of the Scientific Manpower Commission, concludes that "I cannot emphasize too strongly that where educational institutions have done their job well in the first place in educating young scientists and engineers, there is very little need to bring these people back to school to teach them how to apply their teaching in a new area." Any new employment situation, whether it be for a new graduate or an experienced engineer, must involve an on-the-job training period.† Educators understand this, but industry and the public sector at times do not. However, it must also be acknowledged that situations do arise, even with the most experienced engineer, in which he may encounter some process requiring a more formal approach than can be given by on-the-job training.

Engineering programs have been sensitive to changing demands; in particular in the last decade there has been considerable effort to broaden them by including more subjects in the humanities and social science areas. In this way students are more likely to attain a better awareness of the importance of their social environment. It is by no means a foregone conclusion that this happens automatically through the taking of humanities and social science courses. Poor teaching does, in some cases, merely produce a group of students disenchanted with the area of study. The remark has been made†† that engineering programs at the undergraduate level should be made more acces-

*Millaney, Appendix I-13.
**Report on Workshop IV, Appendix II-15.
***Vetter, Appendix II-7.
†Report on Workshop IV, Appendix II-15.
††Report on Workshop I, Appendix II-12.

sible to the undergraduate student in the social sciences, so that they in their turn can be made aware of the technological approach to problems. This kind of effort is seen as a necessary spur in improving communications between different sectors. Difficulties encountered in this area range from the quest of acceptance by faculty members in their own disciplines, which militates against venturing in the *terra incognita* of another discipline, to traditional curricula strongly centered on disciplines and leaving little room for interfacial concerns.

Given these difficulties in cross-disciplines communication that must be overcome for an effective involvement of engineers in a broad spectrum of social problems, the question legitimately arises: Should indeed engineers be involved at all? A convincing argument for involvement has been given by Gregg.*: "Urban problams have to be attacked by some one. Should it be the social and behavioral scientist or the engineer? Should an effort be made to develop broader quantitative skills and methodology in the social sciences or should engineering with an already sophisticated quantitative orientation be encouraged to move in on social problems?" Gregg argues that "if there is anything all social problems have in common, it is that they have many variables, and engineering schools are the only ones that are producing students, consistently and in high volume, that are not afraid of five or more variables. In fact, their interest is proportional to the number of variables involved in a problem."

Another significant conclusion is that "The whole question boiled down to where can one get the kind of manpower output that has the greater likelihood of looking at the number of variables that are involved in many of the social problems of today. In fact, the number of variables may indeed be the reason why the problems evolved in the first place. City and regional governments lack the competence to handle these complex matters in any way except for one or two variables at a time."**

Undegraduate programs are influenced to a large extent by accreditation procedures. If faculty awareness of, and interest in, the needs of the public sector increase, then doubtless this will also have some influence on undergraduate programs.** Engineering design is expected to have an important role in most undergraduate programs, providing, as it does, a sound problem-solving orientation. However,

*Report on Workshop I, Appendix II-12.
**Ibid.
***Ibid.

most faculty would concur with the views expressed by Vetter*, favoring a broad presentation and development of basic fundamentals as the core of any engineering program.

Graduate Education

The report of Committee V of the Illinois State Board of Higher Education[8] states that "the purpose of a good graduate program is to develop great flexibility of mind and a broad background; consequently, when supply temporarily exceeds demands in some specialty, the loss is not as great as it first seems." The same report acknowledges that in the foreseeable future there is likely to be less need for aerospace and defense related activities but a greater demand for talent in solving problems in the areas of ecology, housing, urban renewal, transportation, and health care. In this regard Vetter identifies what appears to be the key issue:

"It is vital that we differentiate between 'need' and 'demand.' We need all the trained specialists we can get just to clean up the environment, produce abundant power, reconstitute our molding cities, and a host of other desirable things. But until the decision is made at all levels to budget one or more of these activities at the level needed to solve the problem, we have not translated our 'need' for specialists into 'demand,' which means jobs."

As we have seen, a comparison of the engineering graduate output in Illinois with national levels shows that the State is underrepresented on a per capita basis in faculty, degrees granted below the Ph.D. level, and Federal support of research and development per Ph.D. in science and engineering.

In the regional context the question is: To what extent should indeed a state pay heed to imbalances of this kind? How should these imbalances be examined? What effect do they have on the economy of the state? (For instance, comparisons between this country and the USSR show that in the latter there is a much larger per capita number of qualified engineers in spite of the fact that the GNP is lower.[29 30]) The problem is an extremely complex one and still far from being solved at the national level, where the question is intrinsically simpler, let alone at the regional level. At the national level one deals with aggregates, and the boundaries in terms of manpower

*Vetter, Appendix II-7.

movement are far more rigid and meaningful than those for states or local regions.

At the local level, the statistics for engineering per capita degrees and supporting R & D funds are clearly affected by the interplay of two factors:

1. The extent of the manufacturing and production capabilities in the region.
2. The extent of service activities in the region.

In many cases the former activities could be carried out somewhere else in the nation, while the latter inevitably have a strong local component that must be carried out *in situ.* The question concerning production is one of the decision of a given region to attract or not to attract industry and of the relation of engineering manpower to industrial development. According to the outcome of these decisions and issues, the engineering manpower involved in industry in a region may fluctuate widely on a per capita basis.

On the other hand, in the case of services one can expect the engineering manpower to be more closely related to population. Particularly, as services come to represent an increasing share of our economy and at the same time are characterized by increasing technological complexity, they require more technological manpower. Finally, one must also consider the coupling between production and services. The growth of industry increases the growth of services and population—and vice versa.

Other Education

With regard to extra-mural classes, either viewed as continuing education or for degree granting purposes, it is difficult from the available information to assess the demand in this area. It is interesting, however, to observe from Casper's findings[19], mentioned in the previous chapter, that the distance people travel to such classes remains about 8 to 10 miles almost regardless of where the course is offered. One can infer from this that courses should be offered at as many locations as possible in the state, as people are not prepared to travel very far from their homes or places of work for such purposes. Such a conclusion would not be unexpected, but a more detailed study is in order, as available forms of transport, distribution of industry and the demography of the state must surely be important. The use of telecommunication devices also offers extensive possibilities. In general, the principle of reaching out into the community is important, though again a substantial commitment to the idea is required at several levels.

While students are in the process of receiving their formal education at the university, one area where they have an impact on the surrounding community is through cooperative education programs. There is little doubt that cooperative education can be advantageous, both to the student and to the participating industry, if it is well organized. Students gain experience, motivation and a feeling for reality. The industry can foster its connection with the university and with individuals who may have considerable worth to them in the future. However, it is not a foregone conclusion that a cooperative program is necessarily successful.

Before students enroll, the university can reach out to them in several ways, as mentioned in the previous chapter. This kind of contact is usually quite beneficial. In fact, poor high school counseling is blamed in part for the conclusion that engineering students are the least creative.* If true, improved communication between the high schools and the engineering colleges is needed to rectify this.

Both Haga** and Trow[13] conclude that engineering students tend to come from the lower economic strata. This fact offers an opportunity for the colleges of the central city, such as IIT and UICC in Chicago, provided means can be evolved for assisting bright students handicapped by their poor schooling to attain entrance requirements and to overcome the difficulties of the rigorous engineering curriculum. Examples of mechanisms to attack this problem are the Educational Assistance Program and the Minority Tutoring Laboratory at UICC.

The Educational Assistance Program is a campus wide endeavor that provides a spectrum of essential services, ranging from orientation to university life to elementary courses and tutoring, for bridging the gap between inadequate schooling and the demands of various curricula. Typically, in the Fall Quarter of 1972, it assisted over 1000 students.

Establishment of the Minority Tutoring Laboratory in 1971 was prompted specifically by the problem of providing a mechanism for encouraging a much larger number of minority students into en-

*Haga, Appendix I-3.
**Ibid.

gineering; it subsequently evolved into a mechanism for assistance to all engineering students with handicapped backgrounds or having difficulty with the curriculum. The Tutoring Laboratory offers orientation, tutoring, encouragement, and (a very important factor) identity to the students. The salient feature of its success (which in two years led to an almost fourfold increase in minority students in the College of Engineering, from approximately fifty to almost 200) is the self-help feature, whereby a student helped by the Tutoring Laboratory "repays" for the assistance by committing himself or herself in turn to work for the Laboratory as tutor, advisor or recruiter.

INFORMATION RESOURCE DEVELOPMENT

The discussion of bottlenecks and opportunities in the development of an engineering college as a regional information resource is conveniently carried out according to the problem levels introduced in the previous chapter.

Information Extraction (Problem Level I)

Data on the dimensions and effectiveness of this level of operation are virtually impossible to obtain. Most faculty in an engineering college will report that they receive simple questions from time to time, from a variety of sources. The person initiating the question will often make many phone calls before the right person is located, and in most cases, if the question is straight-forward, the entire contact may only last a few minutes. If an enquirer is sufficiently interested, he will persist until the right person is located. There are, usually, no formal procedures for handling such requests. Nor does it appear at first sight that any special effort is needed to facilitate these contacts. However, in general this is a process that discourages many people that do require assistance. The barrier presented by the lack of a readily visible and accessible channel between a potential user and a faculty member has the positive effect of sheltering the faculty from what could become a barrage of time consuming inconsequential questions. Still, there is clearly a need and an opportunity to open the engineering college, in a controlled way, to the citizens of a community, thereby making a substantial contribution to the welfare of the community. A possible approach is discussed in the next chapter.

Consulting (Problem Level II)

On the average, engineering faculty members spend 6% of their time consulting. Universities usually encourage this kind of activity provided it is carried out in moderation. Not only does it keep the faculty in contact with the real world, but it has the incentive of increased remuneration. Most problems at this level, which originate in the community, are handled on a consulting basis, but not all. Simple research efforts that call for moderate use of resources are often carried out as undergraduate and sometimes as graduate projects, as part of the degree requirements of a college, with the faculty member often contributing time, effort and ideas without recompense.

As consultants it might be felt that the engineering faculty are in competition with engineering consulting firms that obtain their livelihood this way. However, the extent of the two operations is usually quite different. The faculty certainly do not possess the resources of staff or time to work on the extensive applied projects that typify the activities of consulting firms.

Problem Level II offers the university a good opportunity for public service, as many problems encountered by industry and public officials are of this type. While consulting with industry and other private activities is a well established activity of faculty members, interaction with the public sector, specifically at the regional and local levels, is at present very limited.* It is the expressed opinion of both state and local governments[2], that the major obstacle to such interchange is the "lack of information exchange mechanism between levels of government regarding scientific advice and application." Two important contributing factors are:

- The lack of awareness by local governments of the scientific and technical resources within the colleges and universities. There is some feeling that academic institutions seem immobilized in this respect. Cheston** expresses the view, in the present study, that it is up to the universities to develop "agencies for cooperation."
- The lack of arrangements in the universities for quick response that would facilitate short term consultation when university facilities are involved and when the usual private consulting arrangements with a faculty member are not sufficient. The problem is how to streamline for this purpose the machinery for the processing of

*Granda, Appendix, I-11.
**Cheston, Appendix II-3.

grants. The solution may involve standing contracts or agreements for the use of facilities on a short range basis without ad hoc clearance. The agreement in the university may be either with a local government, and for that matter any other interested user of university consulting, or with a faculty member.

Research (Problem Level III)

Universities are in the main committed to a spirit of open inquiry; thus it is not surprising that basic research finds a natural place in the academic community. As stated by Coberley*, "[The university] is the main place where people can presumably sit back without too much vested interest in any particular proposition and hopefully evolve concepts that will impact on society years hence." The engineering college performs a positive service through this generation of new information, apart from its irreplaceable contribution to the educational process at both the undergraduate and graduate levels. Again referring to the NSF National Science Board review[23], the faculty, on the average, devote about 23% of their time to research. The question to be answered is whether or not it is reasonable to suppose that this activity, absorbing as it does considerable resources in money, skill and time, can play a greater role in the public service.

The first point to recognize is that universities seldom undertake research where the sponsors are not willing to tolerate the deliberate pace at which it can be executed.** After all it is only a part-time activity for the faculty and becomes full-time for graduate students during limited periods. But there are further obstacles than this. Most research projects that may be expected to emanate from the public sector have a strong interdisciplinary flavor. Bugliarello*** comments in the present study that if an engineer becomes involved in problems that have a strong social interface, he may not know where to publish his findings. However, publications play a leading role in the judging of faculty performance for advancement. As a result, the peer group tends to take a somewhat quizzical view of interdisciplinary activity.

An additional obstacle to research sponsored by both the private and public sectors is often to be found in university patent policies[31] and in publication policies.

*Coberley et al, Appendix II-8.

**Ibid.

***Robert G. Granda and Samuel Gove, "Report on Workshop II—Methodology for Stimulating the Regional Role of an Engineering College: Political Mechanisms." Appendix II-13.

A university patent policy has the aim to provide equitable compensation for the investment that a university and its faculty members have made in the development of a patentable invention. The interpretation of what constitutes an equitable arrangement varies from university to university. One of the major problems concerns the interest of outside sponsors of research. Most government agencies have their own patent policies, which generally cannot be altered by the university. On the other hand, the university has latitude in dealing with industries and other entities which have not formulated such policies with respect to the university.

The problem is a subtle one, as a patent policy must provide a compromise between protection and maximum profit for the university on the one hand, and the inducement of the outside entity to support research at the university on the other. An unduly restrictive patent policy may force the university to forego a considerable amount of cooperation with and support from outside entities, while an excessively permissive policy clearly can impair the ability of the university (and of the taxpayers in the case of public universities) to properly benefit from a university invention.

The same kind of consideration obtains in the case of publications arising from research. A number of both public and private bodies, for a variety of reasons, need to protect the diffusion—or the timing of the diffusion—of research results. For instance, studies of land value and forecasts of land occupancy that are needed for the establishment of a rapid transit system, if made public prematurely, can lead to real estate speculation. What should the university with its policy of open disclosure of research results do? Not engage in the research—and thus fail to perform a public service? Clearly, the issue is a critical one to the ability of the university to provide assistance to public needs.

In brief, at the research level the resource in terms of time and skill is there, but it is clear that its direction—its involvement in direct public service—is strongly influenced by the available sources of sponsorship, by prevailing faculty opinion, and by legal considerations concerning the disposition of the outcome of the research.

CHAPTER 4

Toward a Greater Regional Involvement: Mechanisms

The problem, as already developed in the previous chapters, is quite clear. Beginning with an assessment of the resources available to the engineering colleges, one must answer the question:

• Are these resources providing services to the community in the best possible way?

Furthermore, if there are areas where improved service can be fostered:

• Are there additional mechanisms that should be introduced for this purpose?

In this regard, a university may find itself in one of three possible states:

• The potential for service is well satisfied, primarily due to the location of a college and the needs of its community, whether by accident or design. In these cases the rightness of the situation has led to mutual growth. Examples are provided by the Aeronautics and Astronautics Department at Stanford and its relationship with the NASA Ames Laboratory, and by the strong public orientation of the University of Wisconsin, which is attributed in part to the fact that the University is located in the State capital.*

• The mechanisms and the potential for service exist but are not being fully utilized due to many of the obstacles discussed earlier.

• New mechanisms need to be created to permit existing potentials for service to be satisfied. Hudson**, in this study, speaks of trigger-

*Report on Workshop I, Appendix II-12.
**Hudson, Appendix II-5.

ing mechanisms where one may assume that everything points to a potentially strong interaction, which merely needs a trigger to activate it.

If we focus on the last two states, it is convenient to center our consideration of the mechanisms that need to be activated or developed for greater regional involvement in terms of the two components of the problems: human resources and information resources.

HUMAN RESOURCES

Undergraduate and Graduate Education

As we have stressed earlier, the major service that a university provides to its environment is the education of students. At the undergraduate level, the formal structure of engineering curricula is generally well established. There is, for instance, fairly substantial uniformity—codified by accreditation requirements—in the length of the curriculum, the amount of mathematics and other basic sciences, the need for humanities, and the disciplinary categorization of specializations, such as civil engineering, mechanical engineering or chemical engineering. However, new thrusts, such as bioengineering or environmental engineering, continuously emerge and become either absorbed in the existing disciplinary categorization, or embodied in new ones—at times also backed by new departments.

These new thrusts represent an important mechanism by which the university can respond to the needs of its environment. Current thinking is that interdisciplinary areas, and especially those with more emphasis on social values, are likely to receive greater accent in the future. However, one can conclude from the discussion in the previous chapter that no matter how pressing an outside need, only when a new area has achieved recognition of what is truly fundamental to its study, has placed such a foundation in well defined terms, and has developed an intellectually satisfying methodology, is it likely to have substantial impact on the undergraduate program. This is certainly one of the major reasons, for instance, why until very recently curricula in areas such as packaging engineering or safety engineering, important as they are to our society, have not acquired academic recognition. If changes do occur it is more likely that they will originate from new graduate programs and from increased faculty involvements in new areas.

Both graduate and undergraduate programs are sensitive to the job market for graduate students in the university and for graduates outside the university. If there are substantial areas of need, these must be translated into demand. Until meaningful commitments in terms of funds are made to specific problem areas, the reservoir of human resources will remain untapped in those areas.

New graduate programs can be created and supported initially through fellowships and traineeships, as well as research projects or endowed chairs; but unless they answer a demand manifested on the job market, the programs will disappear from lack of support. The states and local governments have largely neglected the opportunity, through these mechanisms, to stimulate the development of graduate programs more responsive to their needs. The argument can be extended to undergraduate education. Whether curricula responsive to desiderata such as these will in effect be developed will probably remain open to question, unless local governments take realistic steps in their support. The same can be said of industry.

Other Education

The needs of extension, cooperative or orientation educational programs can be satisfied by existing mechanisms. These mechanisms may be improved by working on the linkages already in existence. The requirements are discussed in detail in the next section. Essentially they include a better two-way understanding and awareness between the engineering colleges and the various community groups which have a need for this extended educational activity. For example, the need for better understanding arises at times in connection with research and independent study courses which require a student to seek cooperation from outside the college. These courses are common and usually encouraged by the faculty, as they tend to be highly motivational because of their accent on "real world" problems. The cooperating industry which talks to the students and supplies them with information, in all likelihood will also receive some sort of final report.

There is a danger that the cooperating institution fails to understand the purpose of such student research and that it confuses the student activity with bona fide university consulting. The university's credibility as a source of information may suffer if this is not realized. In turn, the university may fail to understand the outlook of the outside institution and assume an educational motivation where there is a problem solving goal or a recruiting goal. Very often contacts do result in the educating of students, with great effort and very little return.

INFORMATION RESOURCES

As this study has progressed, certain major issues have surfaced. Important among these is the realization of what comprises an effective communication channel between the university and the institutions in its environment. For example, it is not sufficient to have a telephone with a person at either end. These two persons must also understand each other, desire to talk and feel that some benefit is forthcoming. Also, it is desirable that there be a sense of familiarity gained from previous conversations.

Many people view universities as distant and almost unapproachable ivory towers. The popular caricature of the university professor has done little to dispell these misgivings. The artificial barriers between the university and the community will not be removed, unless the university efforts to be of service to the community are guided by the need to achieve:

- greater visibility
- greater credibility
- greater familiarity

Unless this can be accomplished in each of the areas involved in information transfer, no continuing exchange is to be expected.

The visibility aspect can only be developed by a substantial effort in public relations. Most universities and particularly engineering colleges could well do with an improved and positive press coverage, particularly in today's climate, but this is very difficult because usually faculty do not take on public relations functions. Much of what faculty consider to be important is not newsworthy, while newsworthy things are often trivial to them. Credibility depends on a record of successful performance and can be improved by insuring that positive achievements are adequately aired. Finally, familiarity of the community with the university is not achieved overnight. It depends on the development of all sorts of linkages and joint efforts—from a sustained set of conferences, to constant involvement of the community in advisory committees and in the planning of cooperative activities, to the making available to the community physical facilities and services, ranging from swimming pools, to libraries, engineering testing laboratories, computers, and self-teaching laboratories. Table 17 lists several of these linkages to the community. Few universities, if any, have an involvement with the community that covers all the items in the table and that could not be strengthened.

Of course, many of these activities do not constitute by themselves development of information resources—but are highly desirable catalysts for the development of rapport between the university and the community, without which the effective utilization of the university as an information resource is not likely to occur. The community must feel comfortable that the university is available to them.

Extraction of Codified Information

As already described, problems at this first and simpler level of university involvement in providing the community with information require only access to a reasonable technical library. The problem is mostly one of knowing where to look or whom to ask. Even with problems at the higher levels, the initial contact will need to locate the correct person. This task can be greatly facilitated if the university or the college provides a catalogue of skills potentially useful to the community. If this is done college by college, since the spectrum of expertise varies from one college to another, each college should be aware of other colleges' coverage.

Is it desirable to establish a mechanism to enable the community to abstract information at this rather shallow level? The answer is yes, on two counts: (1) If well done, the community's awareness of the university can be enhanced, and vice versa, and (2) the same mechanism can obviously be used for locating the right person in matters requiring more than the use of well codified information.

Present attitudes toward the establishment of mechanisms of this kind range from the creating of formal information-dispensing structures, to the better use of existing mechanisms. Generally speaking, what is done will depend on the demand that exists or can be created. A single individual at an engineering college, armed with a catalog of skills and some accumulated experience, can easily cope with inquiries at this first level; but a centralized input point is desirable if this approach is to be effective. This centralized point could be simply an individual located in the library—at the very least employed part-time on other matters. His or her actions would be either to locate information in handbooks, on request, or to locate invididuals by using the catalogue of skills. In a sense such a person acts as a central switch-

Table 17 Examples of Mechanisms Giving Greater Visibility to the University in its Community.

- Advisory Committees Involving Community (Industrial, Government, Citizens at Large, etc.)
  - At Course Level
  - At Department Level
  - At College Level
  - At University Level

  (Multiple Advisory Committees from Different Segments of Community)
- Public Availability of:
  - Libraries
  - Self-Teaching Laboratories
  - Computer Center
  - Testing Facilities
  - Recreational Facilities
    - Theaters
    - Tennis Courts
    - Swimming Pools
    - Craft Shops
    - Etc.
- Joint Sponsorship of Events with Various Community Groups
  - Debates
  - Plays
  - Symphony
  - Professional Conferences
  - Sports
  - Etc.
- Other Information Services Made Available to Community
  - Technological Advisory Services
  - Agricultural Extension
  - Etc.
- Community Oriented or Community Sensitive
  - University Publications
  - University Broadcasting
- University Prticipation in
  - Community Publications (e.g. through Columns)
  - Community Broadcasting

board for his own college, calling on other colleges when necessary.

Visibility for such an effort would need to be established. An informational campaign would be useful here. Relevant people in the community, such as librarians and legislative services in state legislatures and city councils, key departments in state and local governments, and community organizations, should be informed (and be placed on a limited mailing list for announcements describing the service). Local journals and newsletters could carry an insert. Furthermore, an entry in the local telephone book under the university's heading, such as "Scientific and Technical Information Service," would help.

The establishment of this central contact point is not a trivial matter as it may have a bearing on several other levels of contact. However, as depicted, it is strictly passive in its role and responds to incoming requests.

Consulting

This second level of information assistance to the community includes not only paid consulting but also the offer of voluntary services in giving advice to state boards, planning groups and so on. Most faculty are active in this type of work and are able as individuals to control the extent of their involvement. This level, as discussed in Chapter 2, mostly involves the manipulation of information, available in the literature or from past experience. The time required is likely to run from a few days to a few weeks.

For this level of public service to be effective, several difficulties must be overcome. The first involves the matter of faculty incentive. There are always some faculty who are only too eager to offer their time and energies where public service is involved. Unfortunately, service of this kind is not always included in the university's evaluation of faculty performance for salary and promotion purposes. Hence, considerable reluctance in getting involved may be encountered. Since the activities envisioned are professional in nature, there would seem to be no valid reason for not acknowledging their worth in a positive manner. Conversely, at times faculty members tend to become so involved in their consulting activities outside of the university as to become divorced from many important normal university functions. Clearly a balance must be struck between these two extremes—a balance that will vary considerably according to institutional objectives.

The second difficulty comes back to the central issue of visibility. Though problems at this level call on relatively small resources in time or money, it is not sufficient to rely merely on an entry in the telephone book to satisfy the casual enquirer. If a positive result is to be achieved, some more direct action needs to be taken in striving for visibility. The contact mechanisms proposed in the previous section for the extraction of codified information can also be effective at this level, but need to be strengthened and complemented by well identified contact points located with potential users—for example, the state legislature. Furthermore, it appears desirable that members of a legislature or a city council, particularly if they are engineers, be informed about the availability of the university mechanism for information transfer.

The level of the operation so far discussed is strictly low-keyed. For the specific case of industrial interactions, a number of models for appropriate mechanisms have been discussed by Roy.[32]

Research

The most substantial generation of new information that takes place within the university falls under the heading of research. It is implicit in the goals of this study that this activity can somehow be directed, particularly in the technological areas, in a way which will be more supportive of the public and private sectors of the community than it is at present. The discussion that follows endeavors to identify both the process and the structure that must be instituted if the university's role is to be broadened in this way.

First, it must be realized that there is a distinct process and a unique set of talents representative of university faculty that are currently not available, in the fullest sense, for the enhancement of progress at the state or national levels. The engineering colleges do not speak with one voice and individual colleges are not heard even when they feel they have something of importance to say. Furthermore, the professional engineering societies are often ill-equipped to deal with the interdisciplinary problems. It must also not be forgotten that the process requires a realization of need by government and industry that would create a demand.

What is this unique capability that emerges from Chapters 1 and 2 of this study? It is a capability with two main facets.

The first arises out of the nature of the educational process itself. Not only does the university extend the high school system, but it supplies, as it were, educated and trained people to the community, and then reaches out in an endeavor to continue their education. Societal survival depends on sound education, and the universities have, over the years, struggled to distill the fundamentals upon which educational disciplines can be built. This factor cannot be underestimated, as it is at the very core of the university idea. Today's campuses provide a broad spectrum of disciplines, and hence talents, sufficient to encompass most demands of society.

The second facet centers around the lack of direct political vested interest in many issues (other than that of its own welfare) which distinguishes the university. University faculty usually do not conduct their affairs with political ends in view. The traditional ideal is one of unfettered and scholarly inquiry and response.

Little extrapolation is required to identify a future direction, commensurate with the university ideal, but responsive to the new generation of problems which beset the highly industrialized nations. Interdisciplinary problems—from bioengineering to urban sciences—do indeed require the attention of the combined talents evident on the university campuses. However, this new generation of problems often still awaits a clear identification of new concepts, fundamental to the creation of a new science. Endeavors directed to these ends still fall under the general rubric of education. However, in this case it is the university community itself which needs to define its own education, so that it may pass it on and thus educate the external community at all levels. This kind of group educational process would seem to be inevitable if new progress is to be made.

It is taken for granted that university faculty be involved in the traditional kinds of teaching exemplified by classroom instruction. It should be no less taken for granted that faculty be associated, through whatever mechanisms exist, be they centers, research institutes and so on, with the new educational process. This involves a change of direction and a different application of resources.

It is in these terms that the educational character of research must be recognized. Accepting that teaching faculty perform research as a part-time activity, as does the skilled labor pool provided by graduate students, research, in most cases, proceeds at a fairly leisurely pace. If the pressing problems of the community are to be addressed in the university community, considerable restructuring may be necessary. Initially this requires a recognition and commitment by the university to reallocate resources to promote the new direction. This may involve a restructuring reflected at all levels, down to the individual faculty member who will be pleased to spend some reasonable portion of his time performing that interdisciplinary research essential to the new education. In time, the new direction would be sure to influence undergraduate curricula too, although the current lack of definition in the area makes it difficult to predict how this direction would reveal itself. But new directions should not be interpreted as implying a cessation of fundamental research within the traditional disciplines. Rather, they inevitably tend to feed new problems to such disciplines and to provide new perspectives.

Key Factors. Mechanisms for making the university's research skills more widely and effectively available to its community have to evolve in terms of specific situations. However, one can identify important factors that must be considered:

1. Any reasonable structure must not only be sensitive to its environment but must also incorporate many options for future development.
2. In terms of the use and development of human resources, the teaching function—more specifically classroom teaching—is a central and continuing activity of engineering colleges.
3. The generation and transfer of information is a function largely of the research activity. Thus far economic factors have restricted university success in this regard to the national level, at the exclusion of the regional level.
4. The universities possess certain unique characteristics not found in other organizations. In the main these grow out of lack of vested interests and out of the broad spectrum of expertise present in the university. As a result they have, inherently, the potential to cut across both disciplinary and political boundaries.
5. Attitudes persist in universities, largely due to existing promotion policies, that tend to militate against any sizeable involvement in public service activities outside of the traditional teaching role—particularly for the younger faculty members. These attitudes are probably more the result of peer-group attitudes than of administration policies. Thus the task of revising them is a complex and slow one.
6. Attitudes exist in public bodies, such as state legislatures, that tend to ignore the institutional potential of the university as a meaningful resource in the formulation of policies.

7. Attitudes have developed in industry that the engineering colleges, apart from teaching and individual consulting connections, are of limited use in a broader sense. An important reason is undoubtedly the fact that university and industry, although in many senses complementary, have different goals and different time horizons for their achievements. But many of the attitudes of industry can also be traced to the well recognized problems inherent in the industry needs for confidentiality and quick response.

8. The structure of engineering colleges continues to be geared in the main to a production-oriented industry and has not yet responded in a commensurate way to the shift to a service economy.

The university appears to be the natural place for addressing the kinds of problems that characterize many of today's most pressing issues. It is the only organization that could command the kind of expertise and intellectual sophistication necessary for the new fundamental areas of inquiry that need to be developed to solve these issues. Some aspects of this theme are further developed by Bugliarello[31] who identifies a new phase of engineering education, the "Socio-Technology Phase," currently being accelerated by the environmental and energy crises and by budgetary pressures.

It is in the university's own best interests to seek for mechanisms to further its involvement in public service. Certainly such involvement should be in the interest both of the public and private sectors if they are to remain responsive to the true needs of a community. Problems such as the utilization of water resources, the prevention of pollution, the development of sound energy policies cut across established political and geographical boundaries, and extend beyond the time cycle of political appointments. The knowledge generated within the university is unfettered in this respect, and thus invaluable for future progress at the local, state or national level.

Desirable Characteristics of Interaction Mechanisms. The nature of any given mechanisms for the interaction of a university or college with its environment is determined, in the final analysis, by the kind of problems that need to be tackled. However, it is possible to identify some general attributes that any such mechanisms must possess if they are to function effectively:

1. Availability of resources. The main university resources are faculty and students, as well as facilities. Typically there are competing claims for these resources, which are released by money.

2. Effectiveness with which interactive links are established between the different groups involved, both within and outside the university.

3. Executive action or authority by which the mechanism operates.

4. Adaptability to changing circumstances. This attribute extends from the need for a built in sensitivity to what is going on in the environment, to an organizational flexibility which incorporates a learning component and permits structural changes.*

5. Acceptability. This essential ingredient depends on various factors, including speed of action, confidentiality of results, visibility, credibility, and a working familiarity with its processes. It should be added that acceptability refers not only to those who are the potential users of the mechanism but also to the participants. It is certainly desirable to foster positive attitudes in all involved.

Proposed Mechanisms. Essentially an interactive mechanism must be recognized as consisting of two parts, a structure which connects its components, and a process which takes place within the structure.

It is remarkable to what extent NASA was able to tap the resources of the universities in implementing the space program. In that case the mission was clearly identified and centrally focused. It was, however, a mission which involved a very broad range of technological and scientific problems, so that a large number of academic efforts in science and technology were relevant to the mission and could easily be directed toward space needs.

The problems with which states and communities are grappling differ from those of the space program in at least two major respects: In the first place they are much more complex, as they intimately involve not only technological considerations but also political and social ones. In the second place, there are no simple focal points for their solution—no agreed upon set of criteria and no authority that can be designated for organizing their solution.

Thus a NASA-type approach, suggested by many, cannot be expected to provide an effective solution unless the conditions that were the premise for its success are also present. The question is a twofold one. In the first place, it asks whether the approach to civilian economy problems such as transportation or criminal justice can be centralized to the high degree implied in a NASA-like structure. In the second place, it asks whether states, cities, local communities can build their own NASA-like authorities, to give them the capability to

*Herman Sievering, "Regional Role: A Simple but Human View." Appendix III-4.

solve these problems at the regional level and to represent effectively the regional viewpoints and interests in an interaction with a national problem-solving effort. Also, the problems surface locally in varying forms and degrees depending on local conditions.

A series of simpler models—most of them visible in current practices—that are, or could be, used by state governments for capturing science and technology for public uses have been identified by the Council of State Governments[1]: consultative, managerial, research, mission-agency, service, and network models. Regardless of whether or not large-scale efforts along these lines are viable or indeed desirable, there is an urgent need to devise mechanisms that can facilitate the interaction of the university with the region. In this section we suggest mechanisms that address themselves to three facets of the question:

- Establishment of a network of university-based State-Federal centers to achieve a more organic interaction between the three entities.
- Establishment of an interface mechanism between the state and the resources in the university, to facilitate both the coordination between state needs and the State-Federal centers, and the identification and assembly of any other pertinent resource in the university.
- Establishment of a more effective university-community interaction through the establishment of Information, Monitoring and Innovation centers.

State-Federal Centers. The first mechanism we suggest envisages a network of resource centers at the state level, attached to the universities, and identified in terms of problem areas. For example a state may have a university-based transportation center, a housing center, a water resource center, an energy center, a human resource center and so on, depending on what is deemed necessary. Adjacent states may have all or none of the same centers. This concept is not new, and many centers already do exist with a strong problem focus—for instance, the federally supported state water resources centers—which merely need to be identified as members of the proposed national network. The Federal government should encourage the systematic creation of such regional centers (i.e., State-Federal centers), and these centers should be integrated in a state, interstate and national network with strong correlating ties.

The essential task of the centers should be the study of regional problems and the development of sound regional policies. Clearly this will require varying degrees of cooperation between centers in a state and often between neighboring states. Ultimately such endeavors will

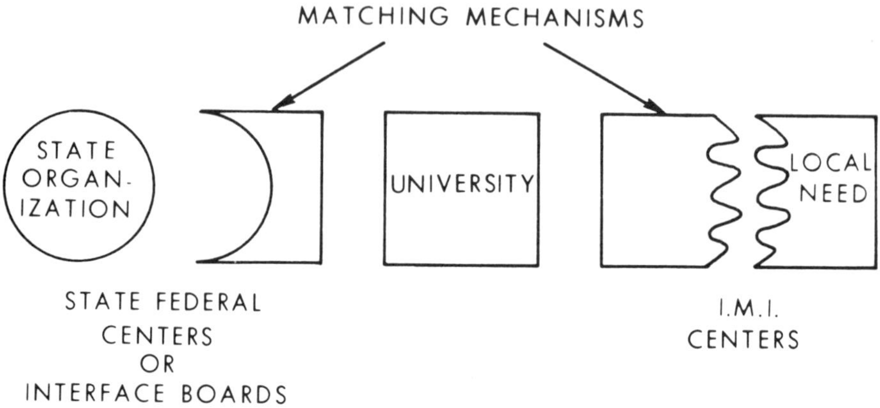

Fig. 9 The matching of the university with state organizations and the local need.*
*Adapted from a suggestion by Merritt A. Williamson, Project Consultant.

provide a vital input to the formulation of national policies.

Interface Boards. The problem of matching the needs of the state with the resources of the university is one of matching entities with very different characteristics with regard to a number of factors, ranging from goals, to typical response times, to executive structure. The nature of the problem is portrayed graphically in Fig. 9, suggested by Williamson.*

To achieve an effective match, coordination at the state level is needed. This can be achieved in a variety of ways, as discussed later in this section. Clearly the centers represent one approach to the problem of how the university can respond to local needs and of how this response can be coordinated with Federal needs. On the other hand, simpler ad hoc arrangements can be devised, if the linkage to Federal policies and needs, and the long range continuity of identified foci in the university and the state, are not overriding considerations in a given state. Under these conditions, for example, the task can be carried out by an "interface board" as proposed by Merritt Williamson and described in detail in Appendix III-5.

The board would consist of persons selected from deans of professional colleges, the directors of government agencies, the state cham-

*Merritt A. Williamson, "An Alternate Interface Mechanism," Appendix III-5.

bers of commerce, state association of manufacturers, and perhaps retired executives with industrial or research experience. Such a knowledgeable group should be able to select a non-political director, who would have a small staff of permanent personnel, supplemented by faculty members from the various universities in the state, on "loan" for limited periods of time. The task of the board would be to reach directly into the universities, identifying and mobilizing the resources that can contribute to a particular problem area or project. The board clearly would increase its effectiveness with time, as it acquires experience with the resources of the universities and the needs of the region. One of the primary ingredients of success of the board would be the intrinsic ability to provide a more rapid response to a need than possibly through the normal academic channels.

Establishment of the board is not mutually exclusive with that of a network of problem-oriented State-Federal centers; in effect their actions could be complementary.

Information, Monitoring and Innovation Centers. As a third major mechanism, complementary to the State-Federal centers, we suggest the establishment of a number of Information, Monitoring and Innovation (IMI) centers at several universities in the state for the purpose of providing a sensitive link with the community, closely attuned to local problems. The centers—which may involve only the colleges of engineering, or be of broader scope—would collect information and store it in an accessible form, monitor technological and other development and community changes, and provide a catalyst for innovation. A view from industry as to the need for innovation is given in Appendix III-6. One of the most important characteristics of the IMI centers should be a rapid time response to problems at the first two levels discussed in Chapter 4. The centers must be interlinked with each other and with the larger resource centers described above.

Innovation in technology is becoming increasingly necessary because of the mounting concerns about the innovative inability of certain segments of our technology in the face of strong competition from other countries. There is also an increasing need for innovation in almost every area of social concern—from health care to education to crime prevention.

Again, the interface between the IMI centers and the community needs to be carefully thought out, using the considerations analogous to those suggested above for the university-state interface.

The need for information, monitoring and innovation clearly extends beyond the technological domain. Thus one can envision the possible establishment of several IMI centers in a university, each geared to a different aspect of community need. Or, from the outset, the first center established should make provisions for accommodating other areas.

The combination of State-Federal centers and Information, Monitoring and Innovation centers would provide a structure highly sensitive to local problems through the IMI centers, but able to perform in-depth research through the State-Federal centers. Science policy at the state or national level stands to benefit from the output of such a network. The concept is illustrated in Fig. 10.

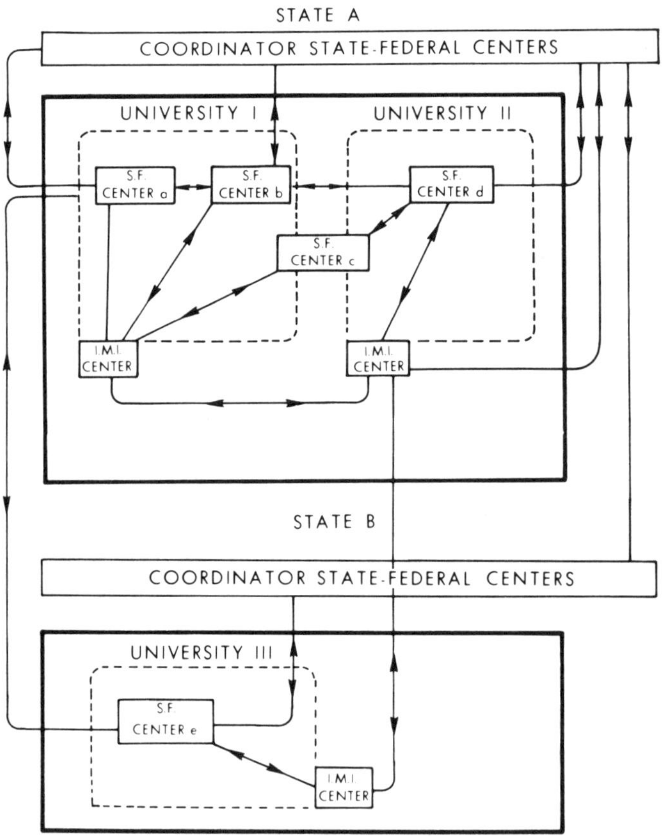

Fig. 10 Linkages among State-Federal (SF) and Information Monitoring and Innovation (IMI) centers.

Criteria for Success.

The structures we have proposed will not function effectively unless the criteria enumerated earlier are built into their operation.

Resources. Engineering colleges have many of the resources required in terms of interdisciplinary expertise, with readily available links to other sectors of the academic community. The money that has been flowing into engineering colleges for research has been mostly in response to Federal needs. In order for the colleges to respond to local needs, appropriate financial support is necessary to make faculty time available. The Federal government has satisfied many of its research needs by using universities, rather than establishing a massive inhouse research establishment. This has provided the Federal government with a great deal of flexibility, as research programs can be initiated without commitment to long-term support and can be allocated among a number of competing institutions. Thus there is no need to establish new groups of employees as each new problem requiring research arises, because the universities, by virtue of their continuing commitment to education, provide a stable faculty resource for this purpose. This arrangement can provide a model for local and state governments, if these levels of government are to benefit from the universities' research capabilities with the same effectiveness as the Federal government. However, some long term commitment is needed by Federal, state and local governments for the establishment of the necessary infra-structures.

The problem is that the continuity of local and state government funding is very uncertain. Probably no state is able today to guarantee even five-year support to the mechanisms proposed in this study, unless such support is written into the budget for higher education (where it will compete with educational programs) or unless incentive funds are provided from Federal sources.

Executive Action and Authority. Universities in general tend to be weak in the area of executive authority and action. Since they are not business organizations, rapid decisions are not easily made; extensive consultation is normal, and the lines of authority are geared to the teaching function. If the universities are to respond more effectively they must evolve administrative structures and procedures that will enable them to act rapidly and assuredly in areas of public service. This cannot be achieved without a strong leadership in the university committed to making public service activities effective.

State-Federal Centers. Though universities have created centers with ease, these centers have not always been popular, as they tend to be administratively awkward; thus they have not always been amply supported. Changes in attitude are necessary if the mechanisms for regional response are to be viable. The director of a center should have a considerable degree of flexibility, so as, for example, to be able to enter into contracts rapidly. As far as possible, the center must be removed from ordinary departmental pressures—but by no means should it compete with the departments. Various types of appointments will be necessary—from the hiring of full-time engineers and other professionals, to joint faculty appointments for varying periods.

To achieve the requisite autonomy, the director of a center should report to as high a level as possible in the university hierarchy, for instance a dean of faculties. There is, however, the danger that this administrative arrangement may create a barrier with the other units of the university. Circumvention of the problem depends on the individual situation and on the persons involved. One possible approach is to aim for a dual structure where, say, the administration of an engineering college and of a center overlap through one or two shared appointments. In any case, furthermore, the director will also have reporting links to the appropriate department in the Federal government, such as the Department of Transportation or the Department of the Interior, which should provide a portion of the center's budget.

At some point the centers must link with the state government apparatus. A prestigious state coordinator should be appointed at the state level to coordinate the activities of all the State-Federal centers. The state coordinator should be a permanent position, linked both to the science advisor for the state (who is normally a political appointee), and to the legislature (possibly through a mechanism such as the Illinois Legislative Council). These links can provide an avenue for authoritative information transfer between the various State-Federal centers and the legislative and executive branches of the state government. The state coordinator should be able to operate with autonomy and should be provided with an independent budget. He should also establish direct linkages with appropriate Federal departments and activities.

The director of a State-University center should be expected to collaborate with the directors of similar centers in the state or in other states, wherever it is pertinent;,similarly, the state coordinator should collaborate wherever appropriate with coordinators in other states. The coordination could be achieved through joint reports and periodic conferences. An overview of the linkages of the State-Federal centers is shown in Fig. 11.

Budget should emanate from the appropriate Federal sources, partly to a State-University center directly, and partly to the state, though clearly labelled for distribution among the centers. As a result, a State-Federal center is forced to think of its relation to state activities. In fact, its yearly report would need to reflect the state of its own problem area in terms of issues in state policy and their relationship to national policies. The report should be directed both to the state and to the Federal government.

Interface Boards. The same considerations about executive action and authority apply to the Interface Boards. The board should have budgetary autonomy, sufficient permanency, and be removed from undue pressures from the political structure of the state and from the academic side.

Information, Monitoring and Innovation Centers. The IMI centers should be offshoots of the engineering college organization, though with a separate budget. Again, coordination among centers, through conferences, newsletters, etc., is essential so that information can be assembled and experiences exchanged. Existing engineering experiment stations or industrial liaison groups already possess many of the attributes required for the IMI centers. They mobilize faculty skills and usually have strong links in the community. They often lack, however, direct experience in entrepreneurship—and so do many of the faculty members—although they possess skills that are indispensable to the innovation process. The IMI center should make a determined effort to assemble staff with entrepreneurial and industrial innovation experience. By so doing it contributes not only to the development of enterprises and the solution of technological problems in the community, but it also enriches the competence of the university. It remains for each university to decide whether the staff of the center would hold long term appointments (an advantage in maintaining continuity but a disadvantage in keeping the staff current in the practice of industry and business), whether to give the staff appointments in academic departments, etc.

In the initial establishment of the IMI centers each university could group various of its existing components together, including certain student activities. A small allotment of Federal or state money could be sufficient to catalyze increased involvement with the local community and with other IMI centers. The IMI centers would be mandated

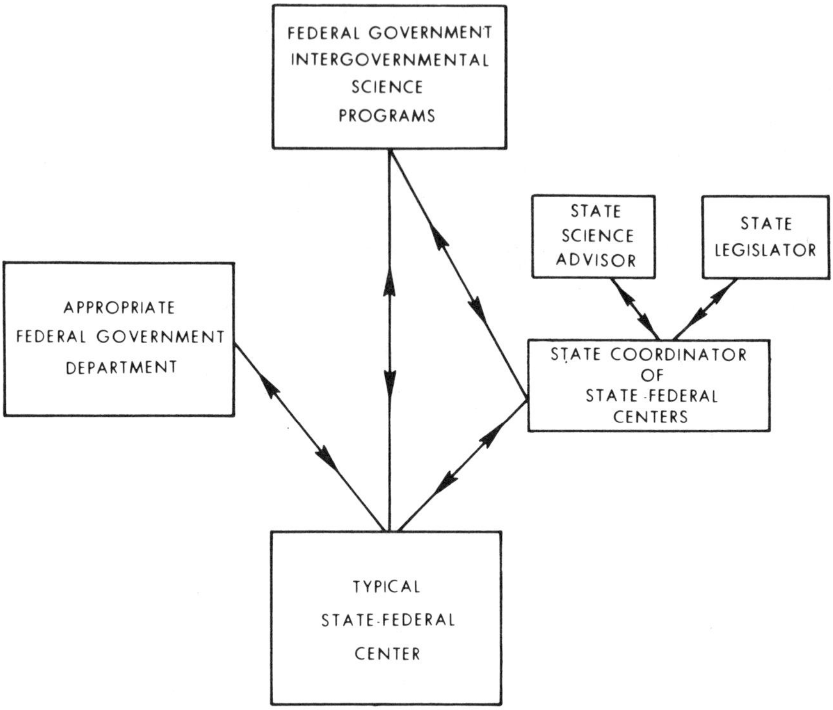

Fig. 11 Linkages and Coordination of State-Federal Centers.

to look at local problems and to assess the extent of assistance that could be provided by the university.

The identification of problems and opportunities is an essential ingredient of the actions of the centers. To achieve this, the center must establish broad contacts and develop appropriate procedures. A simple mechanism for identifying small-scale problems has been proposed by the alumni of the College of Engineering at UICC.*

Adaptability. Inevitably some flexibility is lost as soon as a firm commitment is made to a formal structure. A central requirement of any mechanism for effective university involvement is that sufficient flexibility be retained to make it possible for the mechanism to learn and adapt. Thus, centers should be able to evolve as their roles become

*University of Illinois at Chicago Circle Engineering Alumni Association, "Project Real World," Appendix III-7.

clear. However, their size should be limited in the number of continuing appointments that are made, so that organizational flexibility is retained. The current research activity within a college is a resource that can easily be directed toward particular foci, once the funding is available. To a large extent the mechanisms proposed formalize what already exists. By recognizing a pattern of effort with definite links into state and Federal governments and into the local community, they provide a vehicle for coordination and for a responsible voice of the university on technological matters of importance to the region.

Acceptability. Much of the success of interaction mechanisms depends upon the attitudes of the academic community, of government at all levels, and of the community at large. A concerted effort is required, as well as firm and committed leadership.

The academic community will need to take it for granted that just as teaching and unfettered research are regular on-going activities, so some substantial fraction of time should also be spent in interaction activities with the community. In part this will involve research into fundamental areas, but applied problem focus will clearly carry strong weight. For this to be accepted two immediate changes are necessary. First, the university incentive scheme must take cognizance of participation in interaction activities, through centers, boards, etc. Second, several scholarly journals attuned to the interaction concept will have to be created, with acceptance of publications in interdisciplinary areas as a basic tenet of their operation. The activities of an academic community have considerable inertia but the challenge, if formulated on a nationwide basis, should gradually attract increasing participation.

The attitudes of public and private bodies may also be expected to change if the interaction achieves sufficient visibillty. This is partly a matter of public relations and the repeated use of certain key words in describing the interaction mechanisms, but also depends on a rapid attainment of credibility in the providing of services.

The IMI centers should be the most easily and rapidly accepted. Their function as a sensitive link with the community and as a collector of vital information in accessible form should rapidly prove their worth. Their ability to respond rapidly and to assist segments of the community, which normally, because of limited size (e.g. small companies) or other factors, do not have easy access to expert advice, should quickly develop a clientele. It should be emphasized that the centers should not be in a position of competition with private initiative but should rather concentrate on those services which are not readily available or accessible.

CONCLUSIONS

1. A university constitutes a complex and sophisticated resource for the community in which it is embeded—the city, the state, the region. However, the nature of this resource is understood only in a limited way by the communities. This report focuses primarily on one segment of the university—the school of engineering—for which this issue is particularly urgent if we are to build a system of science and technology responsive to the needs of the states and the communities, just as much as to national needs.

2. Less recognized than the teaching function, but equally important as a resource for a local community, is the information that a university contains or can generate, both institutionally and through the activities and skills of its individual members. Another important resource to the community is represented by the character of the university as an institution which has a peculiar blend of disciplines and attributes and which can play an impartial role in a number of crucial societal issues, such as the establishment of energy policies or the analysis of problems of productivity. Finally, in engineering as well as in other disciplines, the university has a socializing function that may have as far reaching an effect as the teaching and research function.

3. With few exceptions, most noticeably in agriculture, the local communities have been particularly unable to tap the universities' resources, as compared to the success of the Federal government in this respect. The Federal government has been able to utilize effectively universities of great national standing, either institutionally or through their individual faculty members, as pools of resources for a broad spectrum of national needs, from research to advice on foreign policy, from industrial development to arbitration and science policy.

4. There are a number of undesirable consequences of the failure of state and local governments to utilize the science and technology resources of the universities. The most serious consequence has been the inability of the states to develop policies that are based on the kind of sophisticated advice that the Federal government is able to enlist to a considerable measure from the universities. This lack is particularly evident in the areas of technology, environmental and ma-

terial resources, and science in general. The failure has affected both the communities and the universities. The states and other local environments have not benefitted, and the university has found itself operating in isolation from the affairs of the local community. Both have suffered.

5. To solve this impasse, formal mechanisms need to be created to enable the communities—in the first place the states, but also other smaller or larger entities, cities as well as regions—to avail themselves of the resources of the universities. The key to these mechanisms should be the use of the universities as information resource centers in the formulation of local policies and in relating them to Federal policies.

6. The mechanisms should facilitate university action and provide ready access to the university resources. Acceptability in the community should also be fostered, by the careful nurturing of visibility, credibility and familiarity.

7. As a specific mechanism, the formation of university-based State-Federal centers is proposed. These centers, each with a strong problem focus, such as transportation, energy or housing, should be identified as part of broader regional and Federal networks, if already in existence, or created with Federal assistance where pertinent to the entire scheme. The centers will need an overall coordinator to maintain a strong link with the executive and legislative branches of state government. Each center would be responsible to the overall coordinator in the state, and to the Federal agency that relates directly to the particular center. The structure requires inter-center and interstate cooperation.

8. As either an alternative or a complement to university-based State-Federal centers, Interface Boards are also proposed, to facilitate the matching of the different characteristics of universities with the needs of the state and the rest of the community.

9. Linkage between the university and the communities demands, furthermore, the formation of a more capillary network. Information, Monitoring and Innovation (IMI) centers established at a number of universities (or engineering schools) in a state can provide a sensitive interface with community problems locally. Interaction through publications and conferences would serve to knit the network of IMI centers together, as well as to enhance the tie with the State-Federal centers and Interface Boards. Many of the existing institutes and experiment stations can provide a working basis for these centers.

GENERAL REFERENCES

1. The Council of State Governments, *Power to the States,* Lexington, Kentucky, May, 1972.
2. *Public Technology, A Tool for Solving National Problems,* Report of the Committee on Intergovernmental Science Relations to the Federal Council for Science and Technology, Executive Office of the President, May, 1972.
3. Feller, I. et. al., *Developing Science and Technology Capability in State Legislatures,* Institute for Research on Human Resources, Penn State University, January, 1973.
4. Caffrey, J. and H. H. Isaacs, *Estimating the Impact of a College or University on the Local Economy,* American Council on Education, Washington, D. C., 1971.
5. Hartshorne, R., *The Nature of Geography,* Association of American Geographers, Lancaster, Pennsylvania, 1939.
6. Cross, H., *Engineers and Ivory Towers,* 1st. ed., New York: McGraw-Hill Book Co., 1952, p. 67.
7. Johnson, E. L., "Higher Education and National Problem Solving," *A Question of Partnership,* Report and Recommendations of the Conference on Institutions of Higher Education as a Resource in the Solution of National Problems, Washington, D. C., May, 1972.
8. Committee V, "Graduate Education in Engineering," *Master Plan Phase III,* State of Illinois Board of Higher Education, 1970.
9. Jacobson, R. L., "Data Show the Effects of College," *The Chronicle of Higher Education,* April 2, 1973.
10. Keller, J. E., "Higher Education Objectives: Measures of Performance and Effectiveness," Research Program in University Administration, University of California, 1970.
11. Breneman, D. W., "The Stability of Faculty, Input Coefficients in Linear Workload Models of the University of California," Research Program in University Administration, University of California, 1970.
12. Breneman, D. W., "An Economic Theory of Ph.D. Production: The Case at Berkeley," Research Program in University Administration, University of California, June, 1970.
13. Trow, M., "Some Implications of the Social Origins of Engineers," *Scientific Manpower, 1958,* Washington, D. C.: National Science Foundation, 1959.

14. Loeb, J., "Recent Trends in College Retention, Drop, and Graduation Rates,"' Office of School and College Relations, Research Memoranda 72-3 and 72-9, University of Illinois, February and April, 1972.
15. Cohen, I., "Enrollment Policy Formulation," *Proceedings of the 1971 Forum of the Association for Institutional Research.*
16. Hartman, L. M. (Principal Author), *Graduate Education, Parameters for Public Policy,* Report prepared for the National Science Board, NSB 69-2, Washington, D. C.: National Science Foundation, 1969.
17. Bugliarello, G. et al., eds., *Women in Engineering: Bridging the Gap Between Society and Technology,* Proceedings of an Engineering Foundation Conference, July 12-16, 1971, Chicago: University of Illinois at Chicago Circle, 1972.
18. Division of University Extension, "Annual Report for 1970-1971," University of Illinois, 1971, p. 5.
19. Casper, R. F., "Memorandum on Distances Travelled by Students to Extra-mural Classes for Academic Year 1968-1969," Internal Report, Division of University Extension, University of Illinois, 1970.
20. Casper, R. F., "A Study of the Extra-mural Class Program in the Metropolitan Chicago Area," Internal Report, Division of University Extension, University of Illinois, June, 1967.
21. Bitzer, D. L. and R. L. Johnson, "PLATO—A Computer-based System Used in the Engineering of Education," IEEE *Proceedings,* Special issue on Engineering Education, Vol. 59, No. 6, June, 1971, pp. 960-9868.
22. Dekker, T. H. and E. Wilson, "Digest of Community Service Projects," Urban Information Processing System, University of Illinois at Chicago Circle, 1971.
23. *Graduate Education,* National Science Board, Washington, D. C.: National Science Foundation, 1969.
24. Romney, L. C., "Faculty Activity Analysis," T.R. 24, National Center for Higher Education Management Systems at WICHE, December, 1971.
25. Tewkesbury, 'Bright Picture Painted for Young Engineering Graduates,' "The Trend in Engineering," *University of Washington Newsletter,* Vol. 24, No. 3, July, 1972.
26. Lambright, W. H., *Science Policy and the Universities of the Empire State,* Office of Science and Technology Report OST-100, State Education Department, Albany, New York.

27. Underwood, J. E. and A. L. Fritschler, "Science-Technology Advice in Local Governments," *Urban Data Service,* Washington, D.C.: International City Management Association, Vol. 2, No. 11, November, 1971.
28. Krasnow, M. E. and M. K. Blanchard, "Annual Report of Office of University-Industry Relations," University of Illinois at Urbana, 1970-1971.
29. "Stock of Scientific and Engineering Manpower," *Statistical Yearbook 1971,* United Nations Educational, Scientific and Cultural Organization, Paris: 1972, pp. 586, 588.
30. International Monetary Fund and the World Bank Group, *Finance and Development,* Vol. 10, No. 1, March, 1973, pp. 26, 27.
31. Krasnow, M. E. and N. Holonyak, "University Patent Policy," Internal Report, Office of University-Industry Relations, University of Illinois at Urbana, 1969.
32. Roy, R., "University-Industry Interaction Patterns," *Science,* Vol. 178, December, 1972.
33. Bugliarello, G., "Engineering and the New Liberal Education," *Engineering Education,* February, 1973.

APPENDIX I
PROJECT REPORTS

*Study reports presented at the Regional Role of Engineering Colleges Conference.

Impact Identification Model (Boundary Crossing Analysis)

Clarke H. P. Schneider and Joseph L. Schofer

The problem of investigating the impact of an engineering college upon its region may be broken down into several sub-problems. Of these, the most crucial is identifying the possible range of impacts. This is the problem treated briefly here. The strategy which has been adopted is to specify the actors and their actions which constitute an engineering college. Also, regional groups and institutions which might be affected by these activities are categorized. Following this, a three dimensional matrix with actors, activities and affected groups as the axes may be constructed. Each cell of this matrix then serves to represent a rather specific type of potential impact of an engineering college upon its regional environment.

With this as a basis, each cell may be considered with regard to the importance of the particular impact which it represents. Further, study may proceed by investigating the nature of the interaction of a particular cell or of any row, column, or vertical of cells, since each of these vectors represents a set of interactions in which one activity or group is constant throughout.

If this structure adequately represents the process under study, the completeness of effort may be evaluated by observing the proportion of cells which have been studied. Also, this matrix may be utilized as a structural basis for fitting together many of the rather broad range of

Clarke H. P. Schneider is an Assistant Professor in the Center for Urban Studies at the University of Illinois at Chicago Circle.
Joseph L. Schofer is an Associate Professor in the Department of Civil Engineering at Northwestern University in Evanston, Illinois.

activities which have been undertaken in the course of this project. It is necessary to note, however, that this taxonomy is trivial in the sense that the focus of interest is the boundary crossings and not the nature of the processes which have been labelled "a college" and "a region." We proceed by specifying the variables of interest for each dimension of the matrix.

A college is composed of the interaction of students and faculty within the framework provided by the physical facilities of the college. The administration of the college provides the fiscal and managerial control necessary to maintain the satisfactory operation of the college, while the staff maintains and sustains the physical plant and managerial operations. The groups of individuals which are involved in the operation of a college then are the faculty, students, staff, and administration. As noted, the faculty's main preoccupation is ostensibly the instruction of students or, more generally, the transfer of skills and ideas through both teaching and the spread of graduates. Other faculty activities include research, public service, and service to the institution—i.e. administrative and other internal duties, and community service. The students may reside at home, in college provided units, or in apartments within some distance of the college. They attend classes, labs, and recreational activities on campus while residing in the locality of the college.

The activities of the college mainly take place within its physical facilities and are based upon a series of fiscal transfers. Money comes into the college administration from students, subsidies, grants and contracts. A portion is paid in wages and salaries to faculty and staff. Other payments go to purchases of supplies and services.

Based upon this superficial representation of a college, note that the activities of the people of the college constitute the main areas of impact upon the region. Of course, the presence of the physical facilities of the campus also affects the region, but really not in ways substantially different from any other consumer of that amount of land, visited daily by a similar number of people, and with the relatively low levels of pollution typical of the operation of an academic establishment. Some of these facilities do differ (library, gymnasium, laboratories), and to the extent that they are available for utilization by other inhabitants of the region, they do represent impacts of the college. In short, the operation of an engineering college may be modelled as sets of flows between four groups of people, and from this framework, the impact of the college upon its region may be ascertained by studying the impact of the presence of these groups upon the region.

Let us note that the crucial features which serve to distinguish an engineering college from any other establishment with roughly equivalent physical characteristics are first, the technical knowledge and expertise of the faculty and second, the broad range of technical devices which are available for their use. In particular, the breadth of this technical knowledge is crucial. A large manufacturer of electronics may in one specialized area exceed the technical depth of an engineering college, but very few institutions can match the breadth and depth of the pool of expertise available in a typical engineering college.

To reiterate, the activities which we have identified whose impacts cross the boundary of college and region are the transfer of skills and ideas, research, public service, consulting, life styles, economic flows, and community service.

The external agencies or groups which are affected by these activities may be broken down both geographically and by function. Thus at local, regional, and national levels, government, industry and business, and other institutions (including "the general populace") are influenced by the activities of the college. For each of these external groups, a matrix exists like the one below. Note that of the activities

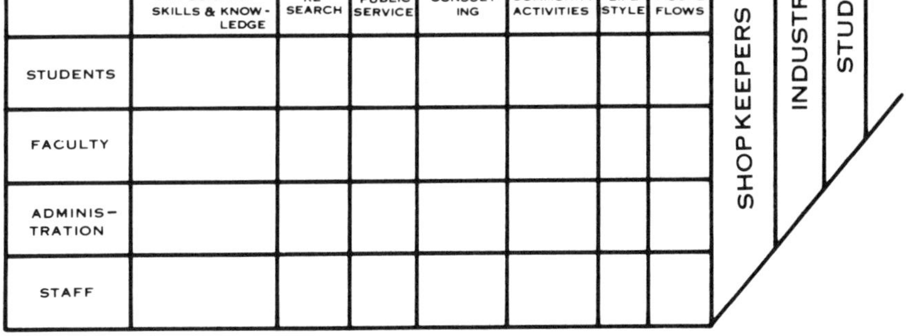

which have been taken as basic to the operation of the college, four are directed primarily beyond the college to entities in the environment. Public service, consulting, life style, and community service are activities which mainly take place beyond the boundaries of the college. Also, the transfer of skills occurs in two distinct ways: first, by on or off campus teaching activities and second, by the actual physical movement of graduates away from this system.

Ensuing reports deal with cells or with portions of the cells noted above. The range of these reports is quite broad, but by considering them in relation to the structure presented here, they may be seen to fairly thoroughly cover the range of impacts anticipated and identified here.

Toward the Development of Models for Studying the Regional Role of Engineering Colleges

Joseph L. Schofer and Amado San Mateo

THE ROLE OF MODELS

Introduction

Conceptual models are useful in the analysis of complex socio-technical systems because they may be helpful in structuring the problem and in suggesting ways in which it might be approached. In both prediction and quantitative evaluation, of course, conceptual models cannot replace mathematical models; yet the former approach to modeling may provide a comprehensive framework within which more responsive mathematical models may be developed.

To evaluate the regional role of an engineering college, for example, quantitative models will be of considerable importance for both evaluating current conditions and for testing alternative policies for operating such educational systems. The formulation of the necessary mathematical models, however, is complicated by the relatively low

Joseph L. Schofer is an Associate Professor in the Department of Civil Engineering at Northwestern University.
Amado San Mateo is a Research Assistant in the Department of Industrial Engineering and Management Science at Northwestern University.

level of quantitative understanding which currently exists regarding the appropriate boundaries for the system to be considered, the elements of that system, and the internal linkages, feed-backs and feed-forwards, which are of significance in policy evaluation.

The early development of one or more conceptual models, perhaps in the form of increasingly more complex flow diagrams, can assist in the identification of these boundaries, elements, and linkages. In particular, such non-quantitative models may serve to formalize and organize the intuitive and experiential knowledge which exists regarding the engineering college and its relationship with its regional environment. Preliminary statements of these formalizations may be tested and evaluated much like mathematical models, by considering the introduction of simple changes in resource inputs, rules, or constraints, and then attempting to determine the extent to which the conceptual model suggests the occurrence of a realistic set of outcomes or consequences.

An initial, and highly simplified, structure for a conceptual model of the university-environmental system was shown in the proposal for this project. This generalized model attempts to demonstrate the nature of the various quantitative models which might be required to evaluate the role of an engineering college. Emphasis was placed on the interactions between the university and its total environment, including: direct and indirect resource inputs, policy and information inputs, the interaction of the outputs of the university with its environment, and finally, the processes by which the consequences of these outputs affect that environment (see Figure 1).

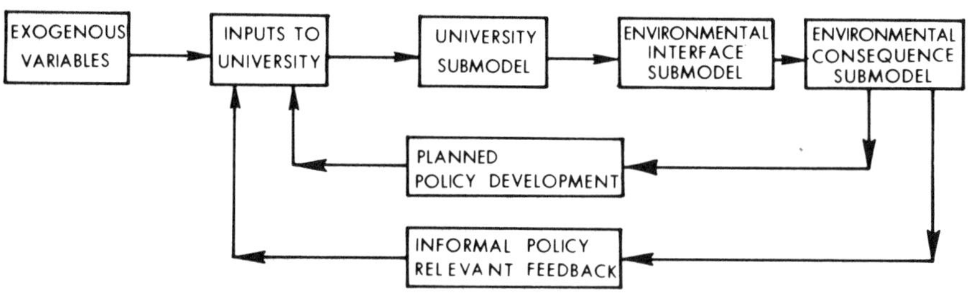

Fig. 1

A more detailed model is shown in Fig. 2. Here, specific categories of inputs and outputs are described. It should be em-

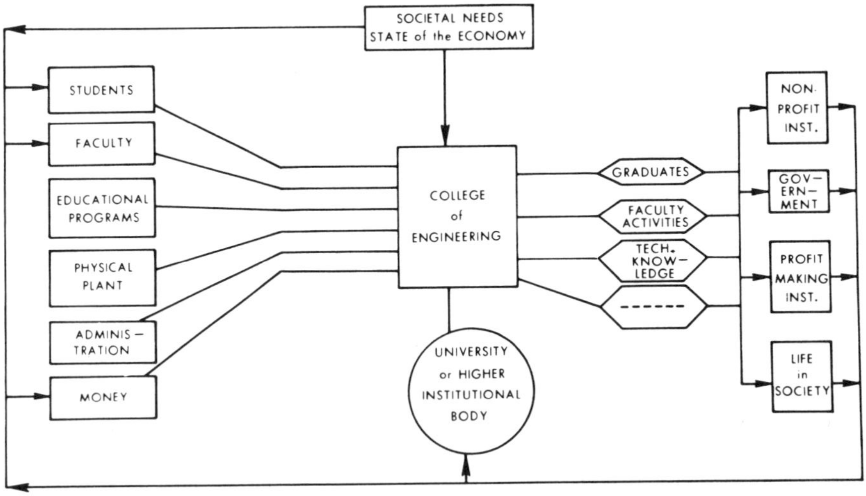

Fig. 2 More Detailed Conceptual Model Identifying Inputs and Outputs

phasized that, at the level of conceptual modeling, it is not necessary to list every specific kind of activity and flow; it should be sufficient to propose a taxonomy or set of taxonomies which spans the space of possible activities and flows. This enables the researcher to avoid the treatment of details, while still requiring that he account in some way for every possibility. A fundamental question to be answered in testing this type of model is: do the taxonomic structures allow us to accommodate specific effect "x" within the model?

It should be evident from the very nature of the problem at hand, as well as from the models in Figs. 1 and 2, that a meaningful approach to the study of the regional roles of engineering colleges must concentrate on the interfaces between the college and its regional environmental environment. The internal operating characteristics of the college are interesting and important, but ultimately, for this project, only in terms of their influence on the set of effects which the college has on the world around it. This suggests a general pattern of emphasis within the research project; it also points out the need for identifying and adopting a number of cost-effective decisions regarding what internal college characteristics should be modeled in detail—because they are important for determining the regional impacts—and what characteristics are best left out of the quantitative models, at least for the time being. For example, it is becoming especially clear that the regional role of the engineering college is strongly affected by the characteristics of the skills, and specialization mixes, of the students who leave the college. This suggests that it may be important to examine the college learning programs in detail, and to model drop-in and drop-out processes, and field switching. On the other hand, to develop quantitative models at this level of detail requires considerable data collection and analysis, which may make it appropriate to lump many of the internal operating characteristics of the college.

Alternative Approaches to Model Construction

One apparently productive path to be explored involves concentrating first on the primary output of the engineering college, trained —or partially trained—students. It may be reasonable to develop a capability to identify and evaluate the regional impact of students leaving the college, and then to proceed to an examination of more complex college/environment interfaces. The nature of departing students and the locations where these primary products of the engineering college function in society are all of critical importance in determining the regional role of an engineering college.

Focusing on graduates of engineering colleges within the State of Illinois, for example, a long list of questions related to student state-trajectories may be formulated, the answers to which will ultimately contribute to an understanding of the role of engineering colleges and to the development of quantitative, policy evaluation models. Typical questions include the following:

Where do students leaving Illinois engineering colleges go?

What are the mixes of levels, skills, and fields of these students?

What are the tasks performed by these students (technical, technical administration, non-technical, education, sales, etc.)?

For what kinds of organizations do they work (locally-oriented, regionally-oriented, state-oriented, nationally-oriented, etc.)?

Are there unfilled engineering jobs in Illinois?

Where do the engineers come from who fill various kinds of Illinois engineering jobs?

Is Illinois a net importer or exporter of engineers? Is this field specific?

If there were an increase of "x" graduates with skill level "y" in field "z", would these persons be employable in Illinois?

What are the labor market areas for Illinois engineering employers?

What are the job location preferences of Illinois engineering students?

Do employers have preferences for in-state or out-of-state engineering students? Is this field specific?

A simplified, student flow model of the engineering college-environment interface is shown in Fig. 3. The feedback linkages which are likely to be of primary importance are suggested, although quantifying their effects will be difficult. The principal components of this model are the following:

Student input vector—including the number and characteristics of new students at all levels, as well as descriptors of continuing students;

Internal college submodel—perhaps including year-to-year transitions in states, drop-ins and drop-outs, field switching, etc.;

Student output vector—including all relevant state descriptors;

Resource consumption-constraint submodel—essentially the production function of the engineering college, to allow the testing of alternative policies in terms of their cost implications;

Initial student output allocation submodel—describing the allocation of engineering students with various skill and level characteristics to particular job types; this model would reflect initial activity choices—the transient characteristics—only,

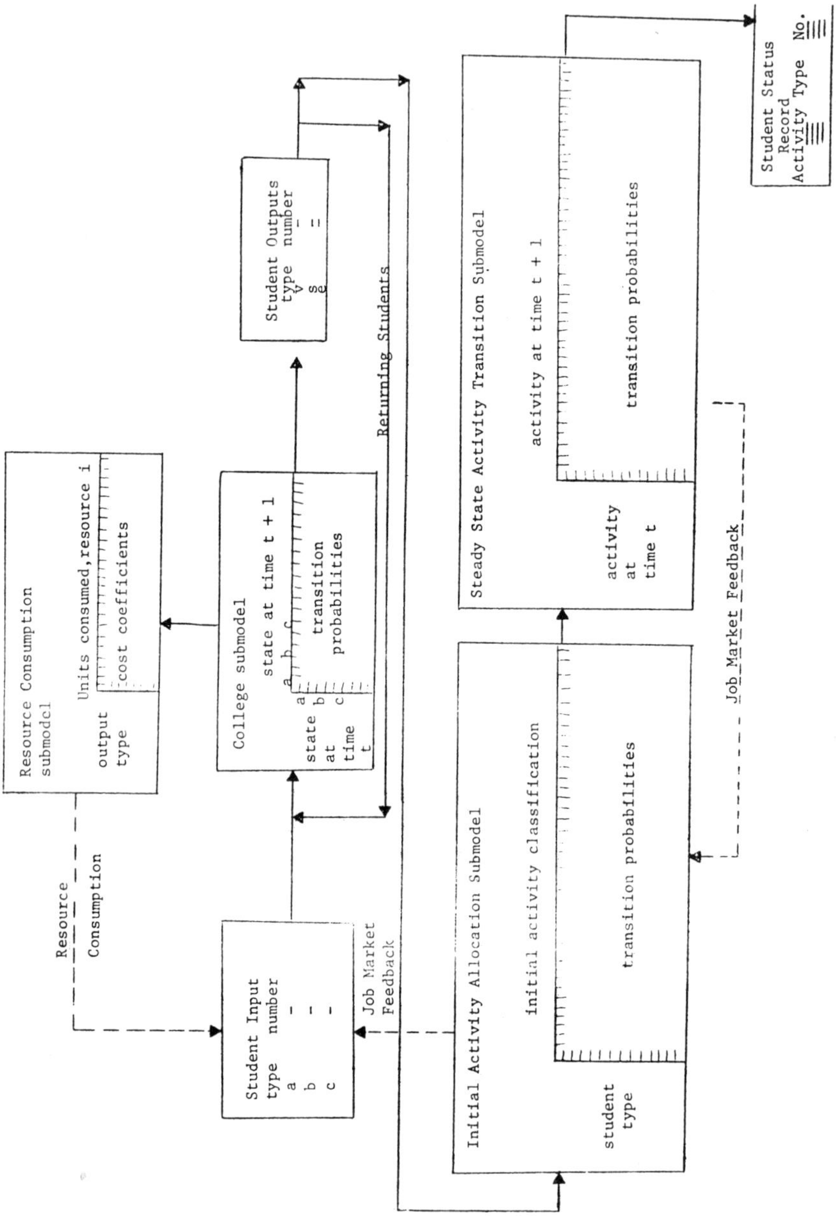

Fig. 3 Simple Student Flow Model

Steady state activity transition submodel—reflecting longer term trends in the student state trajectory.

A natural phasing program for the development of these submodels and their integration into an overall modeling system is suggested in Fig. 4; this program is based on the degree of difficulty of model formulation (easiest first) and the degree to which data are likely to be readily available. It suggests that simple models of internal college student and resource flows might be a logical point of departure because these data should be readily available within the college. Some of the models developed by planners at the University of California might serve as initial bases for further development[1].

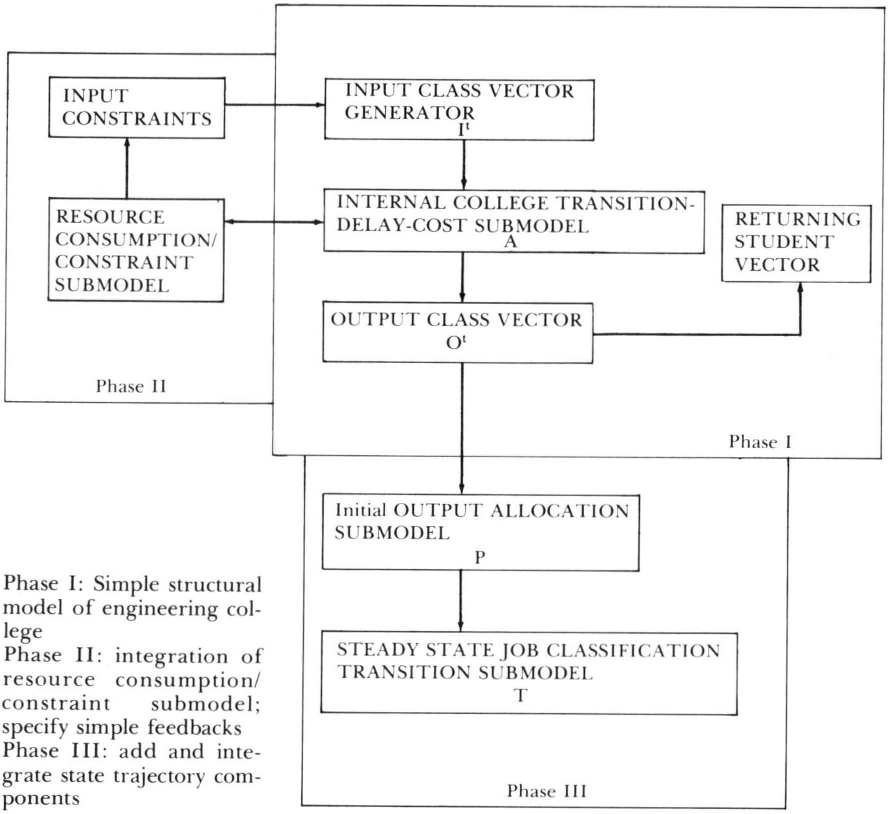

Phase I: Simple structural model of engineering college
Phase II: integration of resource consumption/ constraint submodel; specify simple feedbacks
Phase III: add and integrate state trajectory components

Fig. 4 Phasing Plan for Model Development

A promising approach to the development of such a modeling system might be to consider the student state-trajectory as a first order Markov process, assuming that state changes occur at discreet points in time, and that the probability of a given state change can be developed from available data, depending only upon the most recent student state. For example, the initial student output allocation submodel might be in the form of a transition probability matrix, the elements of which would be derived empirically, as shown in Fig. 5. To formalize such a model it will be necessary to collect and analyze reliable data on the destination of all engineering students who leave the college.

TO
initial employment classification

		public technical	public admin.	public education	private technical	private sales	private admin.
characteristics of student leaving engineering college							
FROM	< BS						
	BS						
	< MS			↑			
	MS		←	probabilities	→		
	<PHD			↓			
	PHD						
	> PHD						

Fig. 5 Initial Student Output Allocation Submodel

Figure 6 provides a further description of the sequence of submodels which might be used to model student trajectories through

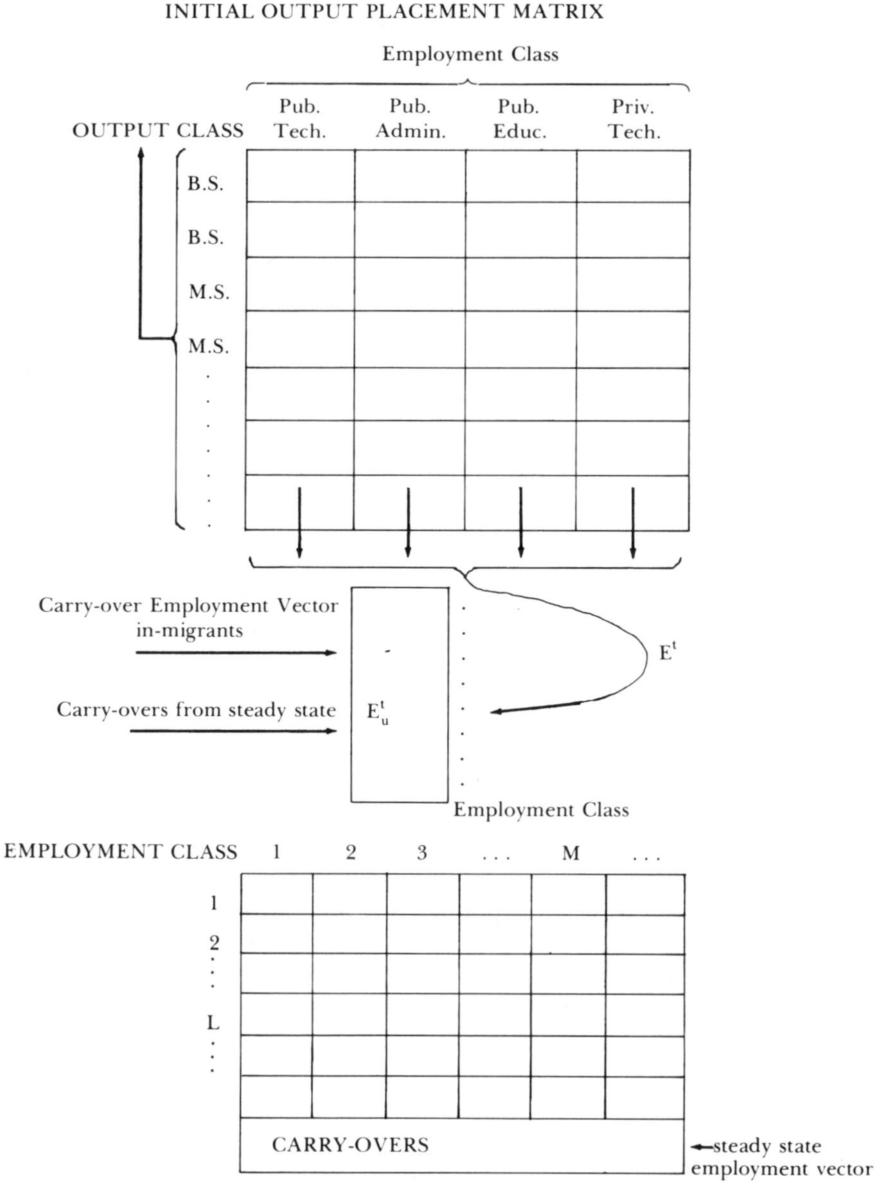

Fig. 6 Sequence of Models for Describing Temporal Trajectory of Engineering Students

time. Operationally, student state trajectories and their response to policy alternatives would be evaluated using a computer simulation approach within a Markovian framework.

Clearly, this general procedure will be limited quite severely by the availability of data. It is possible, however, that a basic model might be calibrated and operationalized using either data from one of the Illinois engineering colleges or by mixing various kinds of data from several colleges. Data for the steady state transition probability matrix might come from national or state sources not associated with any specific engineering college. It would be both desirable and feasible to create a continuous monitoring program focusing on employment status of the graduates of one or more Illinois colleges to provide a valid data base for model calibration and policy evaluation.

This approach has the advantage that it seems feasible, within the data limitations; furthermore, it may be possible to fill initial data gaps using rough estimates, such as provided from experience, or from the use of a delphi process. Over time, more refined data may be utilized to improve the realism of the modeling system. An interesting precedent for this modeling strategy may be found in the work of Blumstein and Larson focusing on criminal justice systems[2]. Eventually, as feedback links are added, and their effects tested and evaluated, this model begins to take on many of the characteristics of the work of Jay Forrester[3].

MODEL DEVELOPMENT STRATEGIES

Modular Approach

Preliminary models should be aimed at identifying key issues, developing methodology and specifying data requirements for a comprehensive model of the interaction of an engineering school with its regional environment. Examples of some key issues of interest in the assessment of the regional role of engineering colleges are listed below.

- What is the contribution of the graduates from the engineering colleges of the region to the economic activities of the region?
- To what extent do engineering schools contribute through their research activities to the technology of the region?
- What role does the presence in the region of the specialized and highly skilled manpower represented by the faculties of the en-

gineering colleges play in state and local strategies for technological and social development?

- What role does an engineering college play as an instrument for unbiased advice on public policy decisions involving an understanding of technology? What role does it play as a sensitive mechanism for forecasting technological trends in a state, in assessing social consequences of technological developments, or in assessing technological needs arising from social developments?
- What is the effectiveness of a College's role as a mechanism for imparting upward social mobility to the first-generation college students that come to it? How is such a role fulfilled for the racial minorities?
- What are the most effective patterns for cooperation between engineering schools and science policy and advisory boards at the state or local levels?
- What is the synergism—actual and potential—of engineering colleges in a region comprehending several states—such as the Midwest, or the Pacific Northwest, or the Southeast, or the Northeast?
- What is the relation between the roles of engineering colleges in the urban, state, and regional communities, and the role that such colleges play in the achievement of national goals?

Data specifications will depend on the key issues identified, and will affect and be affected by the methodology employed in the design of the model. The nature of these data will further evolve as the key issues are crystallized and the models themselves are further structured.

A number of alternative models are being considered in this project, and internal functioning of universities has been the subject of analytic studies in the past. While the relative merits of each approach will not be considered here, as a strategy of model development, it appears highly desirable to encourage the testing of various model structures and relationships among the variables, to insure that the ultimate products are both comprehensive and responsive to policy evaluation needs.

To maximize flexibility in development, and to make it possible to accommodate submodels developed elsewhere, a composite structure made up of building blocks or modules has been proposed. Within such a framework, it should be possible to build and validate modules separately as data becomes available, as well as exercising them together in the form of a modeling system.

Some Potential Modules

An examination of variables which might be considered in the development of preliminary models suggests reasonable groupings which might form the basis of several modules. For example: Internal Student Flow, Faculty Time Allocation, Student Career Trajectories, Resource Allocation, etc. Some variables which might be included in these modules are listed below:

Resource Allocation

For each program—as a function of
its size and characteristics, commitments for:
—faculty at different levels
—capital equipment
—buildings
—administration
—scholarship funds

For each program, mix of learning activities:
—teaching
—laboratory
—research
—cooperative education program
—extension
—cable-TV

Internal Student Flow

For each program:
—number and characteristics of incoming students
—number and characteristics of students in each year (graduate and undergraduate)
—number of undergraduate and graduate degrees
—drop-outs

Student Career Trajectories

For each graduate of each program at each level:
—employment classification
—employment location
—expected sojourn times
—salary

Faculty Time Allocations

For each program and rank, as a function of program characteristics, the time allocated to:

—research
—publication
—consulting
—professional services
—community service
—teaching

Variables Affecting Time Allocation Priorities

—interests
—teaching loads
—university policies
—faculty mix (types, levels)

Relationships within modules may be in the form of mathematical equations, tabular data, graphs, etc., as data will allow. Relationships between modules will emerge as linkages are established among the variables included in different modules. As a comprehensive model evolves, an hierarchical structure among viewpoints, modules, and variables should take form, and validity checks among relationships of groups of variables should suggest themselves.

Approach to Model Composition and Validation

The flexibility and evolutionary properties of the building-block approach may be taken advantage of by using a simple step-wise strategy of model composition. This task of model construction might consist of the following steps:

(1) for the viewpoint adopted, specify the key issues;
e.g., unemployment level of professionals in the region may be a key issue at the regional level; at the college level, a key issue may be the alternative activities to which resources may be committed—teaching, research, community service, etc.; at the student level, a key issue may be the availability of cooperative programs and financial aid, etc.

(2) postulate the variables which may be involved in each particular issue.

(3) hypothesize the relationships among the variables included in the module; these may be in the form of transition matrices, equations, tabular data or charts showing relationships which may or may not include causality as a dimension.

(4) using available or specially collected data, calibrate and validate the relationships within a given module; it may be necessary to "build up" to a higher level to get a meaningful set of variables and relationships which could be validated against available data. It would be desirable to validate at the lowest level (one module at a time) rather than chains of modules, as this would allow verification of internal linkages separately from external linkages among modules.

Candidate Modules for Early Development

In this section, several candidate modules are described to illustrate the proposed strategy of model construction. The examples given here may be appropriate for the early stages of model formulation. These examples illustrate the potential differentiation of viewpoints from which the modules may be formulated, in order to emphasize the flexibility and versatility of this evolutionary approach to model development. Thus, the viewpoints exemplified in these modules cut across the regional, college-wide, faculty, and the student levels; on any of these levels, the key issues may revolve around economic, political, legal, social, etc., questions. Particular attention, however, is focused on policy-sensitive variables appropriate to the viewpoint taken in any given module.

Regional Interface Module. The regional interface module might consider variables which are important to the role of the engineering college in its regional environment. A reasonable viewpoint might focus on the relationships affecting the role of the engineering college in its interactions with other institutions in the region. Of particular concern here would be the present and potential linkages which an engineering college may have with such institutions as:

—legislative bodies
—boards of education
—city government

—state government
—federal government
—professional organizations
—other engineering colleges
—junior colleges and vocational schools
—research laboratories
—industrial establishments
—foundations
—alumni groups

Among the major questions which may be examined here are:

- What role does the presence in the region of the specialized and highly skilled manpower represented by the faculties of the engineering colleges play in state and local strategies for technological and social development?

- What role does an engineering college play as an instrument for unbiased advice on public policy decisions involving an understanding of technology?
  What role does it play as a sensitive mechanism for forecasting technological developments, or in assessing technological needs arising from social developments?

- What are the most effective patterns for cooperation between engineering schools and science and advisory boards at the state or local levels?

- What is the relation between the roles of engineering colleges in the urban, state and regional communities, and the role that such colleges play in the achievement of national goals?

The variables of interest are those which define the nature of the relationships which the engineering college has (or could have) with each of the institutions which could influence its regional role. For example, the variables which define, describe and quantify the information flow between the engineering college and these institutions might suggest the nature of their relationships and the areas where the role of each might be enhanced, shared, or specialized. Potential relationships might be suggested by needs and expectations recognized but not realized.

A variety of structural forms might be proposed regarding the underlying relationships. In any case, it will be important to focus on models with potential for policy evaluation studies. For example, questions along the following lines might reveal policy-sensitive issues affecting the flow of information and communication among the institutions:

- What factors encourage or inhibit the flow of communication among them?

- Are there intended or unintended restrictions in these flows?

- What are the pay-offs or penalties incident to such flows?

- What delays are encountered and how do these delays influence the content of such communication?

- Is there inherent directionality in these flows?

An empirical study of the formal and informal communication nets linking these institutions will be essential for providing insights useful in the development of models which would be suggestive of the behavioral mechanisms sensitive to policy changes. Aside from providing measures of the role of the engineering college in its regional environment, such a study would reveal properties of the system on which responsiveness to change may depend. It may suggest ways in which the presence of the highly specialized manpower represented by the faculties of the engineering colleges could be organized to facilitate a desired level of involvement in the strategies for technological and social development; it may point to alternative and more effective modes of interaction among individuals, teams, offices, etc., through which alternative roles might emerge. As an important channel of expression and realization of institutional role, information nets would merit study in the early stages of model formulation. Preliminary efforts at data collection along these lines conducted as a part of this study suggested that the monitoring of external college communications is both feasible and highly informative. It would be desirable to continue and expand this surveillance, using the results not only for model building but also for on-line policy formulation and evaluation.

A very simplified module structure which could be detailed and expanded as necessary would be:

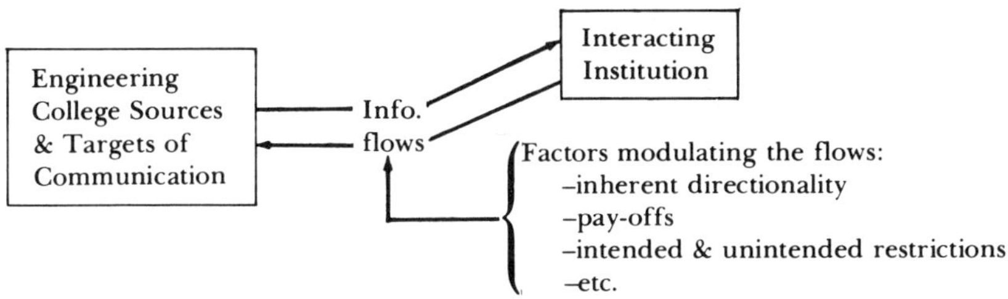

Some examples of communication links meriting early investigation include:

- information concerning research support available from various sources;
- information concerning problem areas which the engineering college might participate in investigating through individual faculty projects or interdisciplinary teams;
- communication channels within the college through which information on opportunities for participation by faculty, students, etc. in issues bearing on technological consequences of regional impact;
- communication links with policy-setting and policy-implementing bodies at the community, regional and national levels;
- communication links with industrial firms, laboratories, and other employers of engineering manpower in the region;
- communication links with professional societies, alumni, etc., which could play a role in the formulation of policies affecting the engineering profession and its role in the region.

Within this module, the existence and significance of such communication links could be established by quantifying their numbers and rating their importance on ordinal scales. Later, behavioral mechanisms underlying the modulation of these flows may be hypothesized and validated against the empirical data.

Student Flow Module. This module might consider flow rates of students entering, staying in, and leaving the college. Flows may be analyzed according to program or field, according to degree levels, or in terms of internal transition rates from year to year. To the extent that policy-sensitive factors may be program-specific, or dependent on degree levels being considered, the choice of level of analysis must be made accordingly.

Initially, it may be desirable to follow the course of previous studies by focusing on the flow rates of graduates in all programs and degree levels as an aggregate. Levels of resource commitments may be studied in relation to these flows. Data on the temporal relationships between levels of resource commitments and student flows should be readily accessible. Strength in these relationships would shed light on internal functioning of the university. Weaknesses might identify the significance of certain exogenous variables.

The student flow module may be represented by the following:

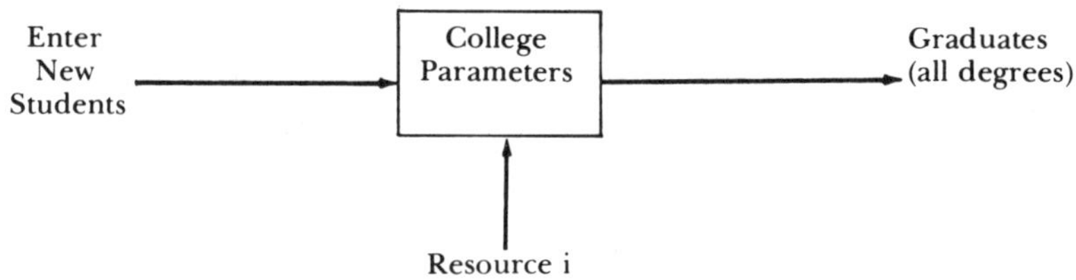

At a more detailed level, the student flow module becomes

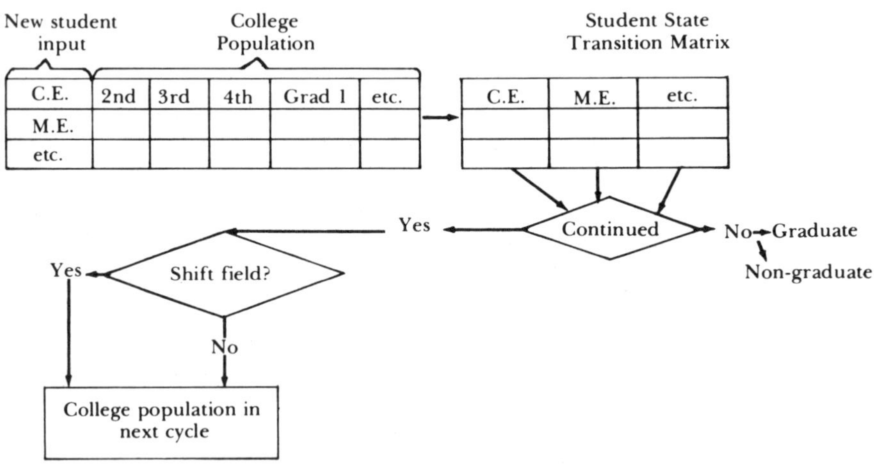

The student state transition matrix may depend on several factors:

Resources allocated
Characteristics of students
Financial aid, cooperative programs, incentives
Mix of learning activities
Faculty time allocation

	C.E.	M.E.	etc.
1st to 2nd			
2nd to 3rd			
3rd to 4th			
etc.			

For purposes of policy design, it will be desirable to trace the influence of separate resources on the state transition probabilities. Data appropriate for this application include:

Faculty hours spent in —teaching
—laboratory instruction
—research guidance
—other activities directly involving students

Laboratory equipment provided
Library services
Research, scholarship funds
Cooperative programs aimed at specific fields and student levels

The specific policies under consideration should influence the emphasis and direction of data collection. For example, given the resources as allocated, how would a change in availability of cooperative programs to incoming students affect the transition from freshman to sophomore stage? How would an increase in research guidance at the senior year influence the transition from B.S. graduate to first year graduate student? Careful sample design and the use of statistical analysis should make it possible to identify these changes.

Student Career Trajectories Module. This module considers the career patterns of engineering graduates and non-graduates in the region.

On a gross level, we may be interested in the "typical" professional life of engineering graduates in this region, if such a profile may be envisaged. This profile may consist of flow characteristics, which would allow prediction of career patterns with respect to mobility after leaving the college, both in terms of changes in status and salary, as well as movement into and away from the region. It should be related to the historical employment patterns for engineers in the region in terms of the number, sizes and nature of engineering employers in the region.

For purposes of designing regional policies, some essential factors relevant to the development of this module include the following:

ZZ {

—impacts of alternative schemes by which the colleges in the region "market" their products
—impacts associated with significant changes in scale of particular program areas (water resources, environmental design, etc.)
—structural changes in the information feedback system by which the college determines the needs in its environment for its products
—structural changes in the processes by which university policy decisions are made
—changes in the human and material import/export balance of the region
—changes in social mobility
—changes in gross regional product
—changes in income
—changes in tax revenue
—regional rounds of effects and multipliers

The inter-relationships between the career trajectory and student flow modules might be described as follows:

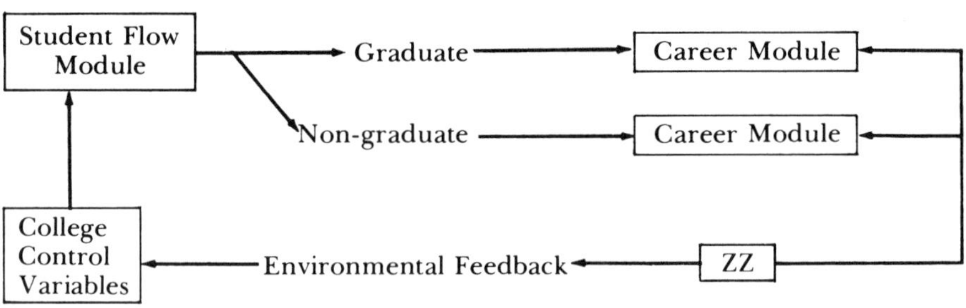

Some specific questions which may be of interest in connection with the Career Module data collection effort include:

—how would we characterize the initial employer(s) of students first leaving the university, based on student characteristics?
—how long does the graduate stay with his first employer?
  —what factors affect his decision to transfer?
—considering the entire career of the B.S., M.S., Ph.D.,
  —what measures of mobility may be used?
  —what regional and college-wide policy decisions could influence the:
    —initial distribution of graduates in various employment categories (government, private, etc.)
    —the mobility of the graduate in his career, in terms of rate of advancement, number of employers, etc.
—what role do professional societies play in the career patterns of engineers in the region?
—considering engineering education as an investment in the lifetime career of the students entering college upon completion of their high school studies, what measures and magnitudes of return are to be expected during the entire career of the B.S., M.S., Ph.D. graduates?
—how may this information influence the policy decisions concerning the encouragement of engineering education at these levels?

The structure of the career trajectory module may be patterned after the Student Flow Module, highlighting the flow characteristics of careers. The underlying factors which influence these flows may be traced from the profile of historical employment patterns for engineers in the region, as well as some or all of the elements listed under ZZ in the discussion on the previous page.

Since the interactions among variables may be difficult to measure, the career trajectory module may be structured along two lines:

(i) the flow model indicated above,
 —operating in the scenario provided by all the data mentioned
 —factor analysis could indicate significant relationships
(ii) specific policy-sensitive variables may be correlated with the flow model in the context of the scenario presented.
 —time series data may be required

The scenario against which the careers evolve might be modeled as a separate career environment module. This module would consider characteristics of the "employers" of engineers in this region, as well as factors which could influence the decision of the engineer to move into or out of this region.

Faculty Time Allocation Module. This module considers the time made available by the faculty for purposes of teaching, research, publication, consulting, professional activities, community service, etc. Time allocation may be analyzed according to faculty rank, field, administrative position held, number of years in the college, etc.

Considered in relation to the student flow module, the variables of interest in the faculty time allocation module must certainly include the time spent in teaching, research guidance, laboratory instruction and other work directly related to student flow.

Policy-sensitive factors affecting the priorities in allocation of time by the faculty may thus be related to the behavior of the student flow module. How and to what extent such a relationship exists will be the focus of time-series data collection for this module. Policies concerning availability of research funds, reduction of teaching loads, committee work, factors in promotion, and the general incentive structure for the faculty are likely to be key issues in this module.

In addition to data collected at a single college, comparative studies may be necessary using data from many engineering colleges in order to determine the effect of college policies on faculty time allocation. In order to trace the impact of faculty activities on the region, data collection should also include time devoted to consulting with government and private clients, mission-oriented research, and public service activities.

A composite picture of the role of the faculties of engineering schools in the region will thus emerge. For purposes of influencing the direction as well as the nature and magnitude of this role, it may be necessary to study the underlying factors which bring about such a role. For instance, on the regional level, how has the availability of government-sponsored research, or the public recognition of the importance of specific research areas (e.g., environmental problems, traffic safety, etc.) influenced the role assumed by the faculty? What would be the effects of role changes on the student flow module? How may the faculty play a different role (new activities, encouragement of professional activities, political action, etc.) to enhance its regional role? How may regional organizations of engineering faculties influence this role?

The faculty time allocation module might be structured as follows:

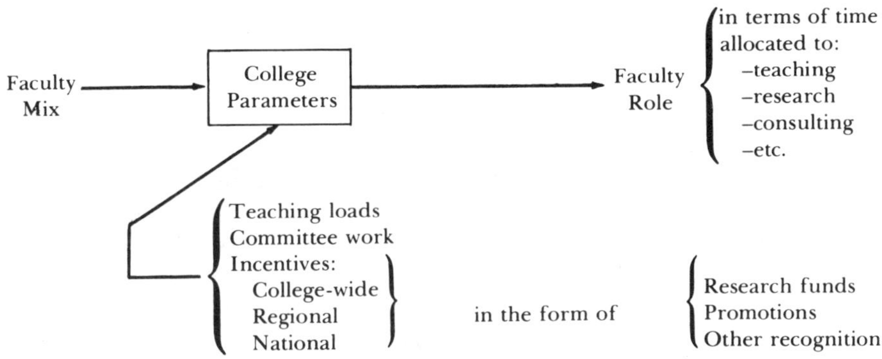

The module structured as above may interact with the student flow module through several links:

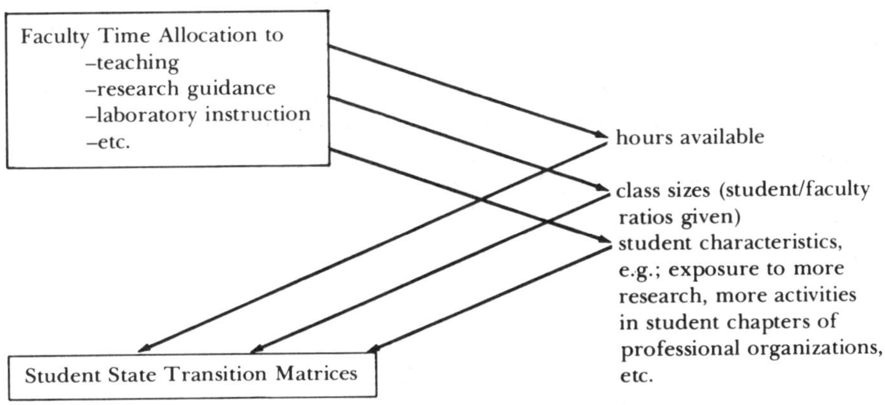

The faculty time allocation module may also be influenced by the incentives structure in the college, regional and national levels:

Incentives:

—research funds from various sources
—promotions policies in the college
—recognition within and outside the college
—faculty time allocation
—availability of other activities, rewards
—etc.

CLOSURE

This discussion has attempted to explore, through examples, the functions which might be performed by conceptual and analytic models, and accompanying data collection efforts, in studies of the regional role of engineering colleges. A generalized, modular strategy for model development and testing has been proposed.

It should be clear that a considerable amount of detailed data will be required to operationalize these models. Some of these data may be routinely collected as a part of on-going college operations. Much will be compiled only through the initiation of new data collection efforts. While on-going surveillance of college activities, the career trajectories of students leaving the college, and the relationships between college and the region will form the basis for model building, it is important to recognize the larger value associated with the availability of such data. The models, themselves, are proposed only as tools for enlarging the understanding of the relationships between engineering colleges and their environments. They may be useful for long range planning, for policy evaluation, and for enhancing the regional role of engineering colleges. Similarly, the availability of relevant data alone will be of direct value in these efforts.

REFERENCES

1. Breneman, D. W., "An Economic Theory of Ph.D. Production: The Case at Berkeley," Research Program in University Administration, University of California, June 1970.
2. Blumstein, A. and R. Larson, "Models of a Total Criminal Justice System," *Operations Research*, Vol. 17, No. 2, March-April 1969.
3. Forrester, J., *Urban Dynamics*, Cambridge: MIT Press, 1969.

Who Goes to Engineering School? Who Graduates? A Cohort Case Study

William J. Haga

THE ISSUES

In examining the regional role of engineering colleges as institutions ostensibly transforming manpower resources toward a socially beneficial end, it immediately becomes necessary to grasp a set of basic processes:

1) What kind of subset of the overall input into undergraduate programs is represented by those freshmen entering the study of engineering?

2) What happens to that frosh subset between entry and exit from the institution?

3) What becomes of an engineering college's formal output (baccalaureate graduates in engineering in the present case)?

Several issues are raised here together: who goes into engineering programs compared to who enters other undergraduate programs? How efficient is the engineering college as a subsystem in society's acculturation and normative socialization system transforming a raw

William J. Haga is a Research Assistant in the Department of Sociology, University of Illinois at Urbana-Champaign.

material (freshmen) into a socially useful product (engineers)? How does the larger system absorb the subsystem's output? What latent or unexpected consequences attend the formal, manifest purposes of this transformation process?

Identifying these issues and raising the relevant questions, while often considered to be a crucial step in any investigation, proved considerably easier in this study than providing the answers. That this effort identifies the several fundamental difficulties in getting at the answers *is* its essential payoff; it was designed to develop a methodological prototype for data collection on the engineering college as a social subsystem. The empirical findings that were revealed and the meanings they suggest are a consolatory by-product of the main methodological thrust. The results presented and discussed here are meant only to be suggestive of areas for further research; they represent but one cohort moving through a specific period within a particular engineering college.

REVIEW OF THE LITERATURE

While practicing engineers are popular subjects for studying professionals in conflict with employing bureaucracies, treatment of the social sources of engineering students have been few. No research known to the author has examined engineering education as a social process. What little has been done in the way of sociological investigations of these areas is conveniently collected in a single publication issued by the National Science Foundation (NSF, 1959): *Scientific Manpower, 1958*. A later number in this series (NSF, 1961), *Scientific Manpower, 1960* also contains some excellent works by sociologists and psychologists although the focus is upon scientists in basic research rather than engineers.

In fact, only the article by Trow (1959) in the NSF series focuses upon engineers alone. Trow's thesis is that the social origins of engineers are quite distinct from those of students who enter the pure science disciplines. Since the Sputnik event, he notes, commentators spoke of the need to build a "pool of scientists and engineers" without reflection upon the differences between the two occupations.

First, Trow indicates that engineers come from lower economic strata than do science or mathematics majors. Indeed, he found that they come from lower economic strata than undergraduates in most fields except education (he was speaking from data circa 1952). A study of winners and runners-up in the 1952 National Merit Scholarship Program found 50% of the future engineers had fathers in manual or lower middle class occupations while the same was true for only 33% of the physics and mathematics students. A later study by Stice (1956) showed even more disparity in these class origins. Trow argues, "Whatever the proportions in different schools or subpopulations, the data support the view that engineering has a special appeal for bright boys of lower and lower middle class origins" (1959:68).

Second, Trow speculates that bright students from lower and lower middle class families do not see fields such as physics presenting a genuine possibility for themselves. His belief is that for boys who have been raised to not want too much nor to aspire too high, physics or mathematics seem too difficult and too prestigious. By contrast, they view engineering as both rewarding and attainable, "a sensible way to pursue their interests in mechanical and scientific things..." (1959:69). In addition, engineering is seen as a way to reach middle class status, thereby improving their family's position in their own generation.

Finally, Trow reveals that while the lower economic strata origins of engineering students tend to be a major impetus to their upwardly mobile aspirations, these same origins work to limit how far they rise on the socio-economic scales. Engineering students are unquestionably among the brightest of all college students according to data from a variety of measures examined by Trow. Yet, they are at once among the least creative of students. His explanation is that future engineers see a college education as purely instrumental in nature. That is, it is a place to learn specific skills in order to attain a "union card" into a respectable career and little more. They go to learn, not to question. Trow's position has been supported by recent data (Elton and Rose, 1971) showing that students who transfer out of engineering are more creative and reflective in their thinking than those who stay. The engineering "persisters" were found to be practical of mind and intent.

Studies cited by Trow repeatedly found that engineering students avoided the liberal arts subjects, especially the arts, literature, and history. Yet, Trow notes, it is a mastery of certain liberal arts disciplines that provide a college graduate with polish and a new life style, enabling him to rise to positions of management and leadership among the upper economic class.

The burden of Trow's work is that high school counseling and curricula tend to guide the bright children of lower social status to the attainable instrumentality of engineering education, while not encouraging them to extend their limited aspirations. At the same time, he notes, many students of a more creative bent, and with broader cultural interests, are invariably steered away from engineering programs in college even though they may possess the intellectual potential for success in the field. Trow makes an insightful analysis and taxonomy of the various emphases placed upon liberal arts subjects and leadership training in different engineering schools. His findings bore important policy implications at the time they were written, aiming, as they did, at increasing and improving the supply of engineering manpower. They are equally relevant today as a foundation for policy decisions aimed at broadening the awareness of engineering students to the environmental and social impact of technology.

Two other notable papers in the NSF series are by Becker (1959) and Parsons (1959). Becker suggests an analytical model of the process by which young people reach a commitment to an academic-occupational pursuit. Parsons discusses the conflicts and roles of the professions within the total society generally and within the hierarchical authority structure of complex organizations in particular. His focus is upon the symbiotic, but unhappy, meshing of the respective roles of administrators and technical personnel. Both are relevant to policy suggestions for modifying the social role of the engineering college as an institutional subsystem.

METHODOLOGY

As was pointed out earlier, the purpose of this study was to develop a prototype of data sources from which to model the subsystem role of the engineering college. This was to be done by exploring the availability of such data as at single school, the College of Engineering at the University of Illinois, Urbana-Champaign. This choice was based largely on the convenience of access by the author to the school's data, along with the belief that it possessed considerable resources of data on its students and graduates. The latter assumption, indeed, turned out to be true. Unfortunately, the quality and ease of mechanized access to this data was quite atypical of the other engineering schools involved in the regional study.

To model the engineering college as social subsystem, data was required for the social dimensions of the subsystem input (the freshmen), the output (graduates), and the thruput process (who started? who finished? and what happened to those that were diverted along the way?). One approach might have been to have gathered cross-sectional data at one point in time. Data on incoming students, students at each progressive stage of the undergraduate development, and alumni in 1971, for example, could have been gathered. This strategy, however, leads to spurious conclusions and is vulnerable to serious challenges from the rival hypothesis of historical effects. Missing too is the time dimension as process factor. In the cross-sectional approach, time is treated as a static factor. Yet all social processes are embedded in a temporal dimension. The results would be descriptive of structure at one time, but not of process through time.

The focus of interest here was in the engineering school as a processual subsystem, not as a static structure. Therefore, the cross-sectional approach was rejected in favor of a cohort strategy. A cohort is a sociological term for a group or statistical aggregate identifiable on some social dimension. The cohort group moves through time, retaining its social identity. By following a cohort across time, as it encounters significant events, one obtains a processual view of its social history.

The cohort studied here was the class of engineering freshmen that began their education at the University of Illinois, Urbana, in 1966. They were followed through four to five years of undergraduate engineering until most had completed or left the program by the spring semester of 1971. This particular group was chosen for two reasons. First, 1966 was the first freshman class with which the university began using social security numbers as identification numbers for students. Access to data on earlier freshmen classes was impaired by the noncomparability of the computerized finder system for individual student data. Using a freshmen group later than 1966 would have been unfeasible because they would not have had sufficient time to complete undergraduate engineering programs that often take up to five years. The second reason for selecting the 1966 freshmen was that they would provide the most recent engineering graduates in a new market situation characterized by the collapse of the West Coast aerospace industry.

Input Data. Information on the social origins of the 1966 incoming engineering freshmen was taken primarily from answers they gave to a questionnaire that was part of their American College Testing

(ACT) examinations. These tests were taken when the freshmen were still seniors in high school. The University of Illinois requires the ACT as an undergraduate admission credential.

The socio-economic questionnaire is designed to be coded and punched for processing on its own single punchcard. Of the total testing data collected by an ACT examination, the social origins information is the fifth card of an individual student's ACT deck. It is thusly named the "ACT-5 data."

The ACT-5 covers a number of areas that are not of direct interest to a study of social origins (the entire set of questions asked and coding format of ACT-5 is included in Appendix I). The university's administrative data management system, however, provided efficent selection and access to those variables dealing with personal aspirations and aptitudes as well as characteristics of each freshman's family, high school and community.

Access to the ACT data was guarded by two levels of organizational gate-keepers in the university. The first was the Bureau of Institutional Research and the second the Office of School and College Relations. Both of these agencies were engaged in systematic studies of the university itself and its effectiveness as an educational subsystem. They were particularly encouraging and cooperative with this study. With their approval, it was then feasible to approach the next level: the university's administrative data processing unit. With proper funding and project identity information, this agency too, was especially cooperative in making the required access and selection runs from the student data files.

The first step in retrieving the social origins data from ACT-5 was the creation of a "finder" deck by identifying all of the social security numbers of freshmen in the various undergraduate engineering programs and curricula. This was done by requesting the administrative data processing (ADP) unit at Urbana to create a deck of social security number cards pulled from the data files of the 1966 freshman group according to program and curriculum codes found in a manual prepared by the university level office of the Bureau of Institutional Research.

The second step was simply to request ADP run the "finder" deck against the ACT-5 data file and retrieve the social origins data, creating a new deck of punch cards representing the engineering college subset of the total 1966 University of Illinois, Urbana, freshman ACT-5 data. This deck was then used for further statistical package programs such as SOUPAC (1971) or SPSS (Nie, Bent and Hull, 1970).

Trend data on engineering freshmen, permitting comparison of the 1966 input with three later inputs, was provided by the Office of the Associate Dean, College of Engineering (1967). Data on earlier classes was not available.

Thruput Data. The "finder" deck, used above to initially identify the 1966 engineering freshmen by social security number starting with curriculum codes, was now used in a reverse procedure to trace the timing and location of the undergraduate fall-out from engineering over a five year period. For each of the ten academic semesters between Fall, 1966 and Spring, 1971, ADP asked their computer to print out the college or program codes of each of the original 1966 engineering freshmen cohort by social security number. Semester by semester, it was possible to trace the path of each member of the beginning group. By the Spring, 1971 semester, the author was able to determine the following about each of the 1966 engineering frosh:

1) Who graduated with a baccalaureate degree in engineering from the Urbana campus. This subset group of the original cohort was labeled "Engineering Graduates, 1971" for later analysis.
2) Who were still in the engineering curriculum, but had not yet completed the baccalaureate degree. This group was labeled "Still in Engineering School, 1971."
3) Who transferred out of an engineering curriculum to another undergraduate curriculum on the Urbana campus and eventually received a baccalaureate degree from that new program or department. The specific program or curriculum to which they transferred was also known, as well as the semester in which the transfer was made. This group was labeled "Other Graduates, 1971."
4) Who transferred to another undergraduate program on the Urbana campus, but were still working on the baccalaureate degree. As with "3)" above, the exact program and timing of the transfer was known. This group was labeled "Others Still in School, 1971."
5) Who left the University of Illinois, Urbana, undergraduate system entirely and which semester they left. On the tables accompanying the analysis section of this report, this group is inaccurately labeled "College Drop Outs." The term "drop out" has connotations that go beyond this data. The reason for these individuals' departure

from the University is not known, nor is their destination. It is not clear whether they were transferees to another institution, students serving a military obligation, or if they, indeed, just dropped out of higher education. In a few cases, those who left in an early semester reappeared later after a personal hiatus. This group might better be labeled "College Leavers" after the British euphemism.

In subsequent analyses of the social dimensions of the 1966 engineering freshmen cohort, the results are grouped under these five categories. It is therefore possible to examine if the social origins of those who persist and succeed in the undergraduate engineering programs vary from those that leave college or transfer to another curriculum. The five groups are also compared to the social origin data of the entire freshman group entering the University of Illinois, Urbana campus, in the Fall, 1966 semester. The latter comparison data was obtained from the University level Office of School and College Relations.

Output Data. The flow of analysis is discontinuous between the 1966 engineering freshman cohort data and the available data for the output end of the subsystem model. The University's computerized data files stop when a student graduates or leaves the institution. Because the focus of this study is engineering colleges, no attempt was made to look at the post-college pursuits of the five subsets of the 1966 engineering cohort except those that graduated as engineers. For this information, the present study was dependent upon the Engineering Placement Office at the Urbana campus. Fortunately, and this was a primary reason for making the study at this school, this agency gathers rather precise information on the immediate plans and later realizations of each class graduating from the College of Engineering. Moreover, this office makes rather detailed analyses of survey responses from certain classes at the five and ten year points beyond graduation. The results presented summarize and highlight points from the extensive data that is available from this source.

Data on the June, 1971, recipients of baccalaureate engineering degrees was taken from a salary survey completed in August, 1971 (Chapman, 1971). Comparisons with the output data from earlier engineering graduating classes were taken from reports published by Pauline Chapman, Director of the Engineering Placement Office (1970a and 1970b) on the 1960 and 1965 graduates.

Percentages calculated for the output results are of the base of the

1971 graduates only, not upon a base of the original 1966 freshman cohort.

FINDINGS

Social Origins of Engineering Freshman

ACT Scores. In looking at Table 1, two observations are apparent. First, that all of the engineering freshmen, regardless of their subsequent path as undergraduates, scored higher than the total class of freshmen entering the University of Illinois in Fall, 1966. Second, those engineering freshmen who ultimately attained baccalaureate degrees, whether in engineering or in another field, scored slightly higher than their peers who either left the university or were still working on their degrees after five years. Those who left undergraduate studies altogether had the lowest scores among the engineering freshmen cohort but still slightly higher than their freshmen class as a whole.

Table 1 Average ACT Scores, University of Illinois-Urbana Freshmen, Fall, 1966.

All entering freshmen, U. of I. Urbana	25.5
Those who later graduated as engineers by June, 1971	27.0
Those who started as engineering frosh but later graduated from other fields	27.0
Those still in engineering school after June, 1971	26.8
Those still in school in other fields after June, 1971	26.1
College drop outs from engineering	25.9

Source: ACT-5 Code Handbook

Table 2 shows the trend of ACT composite scores for the 1966 through 1969 entering classes of freshmen in engineering alone. Across that four year period there is a perceptible upward shift in the modal score. This indicates, in a rough analysis, that the educational preparation of engineering freshmen is improving with time.

Table 2 ACT Score Educational Profile for Engineering Freshmen, University of Illinois-Urbana, Fall Semesters, 1966-69.

ACT SCORE RANGE	1966 COMPOSITE SCORE NO.	%	1967 COMPOSITE SCORE NO.	%	1968 COMPOSITE SCORE NO.	%	1969 COMPOSITE SCORE NO.	%
0 – 15	1	1	0	0	6	1	1	1
16 – 17	2	1	0	0	5	1	2	1
18 – 19	12	2	7	1	4	1	7	1
20 – 21	31	4	17	2	9	1	16	2
22 – 23	73	6	67	7	34	4	38	4
24 – 25	160	18	159	16	135	15	128	14
26 – 27	251	29	249	29	246	28	220	24
28 – 29	227	27	243	28	269	32	280	30
30 – 31	94	10	139	14	137	14	160	17
32 – 33	11	2	23	3	23	3	49	5
34 – 35	0	0	0	0	0	0	1	1
36 – 37	0	0	0	0	0	0	0	0
TOTALS	862	100%	904	100%	868	100%	902	100%

Source: College of Engineering.

Table 3 Class Standing in High School for Engineering Freshmen, University of Illinois-Urbana, Fall Semesters, 1966-69.

PERCENTILE	1966 No.	1966 %	1967 No.	1967 %	1968 No.	1968 %	1969 No.	1969 %
99 – 100%	56		69		78		78	
95 – 98%	134	38%	172	45%	202	50%	203	49%
90 – 94%	152		182		189		191	
85 – 89%	141	30%	126	26%	147	29%	157	28%
80 – 84%	128		113		116		118	
75 – 79%	47	15%	100	18%	76	15%	75	15%
70 – 74%	61		68		55		57	
60 – 69%	97	10%	78	8%	38	4%	60	6%
50 – 59%	37	4%	33	3%	17	2%	19	2%
40 – 49%	10	1%	1	nil	3	nil	6	nil
30 – 39%	3	nil	0	–	1	nil	3	nil
20 – 29%	0	–	0	–	0	–	1	nil
0 – 19%	16	2%	1	nil	4	nil	0	–
TOTAL	909	100%	943	100%	926	100%	968	100%

Source: College of Engineering.

The results from these data were not unexpected. The measures examined in the 1950's by Trow (1959) show the same educational superiority of prospective engineering students. Table 3 supports the ACT scores in terms of high school class standings of the engineering freshmen. The 1966 cohort had 68% of its members in the 80-100 percentiles of their secondary classes. The trend comparison shows a marked increase in high school standing for engineering students in the late '60's.

Vocational and Educational Aspirations. When the ACT-5 questioned the 1966 freshmen class about their future occupational aspirations (they answered while still in high school), the results were as depicted in Table 4. The most popular response was "producer of services" among the entire class although it was less popular among the engineering cohort, especially those that later completed their undergraduate programs. The "researcher" role was the modal choice of the engineering group. Interestingly, the freshmen that eventually received engineering degrees had the highest percentage choosing the "researcher" role. They also had the lowest number of "undecided" responses among the engineering cohort.

Table 4 Vocational Role Preferences of Freshmen, University of Illinois-Urbana, Fall, 1966.

	All 1966 Frosh	Engineering Grads 1971	Other* Grads 1971	Still in Engrg. School 1971	Others* Still in School 1971	College* Drop Outs
Researcher	17.6%	41.0	33.9	33.9	29.4	35.9
Teacher	17.5	0.5	—	1.6	—	0.7
Administrator	6.3	6.9	10.1	4.8	8.8	6.3
Salesman	2.0	4.1	1.8	3.2	2.9	2.3
Producer of Services	24.8	16.1	19.3	12.9	23.5	19.6
None of These	9.3	6.9	9.2	14.5	2.9	10.0
Combinations	4.7	4.6	4.6	6.5	5.9	4.7
Undecided	17.9	19.8	21.1	22.6	26.5	20.6

*Started as engineering frosh, Fall 1966.

Source: ACT-5 Code Handbook

Table 5 Educational Aspirations of Freshmen, University of Illinois-Urbana, Fall, 1966.

	All 1966 Frosh	Engineering Grads 1971	Other* Grads 1971	Still in Engrg. School 1971	Others* Still in School 1971	College* Drop Outs
High School Diploma	0.1%	0.9	—	—	—	0.3
Votech under 2 yrs.	0.2	—	0.9	—	—	—
Jr. College Degree	0.8	—	—	—	—	0.3
Bachelor's	36.0	28.1	31.2	27.4	32.4	38.9
MA, MS, MBA, equiv.	39.7	53.9	52.3	59.7	61.8	47.2
Ph.D.	9.5	13.8	13.8	11.3	5.9	10.0
MD, DDS, equiv.	7.9	0.5	0.9	—	—	1.3
LLB.	3.5	—	—	—	—	0.7
Divinity	0.1	0.5	—	—	—	—
Other	2.2	2.3	0.9	1.6	—	1.3

*Started as engineering frosh, Fall 1966.

Source: ACT-5 Code Handbook

Table 6 Most Important College Goal of Freshmen, University of Illinois-Urbana, Fall, 1966.

	All 1966 Frosh	Engineering Grads 1971	Other* Grads 1971	Still in Engrg. School 1971	Others* Still in School 1971	College* Drop Outs
Enjoy Life	0.3%	—	—	—	—	0.7
Develop Mind	48.1	44.7	40.4	58.1	44.1	39.9
Job Training	44.4	49.3	53.2	32.3	47.1	53.8
Desirable Marriage	0.2	—	—	—	—	—
Raise Income**	3.2	4.1	4.6	8.1	8.8	5.0
Develop Ethics	—	—	—	—	—	—
Develop Culture	1.1	—	—	—	—	0.7
Develop Personality	0.7	—	0.9	—	—	—
Develop Philosophy	0.9	0.5	0.9	—	—	—
Some Other Goal	1.1	1.4	—	1.6	—	—

*Started as Engineering frosh, Fall 1966.

**This item has questionable validity due to demand characteristics of socially desirable responses. Actual most important goal may be revealed in the next question about *second* most important college goal.

Source: ACT-5 Code Handbook

Table 7 Second Most Important College Goal of Freshmen, University of Illinois-Urbana, Fall, 1966.

	All 1966 Frosh	Engineering Grads 1971	Other* Grads 1971	Still in Engrg. School 1971	Others* Still in School 1971	College* Drop Outs
Enjoy Life	2.7%	4.6	3.7	4.8	—	2.0
Develop Mind	28.5	23.5	23.9	29.0	29.4	27.2
Job Training	29.5	23.0	24.8	22.6	35.3	24.3
Desirable Marriage	1.2	0.5	—	—	—	0.3
Raise Income	17.5	32.7	31.2	21.0	29.4	31.9
Develop Ethics	0.5	—	—	—	—	2.0
Develop Culture	6.6	3.7	5.5	6.5	2.9	3.7
Develop Personality	6.6	5.5	5.5	9.7	2.9	4.7
Develop Philosophy	4.5	4.1	2.8	3.2	—	2.3
Some Other Goal	2.4	2.3	2.8	3.2	—	1.7

*Started as engineering frosh, Fall 1966.

Source: ACT-5 Code Handbook

The tiny percentage of the engineering freshmen selecting the "teacher" role, especially in comparison to the overall frosh class, is consistent with Trow's thesis that future engineers do not see themselves in intellectual roles. The data on educational aspirations of this cohort are moderately consistent with this finding: most of the engineers see their education terminating at the master's degree level (see Table 5). The only clear implication from the data on educational aspirations is that the engineering freshmen have higher goals for achievement than their peers throughout the university.

Purpose of College Education. ACT-5 asked the 1966 freshmen group what they saw as the most important purpose of their college education and then it asked for the second most important purpose. The interest of this study is upon the second goal, for the reason that responses to the first question are likely to be distorted by demand characteristics and social desirability effects. The opinion here is that the second question probably elicited responses more honest than the first.

The only area where the engineering cohort differed from all freshmen was that an appreciable percentage of the engineers indicated a major purpose of their education was to raise their income level. The inference might be drawn that these responses confirm the well known materialistic nature of engineers. Or it might show them to simply be more candid than their fellow students. More interesting, however, is the view that these particular data support Trow's (1959) view that engineering students see their education from a rather instrumental perspective. It shows, again, that they see their prospective engineering degree as a means to elevate the socio-economic status of their family in their own generation.

Table 8 Size of Home Community of Freshmen, University of Illinois-Urbana, Fall, 1966.

	All 1966 Frosh	Engineering Grads 1971	Other* Grads 1971	Still in Engrg. School 1971	Others* Still in School 1971	College* Drop Outs
Farm or Open Country	21.3%	27.6	25.7	14.5	20.6	22.6
Suburban in area of:						
over 2 million	15.2	32.7	12.8	17.7	23.5	14.3
500,000 – 2 million	3.2	21.7	3.7	6.5	—	1.3
100,000 – 499,999	4.9	7.8	3.7	4.8	8.8	4.0
Under 100,000	19.1	5.1	18.3	16.1	11.8	21.3
Central City in area of:						
over 2 million	11.0	3.2	6.4	4.8	8.8	11.3
500,000 – 2 million	1.5	0.5	0.9	—	2.9	2.0
100,000 – 499,999	4.1	0.9	1.8	1.6	8.8	3.7
50,000 – 99,999	7.3	0.5	11.0	12.9	—	7.3
Under 50,000	12.4	—	15.6	21.0	14.7	12.3

*Started as engineering frosh, Fall 1966.
Source: ACT-5 Code Handbook

Size of Home Community. Table 8 reveals that the largest social source of engineering freshmen in this cohort was suburban communities, specifically those in the Chicago metropolitan area. The percentage of freshmen who went on to receive engineering degrees is markedly higher from the suburbs than the percentages of engineering freshmen who did not. It is over twice the percentage of 1966 freshmen as a whole that came from suburbs.

While the successful engineering students were heavily from the suburbs they were almost not at all from central cities of any size. More of the engineering cohort came from rural residences, however, than the total 1966 freshmen class. Of those who started as engineering freshmen, the students who later shifted to other fields or left the university tended to come more from the central cities.

The implication of these hometown figures is not clear. Working from the upward mobility hypothesis, one would have expected more of the engineering cohort to be derived from the central cities and the rural areas. That the bulk of those who completed engineering degrees came from the suburbs clouds the expectation. Only more detailed descriptive information on the ethnic and class dimensions of these suburban sources would provide a better test of the mobility thesis. Data on the high school sources of the engineering freshmen cohort was available by community name but the limits of this study did not justify the effort that would have been involved in developing a social profile of each locale.

Table 9 High School Characteristics of Freshmen, University of Illinois-Urbana, Fall, 1966.

	All 1966 Frosh	Engineering Grads 1971	Other* Grads 1971	Still in Engrg. School 1971	Others* Still in School 1971	College* Drop Outs
Size of Graduating Class:						
Under 25	0.9	0.9	—	—	—	1.7
25 – 99	15.9	14.3	20.2	21.0	20.6	19.6
100 – 400	40.1	39.2	43.1	41.9	26.5	38.5
Over 400	43.1	45.6	36.7	37.1	52.9	39.9

*Started as engineering frosh, Fall 1966.
Source: ACT-5 Code Handbook

Size of High School. Tables 9 and 10 show only that the size of the high school graduating class, from which the 1966 freshmen came, is not a significant variable in analyzing their comparative social origins. There are no appreciable differences between the engineering freshmen and the total freshmen class on this measure. Within the engineering cohort, the only notable difference is that those who later transfered to non-engineering fields tended to come from smaller schools. The trend data in Table 10 reveal a slight movement in the

Table 10 Size of High School Class of Engineering Freshmen, University of Illinois-Urbana, 1966-69.

NUMBER OF STUDENTS IN HIGH SCHOOL CLASS	1966 NUMBER OF ENTERING FRESHMEN	PERCENT	1967 NUMBER OF ENTERING FRESHMEN	PERCENT	1968 NUMBER OF ENTERING FRESHMEN	PERCENT	1969 NUMBER OF ENTERING FRESHMEN	PERCENT
0 — 49	81	8	50	5	55	5	39	4
50 — 99	98	11	80	8	87	10	84	9
100 — 199	132	14	140	15	137	15	148	15
200 — 299	117	13	136	14	118	13	94	10
300 — 399	119	13	115	12	158	18	149	15
400 — 499	115	12	118	13	93	10	141	15
500 — 599	78	8	99	11	67	7	89	9
600 — 699	62	7	85	9	80	9	84	9
700 — 799	42	5	39	4	69	7	61	6
800 — 899	25	3	37	4	21	2	24	3
900 — 999	22	3	7	1	12	1	12	1
1,000 — 1,199	17	2	37	4	29	3	43	4
1,200 — 10,000	1	1	0	0	0	0	0	0
TOTAL	909	100%	943	100%	926	100%	968	100%

Source: College of Engineering.

direciton of larger high schools as the source of engineering freshmen. This statistic is not meaningful by itself because of the general trend in the nation toward larger high schools through consolidation, population growth, and certain urban programs that centralize the education of brighter students.

Table 11 Geographical Origins of Engineering Freshmen, University of Illinois-Urbana, 1966-70.

	1966	1967	1968	1969	1970
Major cities and suburbs in Illinois:	52.0%	54.0%	55.0%	53 %	58 %
Chicago	33.0%				
"Downstate" Illinois:	43 %	42.0%	38.0%	43 %	38 %
Out of State:	5.0%	4.0%	7.0%	4 %	4 %
Midwest	1.0%	2.3%	3.1%	0.7%	1.6%
East	2.2%	1.3%	2.7%	2.5%	1.6%
West	0.9%	0.1%	0.3%	0.2%	0.4%
South	0.9%	0.3%	0.9%	0.6%	0.4%

Source: College of Engineering.

Geographical Origins. The trend data on geographical origins are presented here largely as a basis for making later comparisons with the geographical destination data on the output side of the model for the 1966 engineering freshmen cohort. It will then be possible to test the "Illinois is an exporter of engineers" maxim. The only noticeable trends within Table 11 are the declines in both the non-metropolitan areas of Illinois and other states as sources of prospective freshmen for the University of Illinois-Urbana campus.

Family Characteristics. Table 12 displays no clear relationship between the number of younger siblings in the family and any of the categories of the 1966 engineering freshmen or total freshmen. The same is true for the number of dependent relatives in the family.

Table 13, on the other hand, lends support to Trow's (1959) earlier findings that engineering students tend to come from lower economic strata families than other college students. This was most true for this cohort in the cases of those who succeeded in completing engineering degrees. The engineering freshmen who eventually switched to other fields tended to come from more affluent families than those who

Table 12 Family Size of Freshmen, University of Illinois-Urbana, Fall, 1966.

	All 1966 Frosh	Engineering Grads 1971	Other* Grads 1971	Still in Engrg. School 1971	Others* Still in School 1971	College* Drop Outs
Number of Younger Siblings:						
None	30.2	27.6	34.9	22.6	29.4	26.6
1	30.4	32.7	27.5	35.5	26.5	34.2
2	20.3	21.7	17.4	16.1	23.5	19.3
3	10.4	7.8	9.2	12.9	14.7	10.3
4	4.7	5.1	6.4	9.7	2.9	3.3
5	1.9	3.2	2.8	3.2	2.9	3.0
6	0.9	0.5	0.9	—	—	1.0
7	0.7	0.9	0.9	—	—	2.0
8 or more	0:5	0.5	—	—	—	0.3
Number of Other Dependent Relatives:						
None	84.1	87.1	79.8	88.7	88.2	82.7
1	10.9	7.8	16.5	3.2	11.8	11.0
2	3.5	2.8	3.7	6.5	—	4.0
3	0.6	1.8	—	—	—	1.3
4 or more	1.0	0.5	—	—	—	1.0

*Started as engineering frosh, Fall 1966.
Source: ACT-5 Code Handbook

persisted in engineering. The most affluent category of the engineering cohort, however, were those who remained in engineering school beyond the normal five years. Perhaps their longer tenure was made possible by their parents' more ample financial resources or reflects their higher level of aspirations. By contrast, the engineering freshmen who later left the university altogether tended to be from families grouped in the lower income levels.

Marital or Dating Status. The figures in Table 14 about the marital or dating status of the 1966 freshmen reflect their personal situation as of their senior year in high school. The small percentage of freshmen reporting marriages and children reflect admissions of veterans return to school. The initial impression from these data confirms the

Table 13 Estimated Annual Income of Families of Freshmen, University of Illinois-Urbana, Fall, 1966.

	All 1966 Frosh	Engineering Grads 1971	Other* Grads 1971	Still in Engrg. School 1971	Others* Still in School 1971	College* Drop Outs
Under $5,000	5.9%	8.3	6.4	3.2	8.8	7.0
$ 5,000 – $ 7,499	18.0	23.0	22.0	14.5	20.6	26.9
$ 7,500 – $ 9,999	18.1	20.7	14.7	16.1	23.5	23.9
$10,000 – $14,999	21.7	20.3	26.6	30.6	23.5	19.3
$15,000 – $19,999	5.8	4.6	5.5	4.8	5.9	4.7
$20,000 – $24,999	2.4	0.5	—	—	2.9	1.0
$25,000 and over	2.6	0.9	—	—	2.9	0.3
Confidential	4.7	5.5	6.4	11.3	—	4.0
Don't Know	20.8	16.1	18.3	19.4	11.8	13.0

*Started as engineering frosh, Fall 1966.
Source: ACT-5 Code Handbook

Table 14 Marital or Dating Status of Freshmen, University of Illinois-Urbana, Fall, 1966.

	All 1966 Frosh	Engineering Grads 1971	Other* Grads 1971	Still in Engrg. School 1971	Others* Still in School 1971	College* Drop Outs
Married w/children	0.1%	0.1	—	—	—	—
Married w/o children	0.1	0.5	—	—	—	—
Engaged	—	—	—	—	—	—
Pinned, Going Steady	8.6	5.5	7.3	12.9	11.8	9.6
Date Same Person	18.8	22.6	16.5	19.4	20.6	14.6
Date Several People	53.1	47.0	56.0	48.4	50.0	52.2
Don't Date	14.9	18.4	11.9	14.5	11.8	17.6
Divorced, Widowed	—	—	—	—	—	—
Confidential	4.3	5.5	8.3	4.8	5.8	5.6

*Started as engineering frosh, Fall 1966.
Source: ACT-5 Code Handbook

popular notions, as well as those put forth by Trow (1959), that engineers are something less than socially sophisticated in regard to relations with the opposite sex. Yet, the confirmation here is only margi-

Table 15 Undergraduate Educational Paths of Engineering Freshmen, University of Illinois-Urbana, Fall, 1966 to Spring, 1971

				OUTPUT: Graduate Engineers	INPUT: College of Engineering / Engineers in School	Other Schools in the University / Non-Engineers in School	Non-Engineer Graduates	Not in the University / College Drops Outs
Fall	1966	Frosh	I		100%	—		—
Spring	1967	Frosh	II		85%	9%		6%
Fall	1967	Soph	I		60%	15%		25%
Spring	1968	Soph	II		47%	20%		33%
Fall	1968	Junior	I		42%	22%		36%
Spring	1969	Junior	II		41%	23%		36%
Fall	1969	Senior	I		40%	23%		37%
Spring	1970	Senior	II		40%	22%	1%	37%
Fall	1970	Senior	III	14%	26%	10%	10%	40%
Spring	1971	Senior	IV	30%	9%	4%	15%	42%*

*Considering the typically longer tenure of engineering undergraduates, the comparable college drop out rate for all U of I Urbana frosh entering Fall, 1966 was between 39 and 42 per cent. The drop out rate is decreasing each year.
Source: College Coded Finder Deck, Administrative Data Processing.

nal and of doubtful statistical significance. One is led to expect greater disparity between the engineers and the total freshmen percentages than these figures show.

Within the engineering cohort, however, the percentage differences are more telling. Those who ultimately earned their engineering degrees had the smallest percentage of the cohort involved in serious relationships with girls and the correspondingly largest percentage of those who did not date at all.

Educational Paths of Engineering Freshmen: The Throughput Process

In making a demographic analysis of the engineering college throughput, the 1966 incoming engineering freshmen cohort (N=725) is the basis for calculating the percentages shown in the Table 15 flow chart. Of the freshmen who began college in the engineering programs in Fall, 1966, 30% ultimately received baccalaureate degrees in engineering. Another 9% were still enrolled in the engineering curriculum but had not completed their degrees at the end of five years. Fifteen percent of the original engineering cohort eventually received bachelor's degrees in non-engineering fields. Another 4% remained in non-engineering curricula but had not yet received their degrees after five years of undergraduate study. Finally, by the end of 10 semesters, 42% of the original frosh cohort had left the university altogether.

The rate of attrition of undergraduates from the engineering program is initially high (15%) at the end of the first freshman semester. The rate climbs to its peak (25%) at the end of the freshman year. The exodus is still high at the end of the first sophomore semester (13%), but drops off sharply thereafter, finally stabilizing after the junior year.

Table 16 Choices of Other Areas of Study for Transfers from the Engineering Freshmen Cohort, University of Illinois-Urbana, Fall, 1966. Percentages Are Based on Total of Transfers, Not Total of Cohort.

Liberal Arts	40%
Business	32%
Agriculture	10%
Fine Arts	7%
Physical Education	4%
Communication	4%
Education	2%
Aviation	1%

Source: Administrative Data Processing.

In examining the destinations of the students leaving the engineering program, a minority transfer to other undergraduate studies. The rate of transfer is at its highest point immediately after the first freshman semester (9%) and begins to decline at a decreasing rate thereafter. Table 16 records the rank order of the popularity of other fields chosen by students leaving the engineering curriculum. Liberal arts and business administration programs are the preponderant choices.

Returning to Table 15, the majority of the students comprising the attrition out of engineering were those leaving the university. This exit rate peaks at the end of the freshman year at 19%. Information on the destinations of the college leavers is generally incomplete. It is conceivable that some measures could be developed to at least discover whether the leaver transferred to another institution or not. Even the "not" category begs a number of questions: did the student simply forsake education? Did he enter a trade or vocational program? Did he serve in the military? Will he return to the university after some unspecified interval in which he matures or "finds himself?"

Shedding further light on the fate of this "drop out" category is absolutely necessary before drawing any policy implications or suggestions from this data. It is simply not known in this study whether leaving the university was unfortunate or functional for both the individuals concerned and society, or for what proportions of them.

To make some rough statements on the throughput aspect of the engineering college subsystem, the data in Table 15 show that the subsystem is operating at approximately a 30% efficiency rate, that is, 30% of its original human resource raw material is ending up as a finished product. Another 15% of the raw material is used for producing unintended by-products (graduates in other fields). After five years, 11% of the raw material has not been wasted per se, but is delayed in production, either of the intended finished product or the by-product. But 42% of the human resources taken in are essentially "wasted" in the sense that they do not end up as completed products of the intended or unintended variety. The footnote to Table 15, however, indicates that this "scrap rate" is not inordinate but rather typical of all the undergraduate programs in this university.

One would expect, of course, that a higher human scrap rate could be justified by engineering programs in view of their greater intellectual difficulty. They are much less able than other undergraduate programs to certify a shoddy product as their social output. Faulty economists or sociologists may lead to the destruction of an entire civilization in the long run, but no one will identify the source of the intangible errors; the liberal arts schools will not be held accountable for poor programs. Unprepared engineers, however, would be manifest in the short run by very tangible indicators such as collapsing bridges or faulty circuits. The engineering college at the root of the problems would be surely and swiftly known.

WHERE ARE THEY NOW?: ENGINEERING COLLEGE OUTPUT

Types of Employment. Table 17 shows the distribution of the June, 1971 baccalaureate degree graduates in engineering to types of employment that were recorded by the Engineering Placement Office as of August, 1971. The percentages are based upon the size of the graduating class (N = 336) rather than upon the original 1966 engineering freshmen cohort.

Because this group had only been out of school three months, the figures in Table 17 are not conclusive. The immediate impression is that slightly over a third of the engineering graduates were employed as engineers. This statistic would suggest that only one out of nine of the original engineering cohort eventually was employed by the social system as an engineer. In terms of raw material transformation, questions could be raised about the efficiency of the engineering college as a social subsystem. Such a judgment, however, would be premature. Taking into account the graduates that went on to graduate work in engineering, went into the military, those that are still job hunting, and those that were undecided or did not report, there is a potential for up to 88% of the class to eventually be employed as engineers. Nor can it be said that those who pursued graduate work in other professional fields are cases of subsystem dysfunction in terms of society's needs for skilled manpower.

A surprising figure, in view of government's increasing role in environment control, product standards, urban systems, and transportation development, is that so few of the 1971 graduates were employed by state, municipal or federal agencies (only 2.9%). Only one graduate was employed by an environmental control agency.

The small percentage that pursued MBA graduate programs is also interesting. A decade ago, the engineering bachelor's degree, followed by an MBA, was considered the dream combination and the sure

Table 17 Employment of Baccalaureate Engineering Graduates, Class of June, 1971 as of August, 1971, University of Illinois-Urbana.

34% to Engineering Jobs

26 % to manufacturing, processing, chemical, food, communication, data processing, utility and similar industrial positions.

8 % to non-corporate employment
- 4.8% consulting firms
- 2.0% state government
- 0.6% federal government
- 0.3% municipal government
- 0.3% teaching

25% to Graduate School (17.6% in Illinois)
- 19 % to engineering schools (14% in Illinois)
- 2.6% to law (1.5% in Illinois)
- 2.2% to MBA programs (1.2% in Illinois)
- 0.6% to medicine (0.3% in Illinois)
- 0.6% to labor relations, in Illinois.

13% to Military Service

12% Still Job Hunting

9.8% No Report

1.2% Undecided

4.7% to Miscellaneous Pursuits:

peace corps, second bachelor's degree, own business, head start, overseas volunteer work, etc.

Percentages are based on total of June, 1971 graduating class (N = 336) and not upon total of 1966 freshmen cohort.

Source: Engineering Placement Office.

route to the corporate executive suite. The frustrations and disappointments of young men who were grossly overeducated for the roles that they were actually to occupy in large organizations became legend. Apparently, the "word" has filtered down to the undergraduates. Then, too, the 1971 job market exhibited an inverse relationship between graduate education and employability.

Table 18 Geographical Distribution of Baccalaureate Engineering Graduates, in Engineering Jobs, University of Illinois-Urbana, Classes of 1960, 1965, 1971.

	Class of 1960 After 10 Years	Class of 1965 After 5 Years	Class of 1971 After 3 Months
To Chicago and Suburbs:	19 %	25.3%	36 %
To "Downstate" Illinois:	17.8%	14.9%	31 %
To Out of State:	63.2%	59.8%	33 %
Midwest:	19.5%	20.0%	16 %
West:	22.1%	25.8%	5 %
East:	17.0%	9.1%	8 %
South:	4.6%	4.7%	4 %

Source: Engineering Placement Office.

Geographical Distribution. Table 18 portrays the regional locations, not of the entire 1971 class, but of those graduates in engineering employment. Several interesting shifts have taken place in the trend data. The Chicago metropolitan area is increasingly employing more of the engineers from the University of Illinois-Urbana. But then so are the "downstate" areas of Illinois. These two percentage climbs came from the third category: out of state employment. And within that category, the region exhibiting the sharpest decline in employment of Illinois engineers is the West Coast. This is followed by the Eastern Seaboard, and then the other midwestern states.

These figures clearly demonstrate the impact, on the geographical dispersion of new engineers, of the collapse of the aerospace industry. In 1960, almost two thirds of the University of Illinois engineering graduates were heading out of state after graduation. The proportion leaving the state was still a clear majority as late as 1965. Last year, however, two thirds of the engineering graduates remained within Illinois.

In looking back at the geographical origins of this cohort (Table 11) we find the at Illinois still appears to be a net exporter of engineering manpower. Only 5% of the freshmen engineers were from out of state, yet 33% of the graduates left Illinois. In adjusting these percentages to the same base, however, the picture is quite different, showing a slight shift to a net import situation. Approximately one in twenty-seven of the 1966 freshmen cohort took out of state engineering jobs in 1971 while one in twenty of that cohort came from homes outside of Illinois. In terms of a regional role, then, the engineering colleges may now be producing skilled manpower that is more than wholly retained within the locale of its social origins.

CONCLUSION

A concomitant concern in this effort to explore a methodological prototype for studying the engineering college as a social subsystem was that the findings could lead to suggestions for governmental policy decisions regarding the use of technological manpower. The burden of this report is that offering such suggestions at this point would be premature. The empirical findings here are, at best, rather sketchy and incomplete. Moreover, they represent but one college of engineering with its own peculiarities of location, sources of students, and latent social functions. The exercise, however, appears to have succeeded in identifying the nature and sources of the data required to model the engineering college as a social transformation subsystem. It also identified the difficulties that exist in acquiring this information.

The input/thruput/output analysis appears to be the appropriate foundation for examining the system role of the engineering college, especially in exploring its unintended, nonrational, that is, its social consequences and functions. If some part of the population is entering an undergraduate engineering program for outcomes other than providing society with engineering expertise, what are these hidden agendas of rationales or purposes? Is it to raise their family status? Or is it a ticket to California? Or a path to the executive washroom? Knowing where new engineers come from, where they go, and what happened to them during four to five years of college should provide the answers.

Identifying the path of a cohort through its undergraduate years is the key to the model. All schools possess this data in *some* form. The crucial question is whether it exists in an easily retrievable and manipulable state as was the "finder" deck employed here. In some institutions, such data may exist only as manual notations on student file folders.

From impressions gathered during meetings of the steering committee of the larger study, it appears that data on the employment and geographical dispersion of engineering graduates is a particularly weak area. Representatives of several schools evidenced histories of past difficulties in obtaining, from their alumni, the same kinds of data that were examined in this report, such as type of job and especially salary information.

A particularly difficult problem in this study was matching up data on a freshmen group at "time one" with data on the eventual survivors of that cohort at some "time two" point beyond graduation. This problem may have been unique to this institution at this time because of the identification code changeover. Another aspect of this difficulty, however, was that the data required here was not available from a central, standardized source, not even in an administrative system as presumably sophisticated as this university uses. Five distinct administrative agencies were contacted in order to follow a single freshmen cohort from preadmission testing through to graduation. This problem, however, might conceivably be lessened in smaller institutions. If size and diversity leads to computerized data files, as I hypothesized earlier, then it also leads to frustrating and bewildering divisions of the administrative labor.

Given the richness of the data retrieved from the University of Illinois system, much more could be done with it in terms of the sophistication level of the analysis. Multivariate analysis of this type of data would be entirely appropriate. Elton and Rose (1971) factor analyzed similar variables, including ACT test scores, to distinguish engineering students who persisted from those who left the field. Indeed, some of the ground-breaking work in mathematical sociology is coming out of research on students' educational and occupational aspirations (Haller and Woelfel, 1971). The data here, or data similar to it from other institutions, could be similarly treated. The variables antecedent to the undergraduate attrition from engineering programs across time might be studied through path analysis. The same might be done with the antecedents to choices of types of employment taken after graduation.

Finally, this effort suggests that a number of variables not available here should be considered to portray completely the social process and meaning of the undergraduate experience for engineering students. In particular, data is needed on the social psychological impact of a college education upon the attitudes, beliefs, and life styles of future engineers. Moreover, the intergenerational social mobility hypothesis cannot be conclusively tested without further information on the family income and parental occupations and education levels of engineering students. More information is needed, too, on the social psychological origins of these students, including family size, birth order, family authoritarian/permissive climate, and school counseling they have received.

REFERENCES

1. Becker, H.S., "An Analytical Model for Studies of the Recruitment of Scientific Manpower," *Science Manpower, 1958*, Washington, D.C.: National Science Foundation, 1959.
2. Chapman, P., "University of Illinois College of Engineering 1960 Graduates 10 Years Later, Where Are They Now?" University of Illinois, Urbana, 1970a.
3. Chapman, P., "University of Illinois College of Engineering 1965 Graduates, Where Are They Now?" University of Illinois, Urbana, 1970b.
4. Chapman, P., "Salary Survey, June 1971 Graduates," College of Engineering Placement Office, University of Illinois, Urbana, 1971. (Mimeographed)
5. Elton, C. and H. Rose, "Students Who Leave Engineering," *Engineering Education*, Vol. 62, Sept.-Oct., 1971, pp. 30-32.
6. Nie, N. H., D. H. Bent and C. H. Hull, *Statistical Package for the Social Sciences*, New York: McGraw-Hill, 1970.
7. "Educational Profile for Entering Freshmen for the Fall Semesters 1966, 1967, 1968, 1969," College of Engineering, University of Illinois, Urbana, November 1969. (Mimeographed)
8. Parsons, T., "Professional Training and the Role of the Professions in American Society," *Science Manpower, 1958*, Washington, D.C.: National Science Foundation, 1959.
9. SOUPAC, *Soupac Program Descriptions: Statistically Oriented Users Programming and Consulting*, Urbana: Department of Computer Science, University of Illinois, 1971.
10. Stice, G. et al., *Background Factors and College-going Plans Among High Aptitude Public High School Seniors*, Princeton: Educational Testing Service, 1956.
11. Trow, M., "Some Implications of the Social Origins of Engineers," *Scientific Manpower, 1958*, Washington, D.C.: National Science Foundation, 1959.
12. National Science Foundation, *Scientific Manpower, 1958*, Washington, D.C., 1959.

13. National Science Foundation, *Scientific Manpower, 1960*, Washington, D.C., 1961.
14. Woelfel, Joseph, Haller and Archibald, "Significant Others, the Self-Reflexive Act and the Attitude Formation Process," *American Sociological Review*, Vol. 36, February 1971, pp. 74-86.

Appendix

Act-5 Codebook

University of Illinois

CARD LAYOUT

1966 ACT Class Profile Supplementary Punched Card Reports

Class Profile Supplementary Punch Cards are generated at the time Class Profile Service reports are processed. Information for a student is obtained from the student's responses to Student Profile Section (SPS) items.

INFORMATION	*CARD COLUMN*	
Identification	(1-24)	
Social Security No.	1- 9	
Student Name	10-24	Printed (Interpreted)
Academic Potential	(25-26)	ACT Composite
Goals and Aspirations	(27-34)	*SPS Items*
Major Field	27-28	1. From the list of Vocational Fields (attached) find the field that best describes your future educational or major field. Mark the code number for that field on your answer sheet. Indicate only one field. If you are undecided about your major field, mark "oo" on your answer sheet and go on to the next question.
Vocational Choice	29-30	2. From the list of Vocational Fields (attached) find the field that best describes your future *vocation,* and mark its code on your answer sheet. Again, if you are undecided about

INFORMATION	*CARD COLUMN*	*SPS Items*
		your future vocation, mark "00" on your answer sheet. If your future vocation is not included in these fields, mark, "95" on your answer sheet, or if you anticipate your future vocation to be excusively that of housewife, mark "96" on your answer sheet and skip question 3.
Vocational Role	31	3. Which of the following alternatives describes the main role you expect to play in your future vocation? (For example, if you want to be a physicist and work primarily as a researcher, you would mark "1". If you want to be a doctor who specializes in private practice, you would mark "5". An engineering major who plans to become a sales engineer, should mark "4". A teacher who wants to become a principal should mark "3". An art major who plans to become a professional artist should mark "5", etc.).
		Researcher or investigator 1
		Teacher or therapist 2
		Administrator or supervisor ... 3
		Promoter or salesman of services or products 4
		Practitioner, performer, or producer of services or products 5
		None of above 6
		Two or more roles 7
		Don't know or undecided 8

INFORMATION	*CARD COLUMN*	*SPS Items*
Level of Educational Aspiration	32	4. What is the highest level of education you expect to complete? High school diploma . . . . 0 Vocational or technical program (less than two years) . . . . 1 Junior college degree . . . . 2 Bachelor's degree or equivalent . . . . 3 One or two years of graduate or professional study (M.A., M.B.A., etc.) . . . . 4 Doctor of Philosophy (Ph.D.) . . 5 Doctor of Medicine or Dental Surgery (M.D. or D.D.S.) . . . . 6 Bachelor of Laws (LL.B.) . . . . . 7 Bachelor of Divinity (B.D.) Other 9
Most Important College Goal	33	10. What is your most important goal in attending college? To learn how to enjoy life . . . . 0 To develop my mind and intellectual abilities . . . . 1 To secure vocational or professional training . . . . 2 To make a desirable marriage . 3 To earn a higher income . . . . . 4 To develop moral standards . . 5 To become a cultured person . 6 To develop my personality . . . . 7 To develop a satisfying philosophy . . . . 8 None of these 9

INFORMATION	*CARD COLUMN*	*SPS Items*
Second Most Important College Goal	34	11. What is your second most important goal in attending college? To learn how to enjoy life .... 0 To develop my mind and intellectual abilities .... 1 To secure vocational or professional training .... 2 To make a desirable marriage . 3 To earn a higher income ..... 4 To develop moral standards .. 5 To become a cultured person . 6 To develop my personality .... 7 To develop a satisfying philosophy .... 8 None of these 9
Student Personnel Needs	(35-47)	
Housing Expectations	35	9. Where do you expect to live while attending college? College dormitory .... 1 Fraternity or sorority .... 2 College apartment .... 3 Off-campus apartment .... 4 Off-campus room .... 5 At home (or with relatives) 6
Hours of Part-time Work Expected	36	26. About how many hours do you expect to work part-time while attending college? (Exclude summer work) 1- 9 hours per week 1

INFORMATION	*CARD COLUMN*	*SPS Items*
		10-19 hours per week .... 2 20-29 hours per week .... 3 30 or more hours per week ... 4
Extracurricular Intentions		Questions 12-20 list extracurricular activities that may or may not interest you. Use the responses printed immediately below to respond to all the questions in this group. Yes, I do plan to participate .. Y No, I do not plan to participate N Be sure to respond to every question by marking the appropriate alternative on your answer sheet.
	37	12. Athletics — intercollegiate
	38	13. Music
	39	14. Writing
	40	15. Student Government
	41	16. Science Clubs and Projects
	42	17. Debate
	43	18. Acting
	44	19. Departmental Clubs related to my major field
	45	20. Athletics — intramural
Anticipated Financial Need	46	29. Will you need financial aid from either a scholarship or loan?

INFORMATION	*CARD COLUMN*	*SPS Items*
		Yes, during my freshman year and probably thereafter .... 1 Yes, but probably not during my freshman year .... 2 Probably never 3
Automobile Intentions	47	30. Do you expect to take a car to campus? Yes .... 1 No 2
Non-Academic Achievements	(48-53)	This reports the number (0 to 8) of the student's notable achievements in high school in each of six areas. All items refer to achievements outside of the classroom, not as a part of course assignment.
Science	48	
Art	49	
Dramatic Arts	50	
Literature	51	
Leadership	52	
Music	53	
College Attractions	(54-70)	Questions 31-57 deal with factors that may have influenced your choice of a college. The responses printed immediately below are to be used for all the questions in this group. Of no importance .... 1 A minor consideration .... 2 A major consideration 3
Intellectual Reputation	54	31. Good faculty
	55	32. High scholastic standards
	56	37. Special Curriculum I wanted

INFORMATION	*CARD COLUMN*	*SPS Item*
	57	54. Desirable intellectual atmosphere
	58	55. National reputation of the college
Practical Considerations	59	36. Desirable location
	60	41. Low-cost college
	61	43. Close to home
Outside Influences	62	44. Advice of parents
	63	48. Advice of high school teacher(s)
	64	49. Advice of high school counselor or college counselor
	65	50. Talk with admissions counselor from college
	66	53. College offered me a scholarship or other financial aid
Social Factors	67	33. Desirable social climate and activities program
	68	42. Good athletic program
	69	52. Has fraternities and sororities
	70	56. Coeducational
Demographic Data	(71-80)	
Type of Home Community	71	5. Which of the following best describes the community which you think of as your home town during high school days?

INFORMATION	*CARD COLUMN*	*SPS Items*
		Farm or open country 0
		Suburb in a metropolitan area of
		more than 2 million population 1
		500,000 to 2 million 2
		100,000 to 499,999 3
		less than 100,000 4
		Central city in a metropolitan area or city of
		more than 2 million population 5
		500,000 to 2 million 6
		100,000 to 499,999 7
		50,000 to 99,999 8
		less than 50,000 9
Younger Children in Family	72	6. How many younger children in your family will be living at home when you enter college?
		None 0
		One younger child 1
		Two younger children 2
		Three younger children 3
		Four younger children 4
		Five younger children 5
		Six younger children 6
		Seven younger children 7
		Eight or more younger children 8
Other Family Dependents	73	7. How many other dependents (grandparents or other dependent relatives) will be living at home when you enter college?
		No dependent relatives 0

INFORMATION	*CARD COLUMN*	*SPS Items*
		One dependent relative .... 1 Two dependent relatives ...... 2 Three dependent relatives .... 3 Four or more dependent relatives 4
Estimated Family Income	74	8. What do you *estimate* your family's income to be? (Indicate total income before taxes.)
		Less than $5,000 per year ..... 1 $5,000 to $7,499 .... 2 $7,500 to $9,999 .... 3 $10,000 to $14,999 .... 4 $15,000 to $19,999 .... 5 $20,000 to $24,999 .... 6 $25,000 and over .... 7 I consider this information confidential .... 8 I don't know 9
Age	75	22. How old will you be next September first?
		15 years or younger .... 0 16 years .... 1 17 years .... 2 18 years .... 3 19 years .... 4 20 years .... 5 21 years or older 6
Marital, Dating Status	76	23. What is your current marital or dating status?
		Married (children or expecting) 1

INFORMATION	*CARD COLUMN*	*SPS Items*
		Married (no children) .... 2 Engaged .... 3 Pinned or going steady .... 4 Usually date same person ..... 5 Usually date more than one person .... 6 Do not date at all .... 7 Divorced or widowed .... 8 I consider this information confidential 9
Type of High School Attended	77	27. What type of high school did you attend? public .... 1 Private — nondenominational . 2 Private — church-related 3
Size of High School Attended	78	28. About how many students will be (were) in your high school graduating class? fewer than 25 .... 1 25-99 .... 2 100-399 .... 3 400 or more 4
Sex	79	M denotes male; F, female
College Choice	80	In order of preference, give the three institutions or agencies you have chosen to receive your test results. This college or agency is my first choice .... 1 This college or agency is my second choice .... 2 This college or agency is my third choice 3

List of Vocational Fields

For card columns 27-38 (Major Field) and 29-30 (Vocational Choice)

Social, Religious, & Educational Fields

Counseling and Guidance	10
Educ. Administration	11
Elementary Educ.	12
Home Economics	13
Special Education	14
Phys. Education	15
Psychology	16
Secondary Education	17
Social Work	18
Sociology	19
Theology and Religion	20
Social Sciences	21
Area Studies	22
Amer. Civilization	23
American Studies	24
Library and Archival Science	25
General Education	01

Administrative, Political & Persuasive Fields

Advertising	26
Business Admin.	27
Law	28
Public Admin.	29
Industrial Relations	30
Merchandis. and Sales	31
Military	32
Political Science and Government	33
Internat'l Relations	34
History	35
Foreign Services	36
Government	37
Public Relations	38

Business & Finance

Accounting	39
Business and Commerce	40
Economics	41
Secretarial Science	42
Finance	43
Data Processing	02

Scientific Fields

Anatomy	44
Anthropology	45
Archeology	46
Astronomy	47
Biology	48
Botany	49
Chemistry	50
Entomology	51
Geography	52
Geology	53
Genetics	54
Math. and Statistics	55
Meteorology	56
Oceanography	57
Physics	58
Physiology	59
Zoology	60

Agriculture & Forestry

Agriculture	61
Fish and Game Management	63
Forestry	64

Soil Conservation Work	67
Medical Fields	
Dental Hygiene	68
Dentistry	69
Dietetics	70
Medicine	71
Medical Tech.	72
Mortuary Science	73
Optometry	74
Osteopathy	75
Pharmacy	76
Veterinary Medicine	77
X-Ray Tech.	78
Nursing	79
Occupational Therapy	80
Physical Therapy	81
Arts & Humanities	
Art and Sculpture	82
Architecture	83
Creative Writing	84
Arts & Humanities (Cont.)	
Drama and Theater	85
English and Eng. Literature	86
Foreign Lang. & Literature	87
Journalism	88
Radio-TV-Communications	89
Music	90
Philosophy	91
Speech	92
Other Arts and Humanities	93
Liberal Arts	94
Engineering & Engineering Tech.	
Architectural and Civil	03
Electrical and Electronic	04
Mechanical	05
Chemical	06
Industrial Arts	65
Other	07
Trade & Industrial	
Aviation	08
Construction	09
Electricity and Electronics	62
Metal and Machine	66
Mechanics	97
Drafting	98
Other Trade	99
My future field of training is not included in the fields listed above	95

Cooperative Engineering Education: A Case Study

Gary Erenrich

This report reviews the Cooperative Education Program in engineering at Northwestern University. Cooperative education, a new phase of engineering education in America, was conceptualized in 1906 through the efforts of Dean Herman Schneider, the initiator of cooperative engineering programs at the University of Cincinnati. Cooperative education is a program in which the student works in industry while pursuing his undergraduate education. In 1926, Dean Schneider created the Association of Co-operative Colleges. From its beginning, with only sixteen engineering colleges represented, the Association provided a forum for the study of problems inherent in cooperative education and for the interchange of ideas and experiences among colleges.

Since 1926, the number of institutions providing cooperative education has been increasing, with the most significant expansion occurring in the last two decades. The Cooperative Education Association states that in 1950 there were 35 co-op programs in both engineering and technical institutions as compared with 197 in 1970.

There are essentially four variations of cooperative education with each variation emphasizing different phases of the program: (1) industrial work is an integral part of the curriculum, with the instruc-

Gary Erenrich is a graduate student in the Department of Civil Engineering at Northwestern University, Evanston, Illinois.

tion of the student continuing while in industry; (2) industrial work is selected on the basis of the student's field of specialization and the student is treated like the regular employees; (3) industrial facilities are used for research by the student who remains under supervision of his university institution, and (4) part-time industrial employment with little relationship between the two.

In 1939 Walter P. Murphy bequeathed approximately seven million dollars to Northwestern University for the establishment of the Technological Institute. Mr. Murphy had suggested that the cooperative education plan be followed. When the Institute began operating, the cooperative approach was pursued by all students. This policy of mandatory co-op program at Northwestern can be categorized as variation (2) above, with industrial work selected on the basis of the student's field of specialization, and in which the student in industry is treated as a regular employee.

A student in the Cooperative Education Program at Northwestern University graduates after five years, of which 18 months is allocated to industrial employment and the remaining time is spent in academic work required for the Bachelor's degree. The student elects the co-op option after the completion of his freshman year. If admitted into the program, the student pursues a schedule of alternating periods of college and industry experience which typically appears as illustrated in Table 1.

Table 1 Typical Co-op Schedules
QUARTER

YEAR	Fall	Winter	Spring	Summer
Freshman	1	2	3	Vacation
Sophomore	4	5	6	Work
Junior	7	8	Work	Work
Pre-Senior	9	Work	10	Work
Senior	Work	11	12	———

The co-op program in the past has been administered by three individuals who determine admissions to the program and student job placement in industry. The administrators are also concerned with maintaining good relations with the participating industries. While the cooperative program is operated under the auspices of the Technological Institute, the Institute, its faculty, as a whole, and students assume relatively little control over the program.

Furthermore, there are not specific written procedures or requirements for admission to the program. Admission is based on judgmental evaluation of cumulative grade point average and empirical observations of such characteristics as the appearance and poise of the applicant. Students who are more highly evaluated generally receive the most desired jobs while those who might have a lower cumulative grade point receive jobs if positions can be found by the program administrators, or if the student finds his own co-op job.

At the end of each of the six co-op periods the student is required to complete a report form detailing general information about his employer, including working hours, salary, and a brief description of the job and an evaluation of his experiences. This report form is usually the only contact the student has with the administrators unless there is a specific problem such as a layoff, change in assignment, or student disenchantment with his experience on the program.

Studies have been conducted to attempt to determine the academic merits of cooperative engineering programs. One major study was designed by H. Stuart Smith, Director Cooperative Program, Virginia Polytechnic Institute. Smith tested and confirmed two hypotheses: (1) the cooperative engineering student will perform better academically than the non-co-op student; and (2) cooperative students experience a lower rate of attrition than non-co-op students. A similar study was conducted at Northwestern University by the Cooperative Division on the entering classes of 1960 and 1961. However, the results of this study may not be comparable to those of the VPI study because admission to the Northwestern Cooperative Program is based in part on cumulative graduate point average, which means that any academic advancement of co-op students over non-co-ops may be attributable to pre-selected ability and not necessarily the co-op experience. A similar argument can be constructed for the attrition hypothesis. Marginal students who inherently have a high probability of dropping out are not admitted to the program, and thus co-op students can be expected to have a smaller attrition rate regardless of the effect of the co-op experience. In sum, the significance of the academic merits of the cooperative program at Northwestern University cannot be determined without more sophisticated experimental and statistical studies because of the admission procedures of the program.

In addition to the problems associated with the admission policy, data statistically aggregated by engineering discipline (department) and by class failed to exhibit significant trends largely because of the small number of cases within each classification. For example, in 1966, there were eight sophomores in Civil Engineering—two of the students were non-co-op (25%), and six of the students entered the co-op program (75%). One year later, the sophomores were now juniors and the total population had decreased from 8 to 5 students. Two students were non-co-op (40%) and three students were co-op (60%). These data illustrate that, because of the small population, each individual student becomes very significant. In light of this characteristic of the data, correlations between departments and class of students were impossible.

Therefore, statistics were aggregated for a ten-year period, to permit analysis of the percent co-op by each department over a period of nine years, and by class over a period of ten years. These data are presented in Table 2.

Table 2 Percent of Co-op Students by Department, from 1962 to 1970

Department	*Average Percent Co-op**	*Standard Deviation*
Civil Engineering	69.2	4.84
Chemical Engineering	72.7	9.38
Electrical Engineering	74.2	9.97
Industrial Engineering	74.2	6.16
Mechanical Engineering	72.5	4.97
Science Engineering	61.8	6.47

*Freshman were excluded because co-op decision is not made until sophomore year.

There are two observations to be made concerning Table 2. First, no major variations in co-op participation between departments seems to exist. However, it should be noted that average percent co-op in science engineering is lower than the other five departments, and this may stem from the fact that a student majoring in science engineering receives a very broad engineering background with no specific area of expertise; this is not the case in other engineering departments. Science engineering tends to produce candidates for Graduate School in such fields as nuclear engineering and applied math. Thus, the science engineering student may not elect a five year undergraduate program because he would rather spend the fifth year in graduate school; furthermore, finding co-op jobs appropriate to his area of interest may be difficult. The second observation from Table 2 is that the standard deviations appear to be rather high (near 10% in chemical and electrical engineering). Further examination of the data revealed a sharp decrease in percent co-ops starting with the sophomore class of 1968 and continuing to the present. Table 3 (below) illustrates percent co-op by class from 1961 to 1971.

Table 3 Percent of Co-op Students by Class, from 1961 to 1971

		Percent Co-op		
Year	*Total*	*Seniors*	*Juniors*	*Sophomores*
1961	427	64	74	72
1962	434	65	78	71
1963	398	61	73	72
1964	435	66	72	78
1965	425	67	72	73
1966	404	66	68	71
1967	400	71	69	73
1968	392	66	71	57
1969	428	69	52	53
1970	460	55	51	35
1971	486	45	34	25

Note that this decreasing trend in percent co-op as illustrated in Table 3 originates with the sophomore class in 1968 and follows its cohort in 1969 and 1970. This trend appears in more than one class of students, since it continues in the sophomore classes of 1969, 1970 and 1971.

The source of this trend is of interest in attempting to evaluate cooperative engineering education programs. The absence of a comprehensive data set, along with the complex nature of the problem, namely modeling the student's decision-making process, suggests that only alternative causal hypotheses can be formulated. There are two levels of factors that may explain the observed trend. The first level involves national influences upon student decision-making, while the second involves university influences.

In the mid 1960's the nation began to feel the effects of an economic recession which limited the availability of industrial jobs. The students already on the cooperative program began to discover that the jobs were harder to keep, and layoffs were not uncommon. A layoff has rather serious consequences to the co-op student. First he must attempt to find a new job in his area of specialization in an extremely restricted job market. If the student cannot find another job, he discovers that reentry to the university is very difficult because the courses he will need to graduate are not taught out of sequence.

Thus, the jobless co-op student must remain idle during his co-op period. In addition, the new student may not elect co-op because during a period of economic recession, it appears favorable to the student to remain in school.

Another key factor in the student's decision may be the view of industry that seems prevalent among college students today. Students have outwardly expressed disillusionment with industry because of its role in (1) furthering the military-industrial complex and (2) preserving the status quo in society. Along with their disillusionment with industry comes a disillusionment with traditional engineering values and the engineer's role in society. Together, all of these influences might be expected to lead to reduced interest in the co-op program.

Likewise major changes have occurred within Northwestern which could possibly explain this downward trend of enrollment in cooperative education. The first was a major change in the Northwestern financial aid policy in March of 1967. Prior to that date, a student could receive aid only if his cumulative grade-point average was above a designated level. This policy encouraged students with grades below the cutoff to elect cooperative education as a source of income.* However, with the change in the financial aid policy, financial aid may be granted to the student regardless of his grade point average. Thus, the poorer student in financial need would not have to elect co-op to pay for his education.

The second university change occurring at about this time involved a significant increase in tuition. Thus, the student who elected co-op would find it increasingly difficult to pay for his education. The tuition increase along with change in financial aid policy clearly favors the non-co-op student.

Thus, from the above discussion it is evident that there are apparently significant linkages between cooperative engineering education programs and the conditions of the national economy, industry, and cultural attitudes towards institutions. Internal university policies also influence the operations of the cooperative element of engineering education. Further analysis of the effect of Northwestern's financial aid policy is oriented at this point to better characterize the impact of co-op programs on students themselves. This may also shed additional light on the role of cooperative education in contemporary engineering colleges.

*While precise criteria have not been formulated, it appears that the grade point requirements for admission to the Northwestern Co-op Program were somewhat less stringent than the Pre-1967 financial aid requirements.

A major area of controversy in the evaluation of co-op programs is the net financial benefit provided to the student. It is possible to utilize compiled data on the incomes of summer jobs for non-co-op and co-op jobs to evaluate these benefits. While such data are limited in scope, they may reflect the general trends. However, sufficient data are not available to produce precise and reliable benefit computations. Approximate benefit computations are show in Table 4.

Table 4 Typical Financial Gains of Co-op Students vs. Non-Co-op Students

Non-Co-op					
Quarter	Income	Financial Aid**	Quarter	Income	Financial Aid***
1		660	1		660
2		660	2		660
3		660	3		660
Summer	700		Summer	700	
4		660	4		660
5		660	5		660
6		660	6		660
Summer	700		Co-op	1713	
7		660	7		40 or 190
8		660	8		40 or 190
9		660	Co-op	1800	
Summer	700		Co-op	1875	
10		660	9		40 or 190
11		660	Co-op	1950	
12 Graduation		660	10		40 or 190
1*			Co-op	2040	
2*			Co-op	2100	
3*			11		40 or 190
4*	8520		12Graduation		40 or 190
5*			1*		
6*			2*		
7*			3*		
8*	9420		4*	9000	
Total	$20040	$7920		$21178	$4200 or $5100
Total income and financial aid					
	$27,960			$25,378 or $26,278	

* Quarters worked *after* graduation.

** Average financial aid given to student on financial aid.

*** Average financial aid given to co-op student on financial aid with lower figure given to student living at home while co-oping, and larger figure given to student co-oping away from home.

The data in Table 4 illustrates the total financial gain (income plus financial aid) of non-co-op students. Six typical cases are evident from Table 4. It should be noted that all students pay the same tuition, so average levels of financial aid may reasonably be treated as income. Since co-op students delay full time employment for one year, the non-co-op average salary during the 5th year after entry into the engineering program must be included in the computation.

The following 6 cases are helpful for identifying the most advantageous options for students in a monetary sense:

Case A: Typical non-co-op student on financial aid.

Income	$20,040
Financial Aid	7,920
Net Financial Gain	$27,960

Case B: Typical non-co-op student not on financial aid.

Income	$20,040
FIA	00
NFG	$20,040

Case C: Typical co-op student living at home while co-oping and on financial aid.

Income	$21.178
FIA	4,200
NFG	$25,378

Case D: Typical co-op working away from home on financial aid.

Income	$21,178
FIA	5,100
NFG	$26,278

Case E: Typical co-op living at home while co-oping not on financial aid.

Income	$21.178
FIA	00
NFG	$21,178

Case F: Typical co-op away from home not on financial aid.

Income	$21,178
FIA	00
NFG	$21,178

Case A appears to be the most financially desirable for the student, with Case D the second most desirable. The difference between Case A and Case D is $1,682. It should be noted that the largest difference between the cases is between Case A and Case B, with a difference of $7,920. Futhermore, approximately 60% of all Technological Institute students receive financial aid, and for these students Case A (or non-co-op) is financially the best. For the remaining 40% of the students, Case E or F would be the best choice. Thus, for students receiving aid, the co-op program is not financially attractive; if such students follow the co-op path, it must be because the perceived value of the experience outweighs the financial benefits. The response of students to changes in the Northwestern financial aid policy, namely, the precipitous decrease in co-op enrollment, suggests that the co-op program is viewed by the majority more as a source of monetary support than as a valuable education experience.

Two consequences seem to follow: (1) As tuition increases so will the average financial aid and especially the *number* of students on financial aid. Therefore, the cooperative program will become financially still less desirable than the non-co-op program. Thus the proportion of co-op students will tend to further decrease; (2) During an economic recession, co-op jobs are difficult to find and hold, thus making it advantageous to be a full time student.

An analysis of the geographical distribution of students enrolled in the Technological Institute as well as the geographical distribution of cooperative work experience may illustrate, in part, the impact of the co-op program on various industrial employees. For example, the Technological Institute attracts about 70% of its students from the Great Lakes States (Wisconsin, Illinois, Indiana, Michigan, and Ohio), while an estimated 60% of the cooperative students work in the Chicago Metropolitan Area. One possible explanation is that most students on co-op prefer the benefits of being close to the University community. Thus, the co-op students can receive some of the benefits of the University (dormitories, libraries, athletic facilities, social events) while having no academic obligations. As a result of the high concentration of cooperative students in the Chicago Metropolitan Area, university policy affecting cooperative education has its most immediate effect on Chicago area industries—and, similarly, the characteristics of Chicago industry can have important effects on university programs. For instance, the number of students co-oping has drastically decreased from approximately 300 per year to 125 per year in the last few years, thus decreasing any benefits to industry which the cooperative program may offer. (The strongest effect of this decline has been on Chicago area industries.) In any case the local focus of the Northwestern co-op program may represent one of the more significant effects which Northwestern has on its regional environment.

Until this point concern has been centered upon trends in cooperative education and the effects on cooperative education of the state of the economy and of the financial policy of Northwestern. The discussion now turns to the advantages and disadvantages of the Cooperative Education Program with respect to the students, the university and industry. An informal survey of cooperative students identified the advantages to the student of the Cooperative Education Program as (1) financial aid and (2) goal direction. The financial advantage to the student lies in giving him the ability to pay for his education from the salary earned while on co-op, thus reducing the financial burden on his family and the university. The second main advantage to the student is that as a result of his cooperative education experiences, the student can better determine his future directions (both academically while at the University and vocationally when he leaves the University). A secondary advantage of cooperative education is that the school year does not consist of going to school for three straight quarters, but is broken up with co-op work periods, thus eliminating some of the long term rigors of student life.

The disadvantages which the students perceived are (1) the Cooperative Education Program involves an extra year over the four year (non-co-op) program; (2) the co-op students will lose financial aid from the University in proportion to the money he earns by working (co-op earnings are considered in establishing financial aid levels); and (3) sequence course work is sometimes split because of the co-op work periods (e.g. a student might have to wait one year to take the second quarter of a two course sequence).

The University benefits from the Cooperative Education Program are (1) the University saves on financial aid for the cooperative students (however, the university does incur the administrative costs of the program); (2) the university improves its relations with industry, especially in its immediate vicinity, and particularly with respect to placement of its graduates into industrial positions; and (3) the Cooperative Education Program does attract some prospective students to the Technological Institute of Northwestern University. The main disadvantage of the program to the University is its cost. As the number of cooperative education students declines, the financial attractiveness to the University of the program will most likely become marginal. And while the University seems to be supporting the Cooperative Education Program, its financial aid policies appear to put the program at a disadvantage.

The advantages to industry are more apparent than the advantages to the University. By participating in the Cooperative Education Program, industry (1) improves relationships with the university community for purposes such as recruiting of graduates; (2) trains cooperative students to fit into the industrial organization, thus making the co-op student better suited for future employment upon graduation; and (3) fills important jobs that require some skills which may be classified as undesirable by its regular employees or in cases where non-student employees might be more costly to the industry. Industry is investing money in the cooperative student and in return is receiving work done by the students as well as an excellent opportunity to thoroughly evaluate the student for post-graduate employment. In general, the principal disadvantage to an industry arises when its investment does not pay off, either because the student is unsatisfactory or because a good student does not choose to remain with the employer after graduation.

Students in the Northwestern University Cooperative Education Program do not provide links between faculty and employer for research, consulting, nor do they provide feedback on education programs. It is not clear whether this is due to the nature of the Northwestern University program along with lack of general faculty involvement, or whether this pattern reflects the perceived interests of industry.

IMPLICATIONS

The following implications may be drawn from this brief analysis of the Northwestern Co-op Program:

(1) Cooperative education is affected by its environment and the university.

(2) A major function of the Northwestern University program, beyond providing meaningful work experience, is to provide opportunities for students to support themselves while in school, as well as to reduce the financial burden on the University.

(3) The impact on industry from the Cooperative Education Program is largely localized geographically; the program provides industry with an opportunity to carefully evaluate essential permanent em-

ployees and to train students for long term commitments.

(4) A more definitive evaluation of the Cooperative Program would require:

(a) in depth case studies
(b) and/or institution of experimental changes, the effects of which are subjected to monitoring and evaluation.
(c) formal surveys of affected groups, including students and industry representatives.

ACKNOWLEDGEMENTS

The author would like to thank Dean Willian T. Brazelton, Dr. Ray S. Lindenmeyer, and Dean David Mintzer for cooperating with this study. It should be noted that the responsibility for the findings of this report are of the author and are not necessarily representative of the views of those who provided information.

Regional Role of Engineering Education: Data and Preliminary Models

Nancy L. Nihan and William D. O'Neill

1. Introduction

In order to study the impacts of engineering education upon a particular region, we must first isolate the distinguishing characteristics of engineering education with respect to other forms of higher education. Models which have a sound theoretical base are needed to describe past behavior such as past trends in student flows, growth rates, and socio-economic characteristics of engineering students. We are also interested in determining the effects of relevant characteristics of engineering institutions, such as location type (urban vs. rural, geographical region, etc.), engineering education expenditures, age of institution, etc. Once models are developed which adequately explain the above interrelationships, sensitivity analyses may be performed to examine the probable impacts of various political, social, and economic changes on these generalized trends.

As a first step toward developing an understanding of the data patterns emerging from the time series data on higher education which were collected, several diverse analytical methods were applied. The

Nancy L. Nihan is an Assistant Professor in the Department of Systems Engineering at the University of Illinois, Chicago Circle
William D. O'Neill is an Associate Professor in the Department of Systems Engineering at the University of Illinois, Chicago Circle.

descriptive models which resulted from these initial "fishing" expeditions cannot be considered as satisfactory predictors in their present form. They were used, rather, to provide descriptive information about the dynamic properties of the national engineering education system. Models of the student-flow and economic subsystems were tried first. Three types of models were developed initially (Markov, production functions, and simple regression) to examine possible relationships between student enrollment, labor and capital investments, and degrees granted. Poor results were obtained for the first two model types. The straight statistical analysis gave a good fit for the simple time-lagged relationship between student enrollment and degrees granted. The results obtained from this preliminary model may prove useful as an input to the generalized system-flow model which is recommended in Section 4 as an initial starting point for development of a comprehensive regional impact model.

2. Markov Transition Model: Theory and Practice

A Markov transition model applied to time series data of any form seemed a likely starting point for such a diffuse system. Consider a time dependent variable y(t) to be split into discrete states x_p, $p=1,2,\ldots,k$, where k is sufficiently large to encompass existing data. Let $q_{ij}=Pr[x(t+1)=i|\,x(t)=j]$ where i and j range over the k "data states" *and* appended states. Clearly, the q_{ij} are easily calculated for i and j less than k from the existing data. The appended states allow predicted values of Pr[x(t)] and therefore, y(t), to range over arguments less than or greater than the observed data range. It is obvious, however, that no data exists with which to calculate q_{ij} for i and j outside of the range of p. To calculate the q_{ij} for i and j outside of this range, an iterative process was envisioned. Briefly this was to entail the formation of an error function which depended in some monotonic way (mean square error, for example) upon the deviation of Pr[x(t)] from 1. The error function would be formed by estimating the q_{ij} for the non-data states and iterating the model (essentially calculating Pr[x(t)] from $Q^tPr[x(0)]$ where $Q = [q_{ij}]$). Since Pr[x(t)] now depends upon the unknown elements of Q, it is clear that the error function can be minimized over the unknown q_{ij}.

As a tool for describing certain trends in engineering education, the Markov transition model was found to be deficient in several practical aspects. In the first place the long time series data necessary to have confidence in the model did not exist and (now) we see that they probably will not in the near future. Secondly the computation is deceptively large. Consider, for example, a system to display 10 observed data states. A model that can predict the range of probabilities with arguments above and below the data equal to the data excursions would require a 30 state model with 900 parameters. Finally the optimization problem would be prohibitive. Taking the above example it would be necessary to optimize over 900-100=800 parameters, a non-trivial task.

The work on the Markov transition model proceeded with the available data (17 data points on 5 time series) yielding 421 states including both calculated and estimated parameters. At this point a decision was made to investigate other modeling procedures since the envisioned work involved was not warranted by the scanty data per time series which were available.

3. Deterministic Economic Models and A Simple Regression Model

After the Markov transition model it was decided to attempt an input-output, economically based model. The variables which were based on all yearly data for Urbana (total university) for the years 1949 through 1970, were as follows:

I(t)=investment in plant
L(t)=operations and maintenance
E(t)=student enrollment at all degree levels
D(t)=degrees granted at all levels
K(t)=capital as a result of investment

Three types of economic models were tested. The first, an *exogenous operations and maintenance model* is shown in Equation (1) below.

$$L(t+2) + \alpha L(t+1) + \beta L(t) = 0 \tag{1}$$

Fitting the model to the Urbana time series data resulted in a mean square error fit that gave optimal values of $\alpha = -2.25$ and $\beta = 1.2603$. The maximum deviation of the model was 15%.

The second economic model tested was a *production function degree model.* This model is shown in Equation (2) below.

$$D(t) = cE(t\text{-}2)K^{a}(t)L^{(1-a)}(t) \tag{2}$$

The model above was otpimized over c and a using a mean square error fit to yield best values of $c=5.1\times10\text{-}10$ and $a=0.52$. The fit was less than satisfactory with an average error of 45% per data point. There were 17 data points with a correlation coeficient of 0.425.

The third economic model tested was a *production function degree model with no time lags.* The model was functionally the same as the above model except that the error now averages 52% per data point.

Finally a *simple regression model* was tested correlating degrees and enrollment. The model tested is shown in equation 3 below.

$$D(t) = kE(t-T) \tag{3}$$

The parameters k and T were fit optimally using mean square error. In this model T was restricted to be integer. The results obtained were k =0.2669 and T=2 years. The maximum error was 9%.

From the above results it is evident that degrees and enrollment do not follow economic production function models in any neo-classical sense. It would appear from the encouraging results relating enrollment and degrees granted that the "factors of production" in the university reach a critical level above which these factors do not combine in an economic way. The operating and maintenance model seems to be exogenous to the system although future data could change this. Considering operations and maintenance costs exogenous does follow economic thought so that the production function would appear to fail when capital is mixed.

4. Recommended System Model and Data for Calibration Collected to Date

A generalized systems model — which should be extended further — was also developed for the preliminary study. The model was chosen for the following reasons. First, it has some theoretical basis since the major input has been reduced to the demographic level. Thus, one of the key dynamic relationships involves a simple birth-death process at the national level. Secondly, the model is a simulation model so that relationships which do not have such a sound theoretical base or which exhibit a much higher degree of uncertainty can be tested easily through a variety of sensitivity analyses for their probable impact. Also, since the model is in a form which is easy to simulate on either an analog or digital computer, other types of tests may be introduced; for example, the system's sensitivity to exogenous variables, such as changes in socio-economic structure or government policy.

During the preliminary model-testing phase, data for the years 1954 to 1970 were collected for the following variables: national degrees earned (B.S., M.S., and Ph.D) for total university, with breakdowns by subject area, sex, state, and institution.[4] Also trends in engineering graduates were obtained for selected universities for comparison of regional variations due to urban versus rural location. The dynamic characteristics of these data which led to our choice of the generalized systems model for future planning effort is discussed below.

In Figs. 1-6 we see essentially two types of growth exhibited. Figs. 1, 2, and 3 closely resemble exponential growth functions. Figs. 1 and 2, however, exhibit small oscillations from the pure exponential growth curves similar to those oscillations observed in national demographic trends.[1] Because of the relatively large sample size of B.S. and M.S. degrees earned, these data closely follow the growth trends appearing in the national population (with a time lag, of course) whereas the highly filtered small sample of Ph.D degrees earned exhibits almost none of these oscillations.

The small oscillations in Figs. 1 and 2 and the almost pure exponential character of Fig. 3 suggest a common starting point for modeling these curves. The common starting point for the proposed model is the U.S. birth rate system or "people generator." The oscillations observed in Figs. 1 and 2 are probably due to age distribution phenomenon exhibited by the population at large. We would expect the B.S. and M.S. degrees earned to closely reflect such population trends. The almost pure exponential growth of Ph.D. degrees is also to be expected since the oscillations should be damped out by the selection process at this level.

In Figs. 4 and 5, some rate sensitivity phenomenon with B.S. and M.S. degrees is noticeable (especially in Fig. 4). This rate phenomenon suggests a sigmoid or rate saturation model would be appropriate as a cascaded system to follow the common starting point alluded to above. The oscillations in the engineering curve at saturation may also reflect a "birth-death" process where B.S. degree growth ("births") feeds back causing fewer students to select engineering ("deaths"). Fig. 6, which shows engineering Ph.D degrees earned, does not exhibit the sigmoid property. Because of the nature of the filtering process, it is expected that this is a function of the time lag involved. Thus, based on the

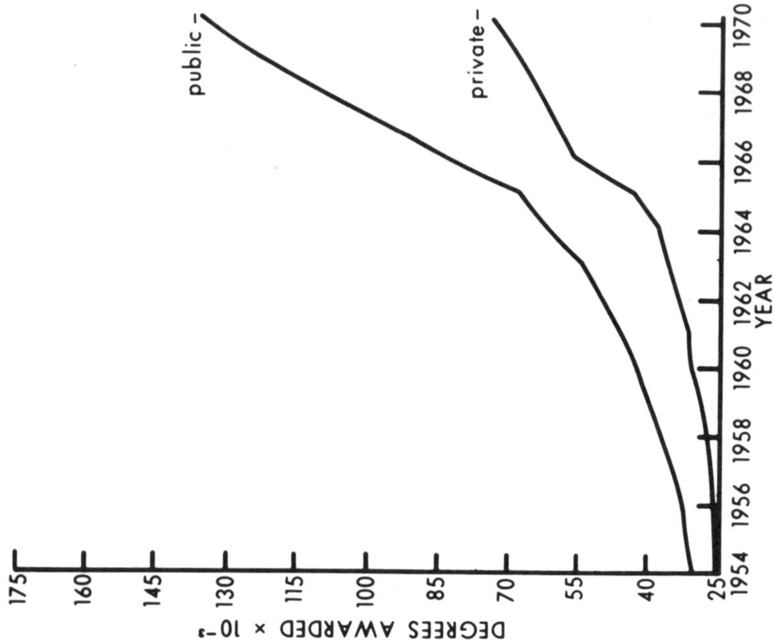

Fig. 2 All University M.S. Degrees from Public and Private Scools.

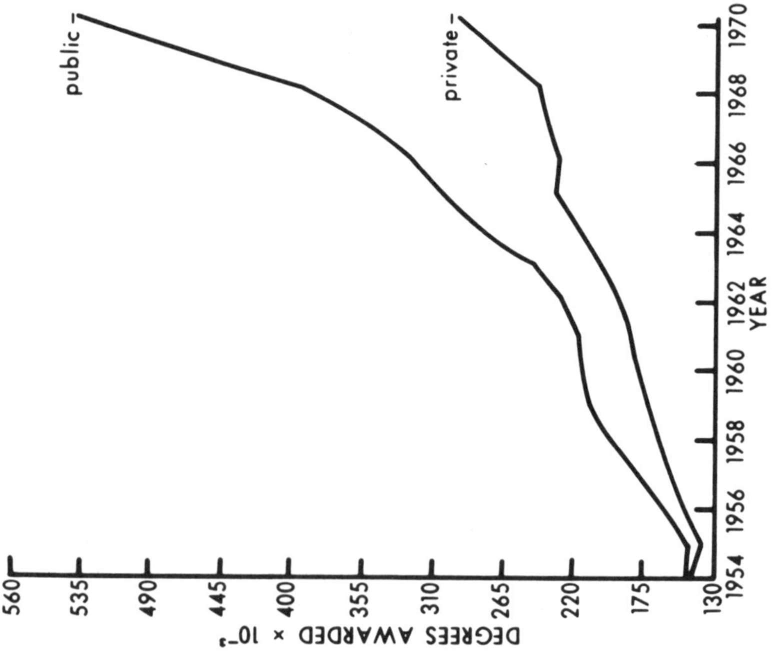

Fig. 1 All University B.S. Degrees from Public and Private Scools.

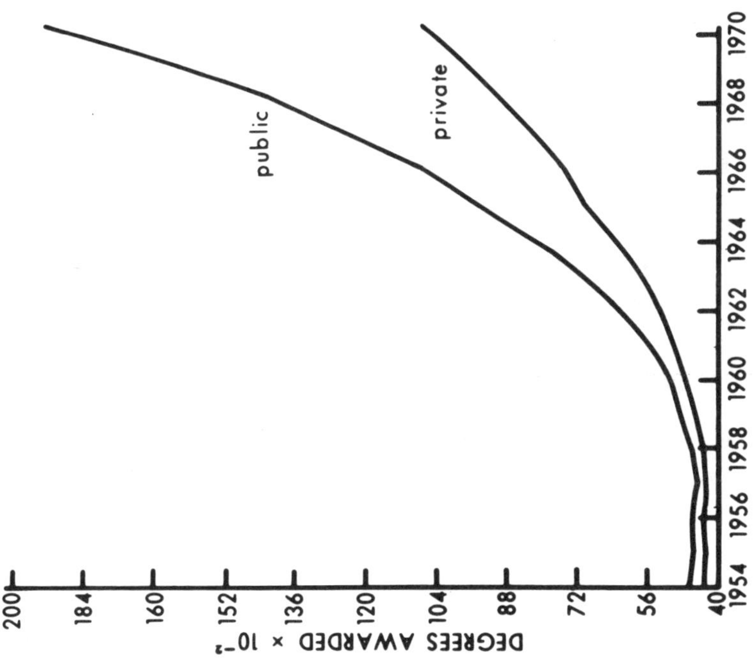

Fig. 3 All University Ph.D. Degrees from Public and Private Scools.

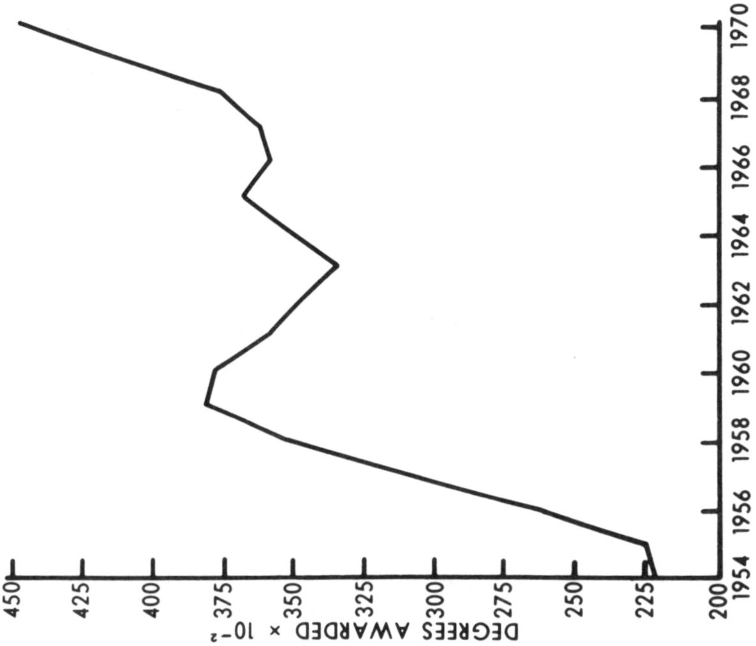

Fig. 4 Engineering B.S. Degrees.

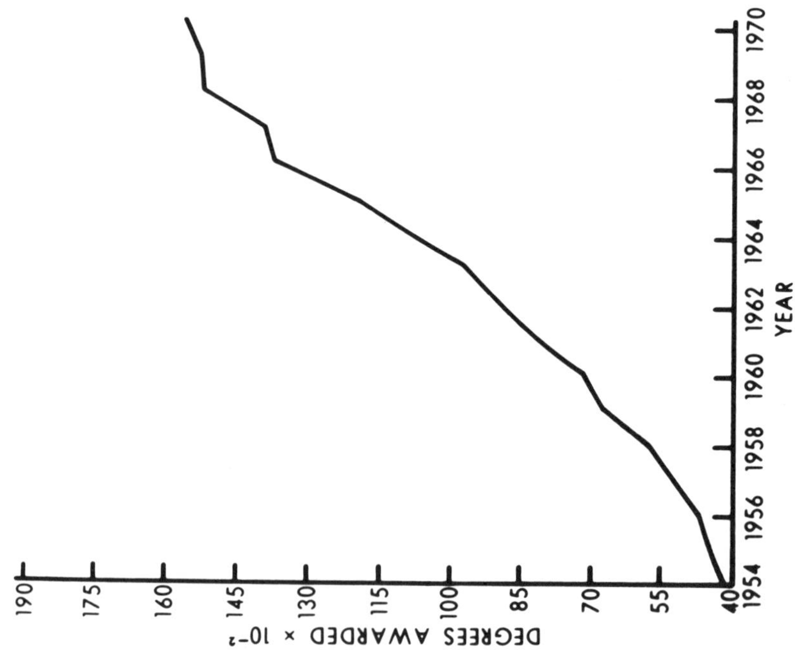

Fig. 5 Engineering M.S. Degrees.

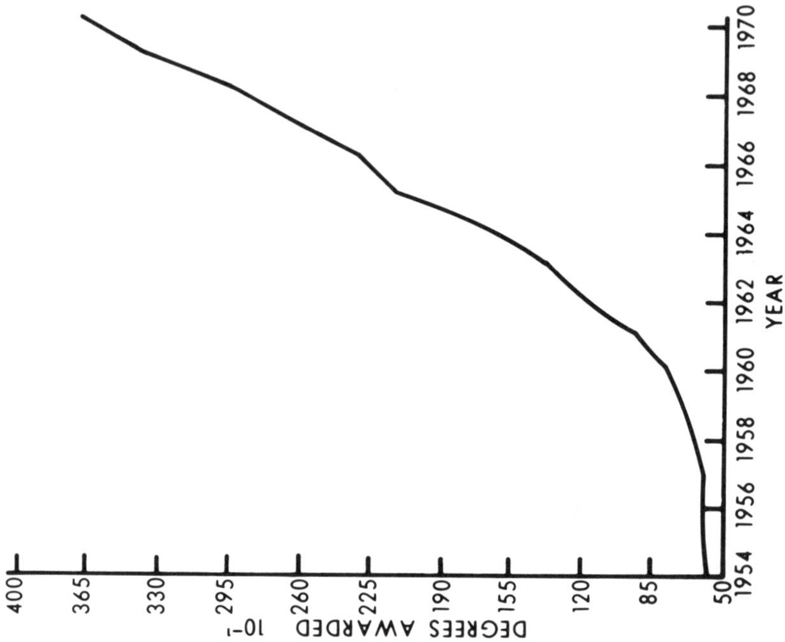

Fig. 6 Engineering Ph.D. Degrees.

trends in B.S. and M.S. degree growth, Ph.D degrees earned can be expected to reach a local saturation in the near future.

The U.S. demographic system which is used as the common starting point for all of the subsequent model systems has an output of births per year r(t). Using a mean square error fit for U.S. population data from 1870 to 1965, it has been shown that the national demographic system can be modeled by the difference equation shown in equation 5 below.[1, 2, 3] The area of model impulse-response was equal to the area of data impulse-response within an error range of plus or minus 4%.

$$r(t+2) + 2(a-1)r(t+1) + (a-1)^2 r(t) = a^2 K r(t-17) \tag{5}$$

where, r(t) = births per year
t = years
a = 1/years
K = net reproduction ratio: number of daughters surviving to age 17 per adult female.

Equation 5 above describes a self-generating feedback system with a feedback of a^2K (no inputs except for the initial 19 states). The systems diagram for Equation 5 is shown to the left in Fig. 7 which represents the preliminary systems diagram for the proposed model. In Equation 5, t takes on values of years. The demographic system model generates the input to a filter with a pure delay of 18 years weighted by the joint probability of surviving to age 18 if born at time 0 and if attending college, Pr(18,C). The output of this filter is the number of people entering into the higher educational system for a particular year t (See fig. 7).

At this point a decision is made as to curriculum major, although there may still be some transfer between majors as the students pass through their early years. However, regardless of the choice, there are yearly lags denoting one year progressions. For our purposes, we will model the path taken by those deciding upon engineering. For example, taking total B.S. degrees earned from Figs. 1 and 4 for the year 1970, we note 830,000 total university B.S. degrees and 46,000 B.S. degrees in engineering respectively. If a successful completion ratio of 1/3 (representative of the University of Illinois) is assumed for all degrees then we can calculate the flow to engineering as attenuated approximately by the ratio of B.S. engineering degrees to total B.S. degrees or 0.055. This attenuator, however, can be subject to the effects of labor market supply and demand. This is particularly important when considering the effect of government spending on engineering-specific programs. (Consider, for example, the impact that government spending on defense and space oriented projects has upon the demand for engineers.) For the preliminary model, therefore, we will call the attenuation k and make it a non-linear function of government spending.

Justification for the above decision can be better explained through use of the following example. Consider the decade from 1955 to 1965, Fig. 4, in particular. With a large expenditure on space programs occurring around the year 1961 we should expect an immediate labor demand for engineers and an increase in B.S. degrees four years later. Thus the 1965-66 engineering upswing was probably stimulated by the space program. In the decade preceding that year, however, we see classical sigmoid saturation as labor demand dwindles. We, therefore, must also make k a function of the total engineering B.S. degrees. This saturation phenomenon can be modeled as shown to the right in Fig. 7. The feedback of the number of B.S. degrees in engineering, B(t), reflects demand saturation while Y(t) represents major governmental spending on engineering-specific projects. The value 0.33 has been selected as a final attenuation in the model to represent successful degree completion of the percent, k(B,Y), deciding upon engineering. This value, as mentioned above, is representative of the successful completion ratio for all degrees for the University of Illinois. The parameter values given above are, of course, subject to revision for different levels of aggregation and are intended only for illustrative purposes.

5. Socio-economic Factors in Engineering Education

Further refinements of the generalized systems model discussed in the previous section should include relevant socio-economic variables. Investigation into the unique social and economic characteristics of engineering students, for example, would improve the model's predictive ability. Attributes such as sex, family background, and costs per student are expected to have a significant impact upon the selection process. Time series data are available for comparing the differences in university populations vs. engineering populations in terms of sex.[5] Since engineering schools differ remarkably from the university popu-

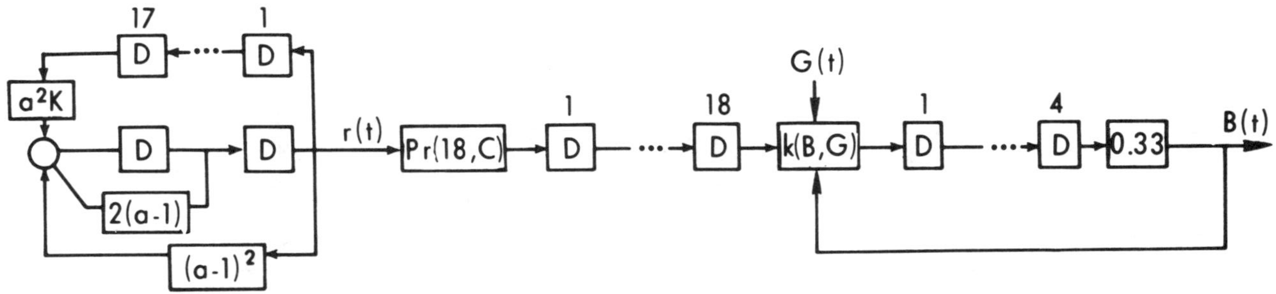

GLOSSARY

a= 1/years
t= years
K= net reproduction ratio
D= unit time delay (one year)
r(t)= national birth rate (millions/year)
B(t)= national B.S. degrees in engineering
G(t)= governmental spending on engineering-specific programs
k(B,G)= percent (absolute) of eligible students selecting engineering careers

Fig. 7 Generalized system model for Engineering Education flows

lation at large the model should be sensitive to policies which encourage a significant change in this social variable.

Figures 8 and 9 show the growth rate in degrees granted for males and females for total university and engineering populations respectively. Although the ratio of females/males receiving engineering B.S. degrees has risen from .0028 in 1955 to .0076 in 1970, it is trivial when compared to the same ratio for total university B.S. degrees which rose slightly from .562 in 1955 to .711 in 1970. This trend is less apparent for advanced degrees earned. Figures A1 through A4 in the appendix show the M.S. and Ph.D. degrees earned by all university students and engineering students with respect to sex of students. The number of Ph.D. degrees in engineering earned by women in the year 1970, for example, was 24 for the entire nation* compared to 3657 engineering Ph.D. degrees earned by male students in that year (ratio = .0066). This contrasts sharply with the figures for total university Ph.D. degrees earned in 1970, which were 3980 and 25892 for female students and male students respectively (ratio = .119). Since the characteristic of sexual bias in engineering enrollment is so pronounced, particularly at the undergraduate level, the generalized engineering systems model should reflect the likely impacts of governmental policies which are aimed toward reducing the deviation in the sexual composition of engineering student bodies with respect to the norm or university population at large.

Other socio-economic attributes of engineering students may also be relevant to the model. At present, limited data have been collected on family characteristics such as father's income and type of employment. The incremental costs per student for engineering vs. other disciplines is also expected to be relevant. Reliable data on costs per student are difficult to obtain. However, some indicators of this variable may be used as surrogates.[6]

In addition to social and economic variables related to the student body, there are corresponding attributes of educational institutions which should also be investigated. The type of institution in terms of public vs. private funding, size, age, primary orientation (technological institute vs. general university) are important factors in the growth patterns of particular engineering schools. Another attribute which may prove influential is the location of the institution in terms of met-

*This is double the value in 1969 which was the next highest year for female Ph.D's in engineering.

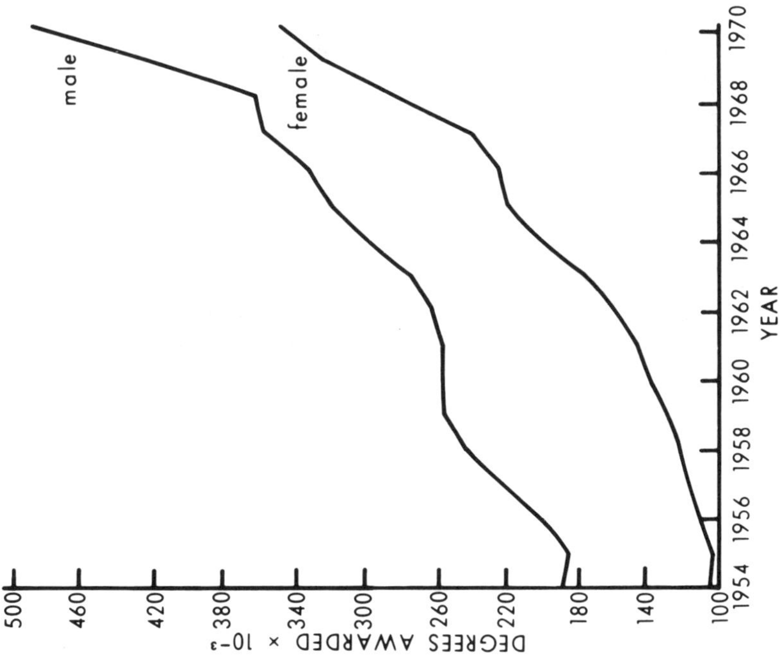

Fig. 8 All University B.S. Degrees, Male and Female.

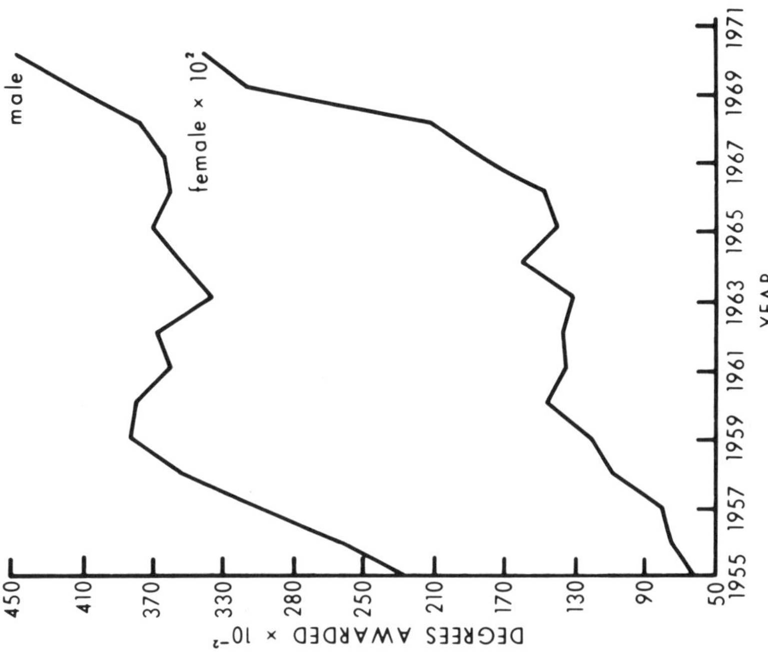

Fig. 9 Engineering B.S. Degrees, Male and Female.

ropolitan vs. small town setting. Figures A5 through A10 in the appendix show past trends in degrees granted for selected institutions. These institutions vary widely in the characteristics outlined above. No conclusions can be drawn from the limited set of time series observations shown. However, these data are available (although in inconvenient form at present) and can be used for future model refinements. In particular, data on the urban dimension (controlled for all other variables) should be investigated for comparison of growth in terms of degrees granted by the type of institution.

Finally the impact of allocation policies for research and development (R and D) funds should be investigated for various universities. Data on R and D money for education institutions are available, but not broken down by major disciplines. Due to time limitations nothing has been collected to date on strategies for State R and D expenditures. However, information on the effects of past policies for distribution of State R and D funds can be obtained for subsequent stages of model development.

6. Does Supply Meet Demand?

The previous pages give nationally, as well as locally, the supply side of educated labor. Using data from the *Survey of Current Business* specific to Illinois, and data from the University of Illinois at Urbana, on where and in what categories the engineers are employed, the question can be partially answered by the graphs in Figs. 10 to 13. (The correspondence between the *Survey of Current Business* categories and those used by the University of Illinois is given in Table I). The figures present:

1. The breakdown by category of the GNP of Illinois for the years 1959 (○) and 1969 (●), and
2. The category of employment of the Urbana engineering graduates staying within the state for employment at the latest date.

Clearly if the graduates were meeting demand in the specific engineering categories the numbers of students employed would be proportional to the dollars earned in those categories. Though some categories reflect such a tendency the results shown in Figs. 10 to 13 are somewhat inconclusive.

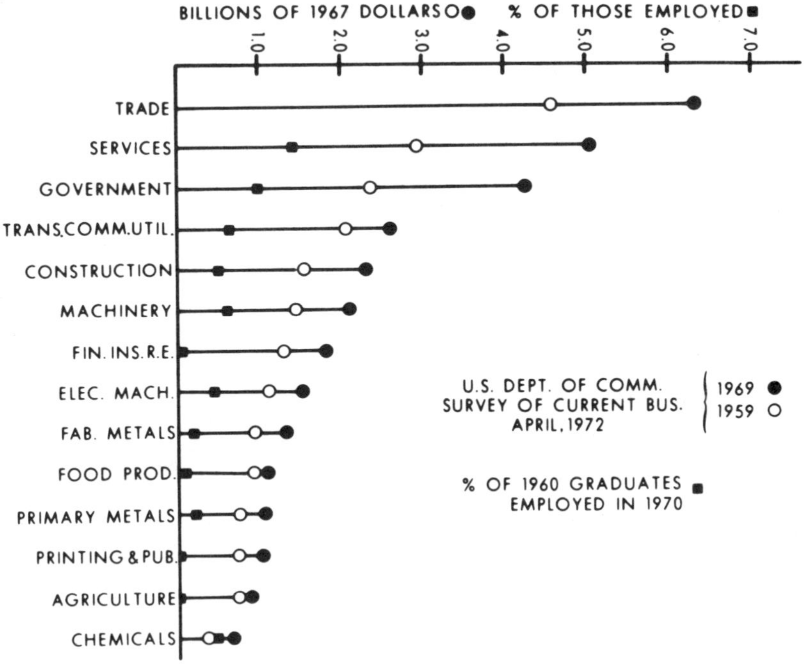

Fig. 10 State of Illinois Earnings in Several Engineering Categories

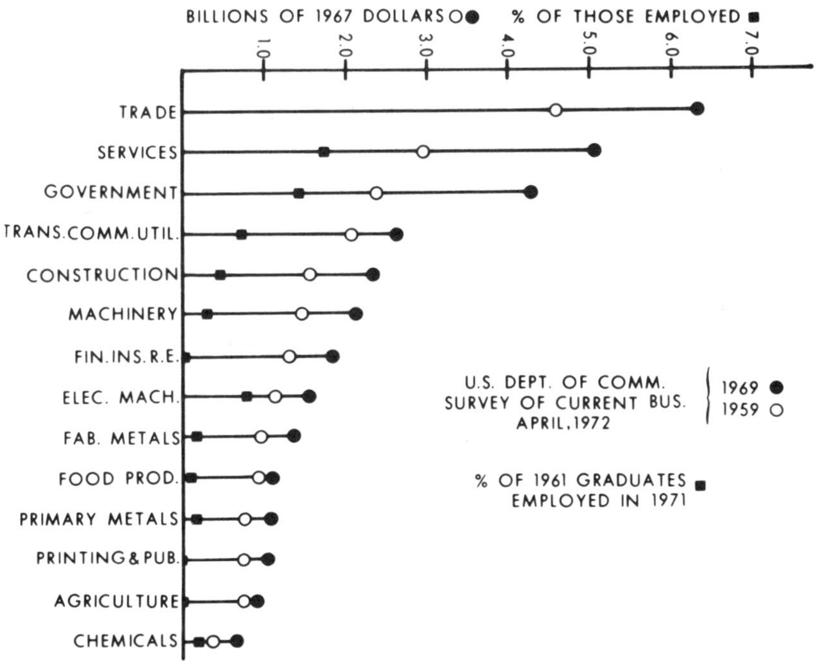

Fig. 11 State of Illinois Earnings in Several Engineering Categories

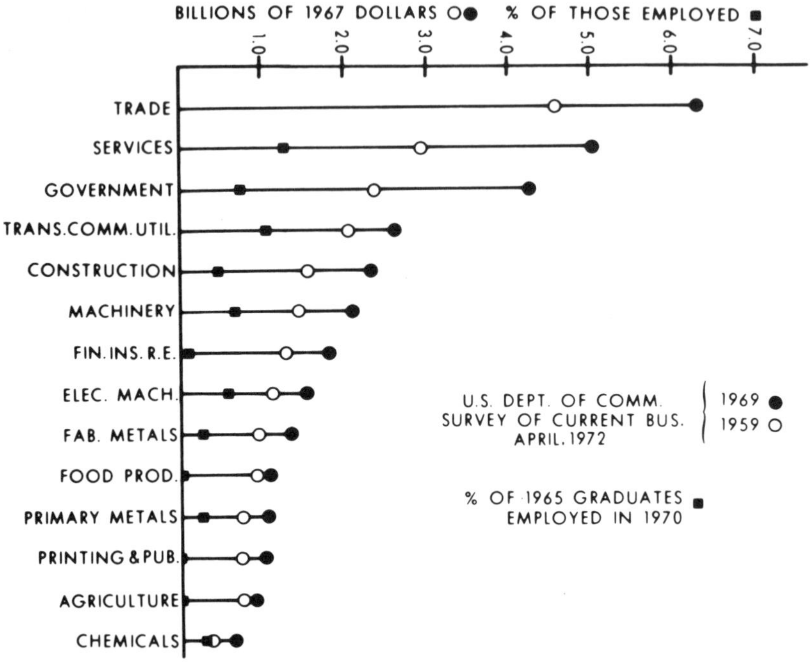

Fig. 12 State of Illinois Earnings in Several Engineering Categories

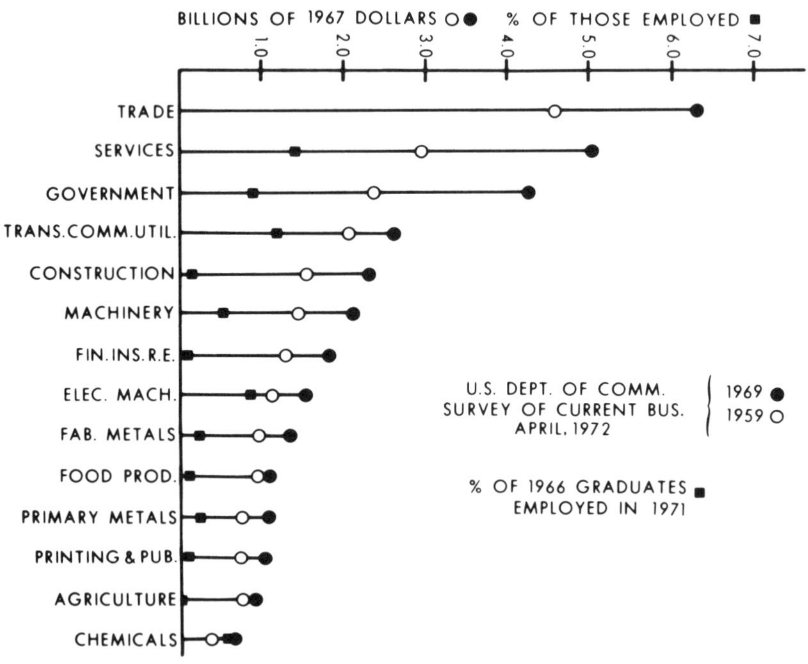

Fig. 13 State of Illinois Earnings in Several Engineering Categories

Table 1 Industrial Category Conversion Table and Definitions

U.S. Dept. of Comm. (Survey of Current Business) Categories	U. of Illinois, Urbana College of Engineering Industrial Categories
Services	Consulting, Research, laboratories, education, law
Government	Federal, State, City and County
Transportation, et al.	Communications equipment, transportation and public utilities
Construction	Construction
Machinery	Heavy equipment, packaging materials and equipment.
Finance, et al.	Investment Banking and Finance
Electrical Machinery	Elec. Equipment, controls and Instrumentation
Fabricated Metals	Metals and Metal Products × ½
Primary Metals	Metals and Metal Products × ½
Food Products	Food
Printing and Publishing	Printing and Publishing
Agriculture	Farming
Chemicals	Chemical and Chemical products

REFERENCES

1. O'Neill, W., "A System Model of the People Generative Process," *5th Annual Princeton Conference on Systems and Information Theory,* March 1971, pp. 79-83.
2. O'Neill, W., "Necessary and Sufficient Conditions for Almost Certain Stability and Asymptotic Stability of a Class of Regenerative Population Systems," *9th Annual Allerton Conference on System and Circuit Theory,* Oct. 1971, pp. 380-84.
3. Bongaarts, J., and W. O'Neill, "A System Analysis of the Population Renewal Process," *Demography,* 1972 (in press).
4. *Earned Degrees Conferred by Higher Educational Institutions,* U.S. Dept. of Health, Education and Welfare, 1954-70.
5. *Ibid.*
6. *College and University Facilities Survey,* U.S. Dept. of Health, Education and Welfare, 1965.

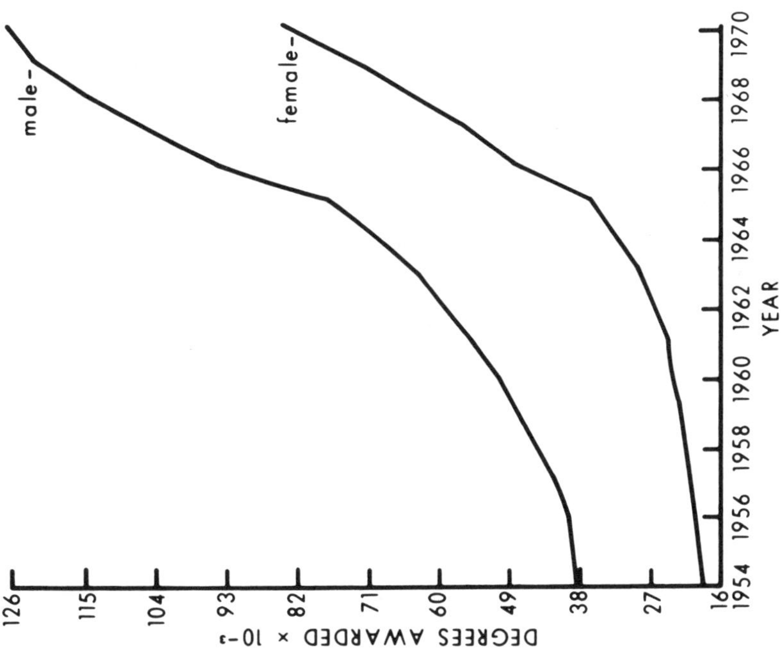

Fig. A1 All University M.S. E Degrees. Male and Female.

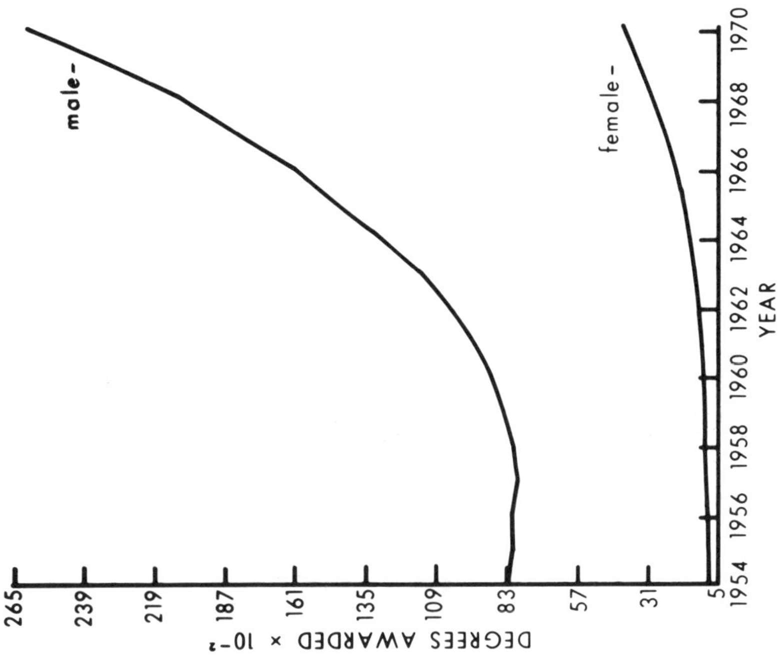

Fig. A2 All University Ph.D. Degrees, Male and Female.

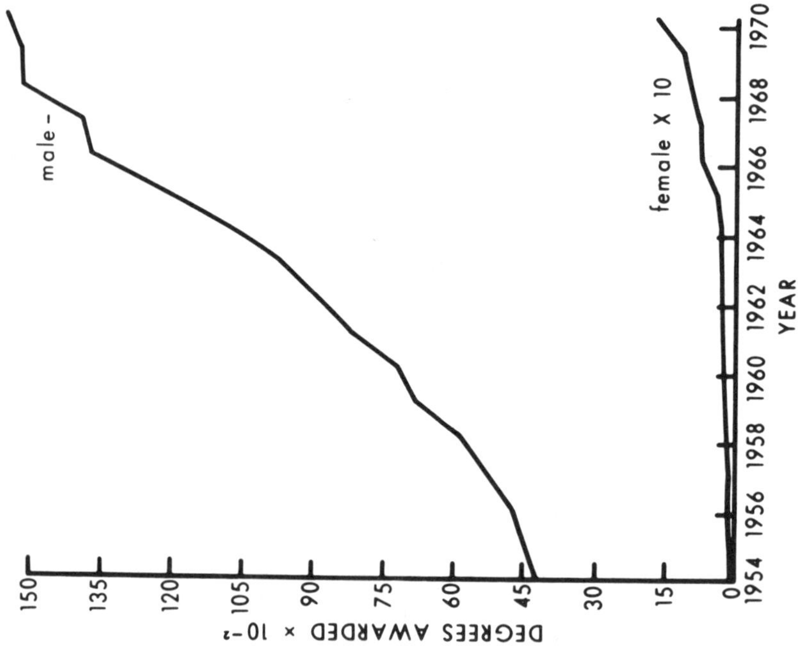

Fig. A3 Engineering M.S. Degrees.

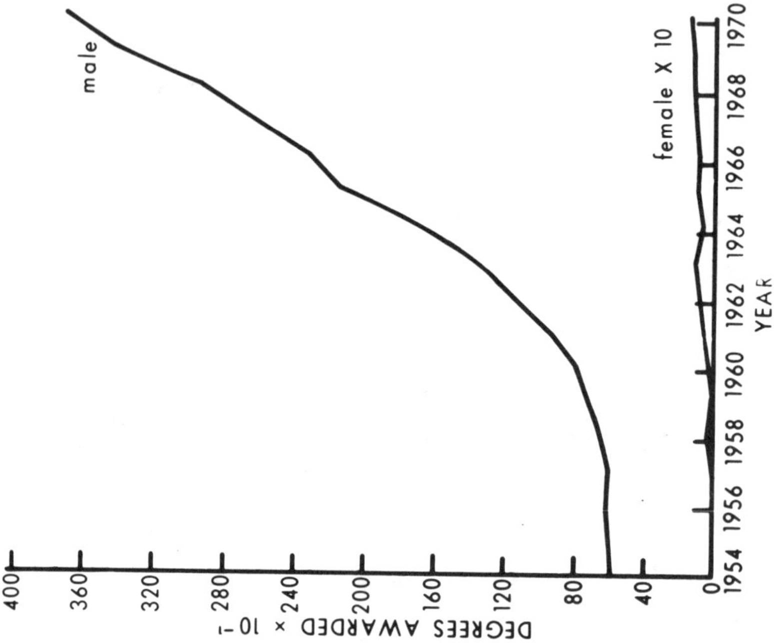

Fig. A4 Engineering Ph.D. Degrees

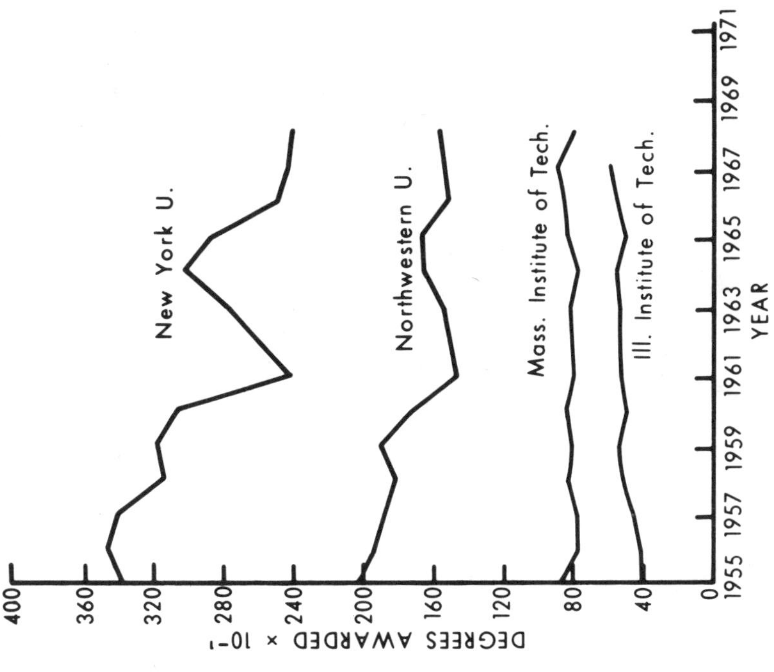

Fig. A5 Total B.S. Degrees

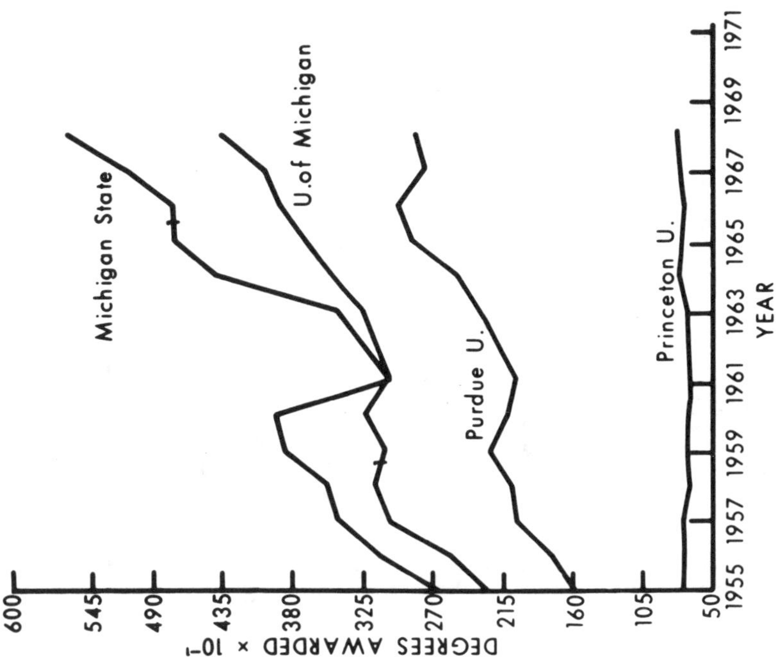

Fig. A6 Total B.S. Degrees.

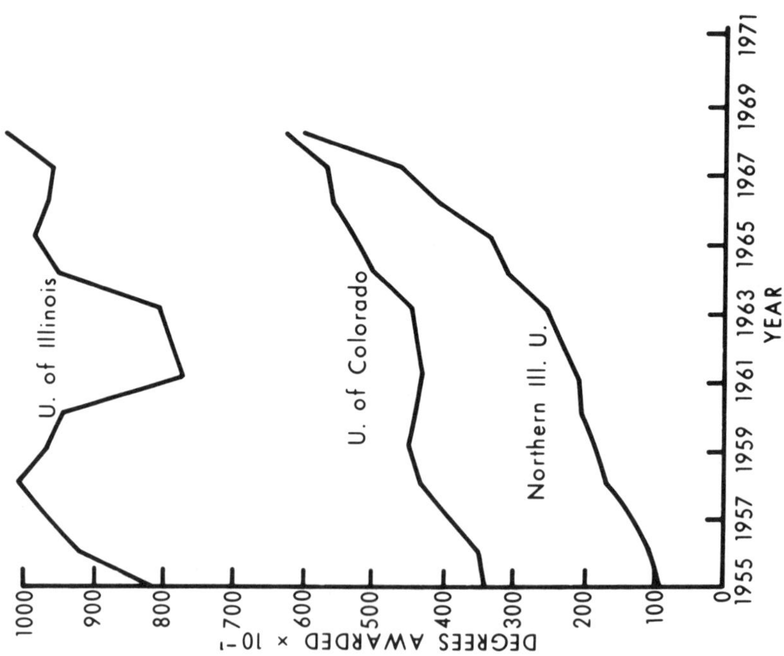

Fig. A7 Total B.S. Degrees.

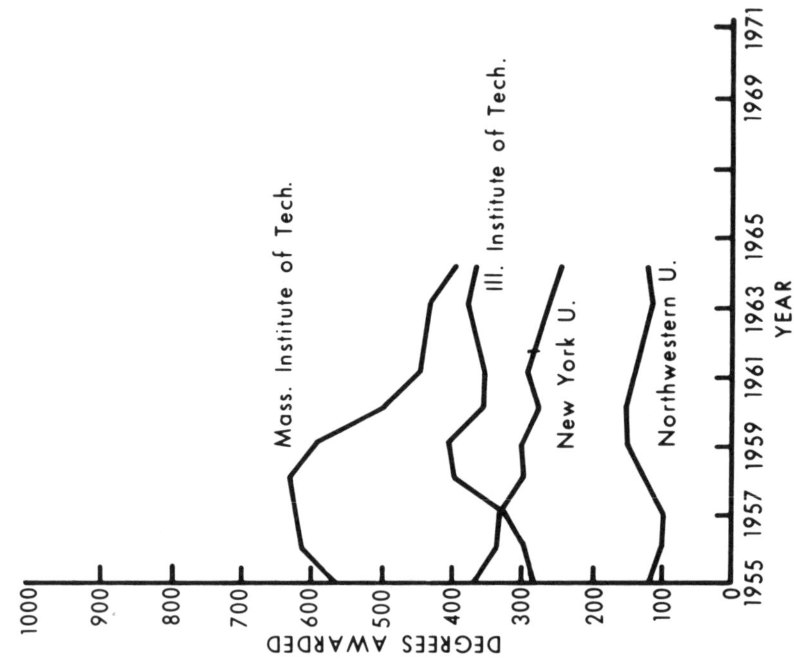

Fig. A8 B.S. Degrees in Engineering.

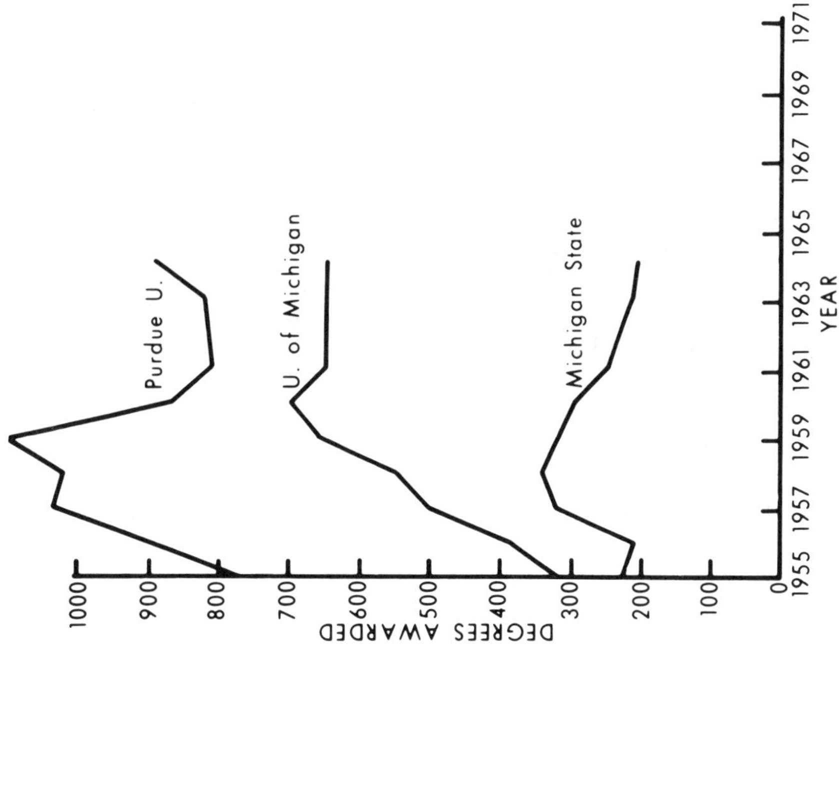

Fig. A9 B.S. Degrees in Engineering.

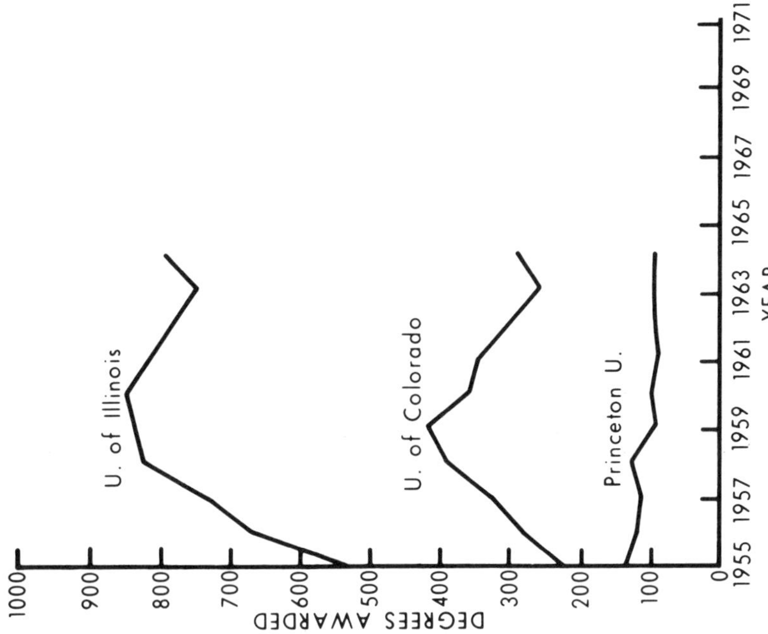

Fig. A10 B.S. Degrees in Engineering.

Graduate Engineering Education in Illinois

Clarke H. P. Schneider

The purpose of this brief report is to review the position of Illinois within the United States with regard to engineering education in terms of a number of factors crucial to higher education and in particular to the production of engineering degrees. The approach adopted will be to first inspect areas which seem to be crucial to the role of graduate engineering education by comparing Illinois to the United States. Two areas in particular may be identified in which Illinois may be suitably compared to the nation. These consist of faculty and enrollment-degree production. The structure of engineering education within Illinois is studied by a more detailed look at the universities in Illinois which grant graduate degrees in engineering.

In 1970, the State of Illinois had 708 full-time faculty members in the three professorial ranks engaged in teaching engineering at graduate and undergraduate levels in the six universities in Illinois which grant graduate engineering degrees. These six institutions are Bradley University, Illinois Institute of Technology, Northwestern University, Southern Illinois University, University of Illinois at Chicago Circle, and University of Illinois at Urbana. These 708 full-time faculty members constitute about 4.7% of engineering faculty members in the three ranks in the United States. The 1970 population of Illinois constituted about 5.5% of the total United States population and so Illinois employment of engineering faculty seems to be

Clarke H. P. Schneider is an Assistant Professor in the Center for Urban Studies, University of Illinois at Chicago Circle.

somewhat less than might be anticipated from the distribution of population within the nation. Note that the device of comparing a particular characteristic of Illinois as a percent of the U.S. to the Illinois population as a percent of U.S. population is merely a comparison method which directly reflects numbers per capita for the U.S. and Illinois without requiring that two new figures be presented for each item to be compared.

Table 1 summarizes the position of Illinois with regard to earned degrees granted in engineering relative to the nation. At all degree levels, Illinois produces a higher proportion of the U.S. female recipients than male degree recipients. However, at all levels below the doctorate, Illinois produces fewer degrees than expected from the population. This, of course, reflects this State's commitment to the advancement of research in engineering through training of skilled researchers.

This trend is perhaps better and more suitably illustrated by the State's own research policies. Specifically, research and development spending by State agencies (and not including expenditures in colleges or universities) places Illinois fourth within the U.S. in terms of State R & D spending, while standing fifth in terms of population. However, in terms of receipt of Federal funds dispersed for research and development, in fiscal year 1965, Illinois stood nineteenth among the states (while ranking fourth in total Federal taxes paid). Similarly Illinois ranked thirtieth in terms of Federal research and development funds paid to industrial firms where, for comparison, Illinois ranks fifth nationally in industrial employment and fourth in value added by industry. Turning to total Federal monies and also to funds for research and development, in FY-1970, Illinois was in the top ten states in support. However, Illinois ranked in the third ten in total dollars per degree granted in 1967-68 and in the fourth ten in R & D per Ph.D. awarded in science and engineering in 1967-68. This figure elucidates somewhat the position of Illinois in engineering degree output summarized in Table 1.

The picture then is that despite the fact that the Engineering College of the University of Illinois at Urbana is in certain respects the nation's largest, the State of Illinois is under-represented on a per capita basis (and also, of course, on the basis of tax dollars contributed to the Federal government) in faculty, degrees granted, and Federal support of research and development in the realm of engineering education.

Table 1 Relative Standing of Illinois with Regard to Three Levels of Graduate Engineering Degrees Granted for Male and Female Recipients.

		Earned Engineering Degrees Conferred 1969-70		
Degree		Number Granted in Illinois by the Six Graduate Engr. Degree Granting Institutions	Total Such Degrees Granted in the U.S.	Degrees Granted in Illinois as % of Total U.S.
Bachelor's	Total	1675	44,772	3.7
	Male	1653	44,434	3.7
	Female	22	338	6.5
Master's	Total	635	15,597	4.1
	Male	626	15,425	4.1
	Female	9	172	5.2
Doctor's	Total	257	3,681	7.0
	Male	255	3,657	7.0
	Female	2	24	8.3
Overall Total		2567	64,050	4.0

Turning to the relative importance of engineering programs within each of the six universities, Table 2 presents degrees granted in 1970 by these institutions in four categories: total degrees granted; masters and doctors degrees granted; and total masters and doctors degrees granted in engineering. Comparison of these figures allows us to inspect within each institution the importance of graduate work, the importance of the engineering program, the importance of graduate work within the engineering program, and the importance of graduate engineering in the graduate program. This is done in the simplest way by presenting the percentage which the subcategory constitutes of the appropriate category.

Inspecting these in order, first, in column 3 the graduate degrees as percent of total degrees granted indicates the relative importance of graduate programs within each institution. A rather clear break exists separating the group into two subgroups with 20 percentage points separating the highest of the low group from the lowest of the high scoring group. Northwestern IIT, and the University of Illinois at Urbana each have graduate programs which grant more than 40% of the degrees given in the institution whereas SIUC, Bradley, and UICC have programs which grant less than 20% of the degrees of their respective institutions.

Column 5 of Table 2 presents total degrees in engineering as percent of total degrees granted by institution. Clearly the institutional importance of the engineering program covers a wide range. In this category, IIT grants twice as many degrees proportionally as the next highest institution—the University of Illinois at Urbana. At the middle range of these institutions, engineering degrees constitute 8 to 12% of total degrees granted, while at SIUC engineering degrees account for only 3.7% of degrees granted. Note that at SIUC the engineering program is very new.

The relative importance of graduate work in engineering is presented in column 7. In all institutions where the graduate program was found important (e.g. about 40% of engineering degrees), graduate work in engineering is found to be stressed, although here the spread is somewhat greater with the University of Illinois at Urbana and Northwestern granting more than 50% of their engineering degrees at the graduate level, while at IIT, these degrees constitute 37.5% of those granted. UICC, which has only very recently begun granting graduate degrees in engineering, has 6%, SIUC,12%, and Bradley, 25% as the proportion of graduate degrees granted within

Table 2 The Relative Importance of Graduate, Engineering and Graduate Engineering Programs in Graduate Engineering Degree Granting Institutions in Illinois in 1970.

	1	2	3	4	5	6	7	8
	Total Degrees Granted	Graduate Degrees Granted	Graduate Degrees as % of Total Degrees	Total Engr. Degrees Granted	Engr. Degrees as % of Total Degrees Granted	Graduate Degrees in Engr. Granted	Graduate Engr. Degrees as % of Total Engr. Degrees Granted	Graduate Engr. Degrees as % of Graduate Degrees Granted
Bradley	1306	187	14.3	156	11.9	39	25	20.8
IIT	1080	443	41.0	464	42.9	174	37.5	39.3
NU	3528	1628	46.1	310	8.9	176	56.8	10.8
SIUC	4434	855	19.3	144	3.2	17	11.8	1.9
UICC	2617	130	5.0	309	11.8	19	6.1	14.6
URBANA	8614	3469	40.3	1184	13.7	467	39.4	13.5
	21579	6712	31.1	2567	11.9	892	34.7	13.3

Table 3 Distribution of Engineering Education within Illinois—1970.

	Number of Faculty in Three Professional Ranks	Faculty as % of Total in these Six Institutions	Numbers of Part and Full-Time Instructors and Others	Graduate Degrees as % of Group Total	Engr. Degrees as % of Total Engr. Degrees Granted in Group	Graduate Engr. Degrees as % of Group Total Graduate Engr. Degrees	Total Degrees as % of Total of Group
Bradley	33	4.7	36	2.8	4.9	2.6	6.0
IIT	77	10.9	93	6.6	14.6	11.6	5.0
NU	115	16.2	6	24.2	9.8	11.8	16.3
SIUC	20	2.8	5	12.7	4.5	1.1	20.5
UICC	62	8.7	4	1.9	9.8	1.3	12.1
URBANA	401	56.7	97	51.7	56.3	71.5	39.9
Chicago Area	254	35.8	103	32.7	34.2	24.7	33.4
TOTAL	962		344				

engineering. Finally, looking at the importance of graduate engineering in the graduate program shows that at IIT, the proportion is greatest and at SUIC, the lowest.

Looking now to the distribution of engineering education within Illinois, Table 3 shows the distribution of faculty among the six engineering graduate degree granting institutions. Geographically, the Chicago area with 62.8% of the State's population encompasses 35.8% of the engineering faculty in the State. While the lack of spatial covariation of the traits is not particularly surprising, it does indicate why attempts to partition the State into regions defined by the equivalence of some engineering trait in order to compare distribution by region of other activities will find extreme disparities for all activities which are linked to urbanization. Looking at percent of graduate engineering degrees granted, the Chicago area's position deteriorates further to 24.7% of these degrees granted within the State. Clearly, programs for engineering education are not regularly distributed with regard to population within Illinois.

Thus, in brief, Illinois compared to the nation is undersupported in terms of engineering education, even though the engineering program of the University of Illinois at Urbana is one of the nation's largest. Much of the State's engineering education capability lies in this one institution. It would be intriguing to ask whether, in terms of the development of a regional system of centers for engineering education, the existence of one preponderant program has the effect of retarding or accelerating the growth of other programs.

Data employed in this report were drawn from the following sources.

1. *County and City Data Book 1967,* U. S. Department of Commerce, Bureau of the Census, Washington, D. C., April 1967.
2. *Digest of Educational Statistics,* Office of Education, U. S. Department of Health, Education and Welfare, Washington, D. C., 1969.
3. *Earned Degrees Conferred 1969-70,* Office of Education, U.S. Department of Health, Education and Welfare, Washington, D. C., 1970.
4. *Federal Support to Universities, Colleges and Selected Non-profit Institutions, Fiscal Year 1970,* National Science Foundation, NSF 71-28, Washington, D. C., 1971.
5. *Geographic Distribution of Federal Funds for Research and Development,* National Science Foundation, Washington, D. C., 1967.

6. Illinois, Advance Report PC(V2)-15, General Population Characteristics, U. S. Department of Commerce, Bureau of the Census, Washington, D. C., February 1971.
7. *Science Resources Highlights,* National Science Foundation, Washington, D. C., June 1970.
8. United States, Advance Report PC(V1)-1, Final Population Counts, U. S. Department of Commerce, Bureau of the Census, Washington, D. C., February 1971.

A Faculty Activities Model

Thomas W. Knowles

INTRODUCTION

The purpose of this model is to present and extrapolate data which has been collected on the nature of faculty activities. The basic premise is that if a specified number of students are enrolled, then a certain number of faculty members will be needed. Also, if sponsored research is being conducted which provides release time for faculty members, then more faculty members will be required. The faculty members provide services other than for teaching and research. The other activities considered were consulting, professional societies, community service, and other professional activities. It was desired to determine the extent of these activities and the relationship between the variables.

A pilot study has been conducted at the Illinois Institute of Technology. The results will be discussed in this paper.

The first section of the paper reports the results of a study to determine the number of faculty members required to support a given enrollment. The study considered the five major engineering departments at Illinois Institute of Technology. It was desired to obtain relationships which would provide staffing requirements by rank within each department. This is the required input which, with the second section of the paper, will give the desired result.

The second section reports the results of a survey of the amount of consulting, community activities, and other professional activities. The

Thomas W. Knowles is an Assistant Professor in the Department of Industrial and Systems Engineering at Illinois Institute of Technology in Chicago, Illinois.

purpose of the survey was to determine the extent to which faculty members make contributions to local, state, and national affairs with those specified activities. Statistical analyses were performed to determine the appropriate levels of aggregation, (e.g., across departments or across ranks) by statistically testing the hypotheses that there were no differences.

FACULTY REQUIREMENTS

The issue of faculty staffing is a very complex one. Staffing decisions are made on a department by department basis and an individual by individual basis. Anticipation and past changes in enrollments, shifting emphasis on subject matter within a department, tenure decisions and many other factors must be considered. The following is not intended to be a guide for staffing decisions by department and by rank, but rather to give an overall estimate. The aggregated estimates are fairly good (based upon statistical measures such as R^2), but a real danger exists in using these departmental results for staffing decisions, because the results do not reflect all relevant factors; similarly, comparisons between schools ignore many of the other important factors.

The statistical studies used historical data and linear regression analysis. It was desired to determine relationships of the form

FACULTY = f (BS, MS and PhD degrees granted, undergraduate and graduate enrollments)

The two potential classes of independent variables considered were degrees granted and enrollments. Given an estimate of the probability of receiving a degree enrollments can be translated into degrees granted. For the purposes of the study of the "Regional Role of an Engineering College," perhaps degrees granted is the variable of major significance. However, for the purposes of this paper, both seem appropriate.

What faculty members should be considered? Clearly, we should include professors (P), associate professors (ASP), and assistant professors (ATP). They clearly have an impact outside of the university, as the survey disclosed. How about instructors, lecturers and teaching assistants? The major regional impact of these employees is within the university. Consequently, they have not been included in the analyses of this paper. Faculty, in this paper, refers only to professors, associate professors and assistant professors. The five basic models considered were:

(1) Faculty = $\alpha + \beta_1$ (Degrees granted)

(2) Faculty = $\alpha + \beta_1$ (BS Degrees granted) + β_2 (Graduate Degrees granted)

(3) Faculty = $\alpha + \beta_1$ (BS Degrees granted) + β_2 (MS Degrees granted) + β_3 (PhD Degrees granted)

(4) Faculty = $\alpha + \beta_1$ (Total enrollment)

(5) Faculty = $\alpha + \beta_1$ (Undergraduate enrollment) + β_2 (Graduate enrollment)

The results of the regression models are summarized in Table 1 by department. The data consisted of nine years of annual data. The model with the highest R^2 was selected, subject to the β coefficients being nonnegative.

The quality of the linear regression models is not sufficiently high that the models should be used for staffing decisions. There are many possible reasons for this. One reason is that the other factors previously mentioned could not be considered in this type of model. Another reason is that the engineering departments were relatively stable over this period of time. That is, enrollments and faculty did not change much and consequently other factors, such as changing emphasis within a department and tenure decisions, were more significant than the response to changes in the number of students. Another reason is the relatively small number of observations, caused by data limitations.

To test the hypothesis that if the variables varied more, the relationships would be stronger, another analysis was done. This considered all departments as indistinguishable with respect to the models (i.e., assume α and all βi the same). The results of the 5 regression models, based upon the 9 X 5 = 45 observations, are presented in Table 2.

As an example, the interpretation of Model #5 is that for any de-

Table 1 "Best" Departmental Faculty Regressions

Department	α Constant	Coefficient	Independent Variable	Coefficient	Independent Variable	Coefficient	Independent Variable	R^2
Chemical Engrg.	5.899	0.0336	Degrees					0.080
Civil Engrg.	4.651	0.0449	BS Degrees	0.0745	MS Degrees	0.1761	PhD Degrees	0.165
Industrial Engrg.	4.796	0.0401	Degrees					0.225
Electrical Engrg.	1.267	0.0502	Degrees					0.617
Mechanics, Mechanical & Aero-space Engrg.	13.489	−0.0235	Undergraudate Enrollment	0.1681	Graduate Enrollment			0.709

Table 2 Aggregated Departmental Faculty Regressions

Model No.	α Constant	Coeffficient	Independent Variable	Coefficient	Independent Variable	Coefficient	Independent Variable	R^2
1	4.296	0.0676	Derees					0.410
2	6.285	0.1409	BS Degrees	−0.1389	Graduate Degrees			0.689
3	6.562	0.1593	BS Degrees	−0.1196	MS Degrees	−0.5296	PhD Degrees	0.711
4	2.884	0.0403	Enrollments					0.573
5	2.508	0.0394	Undergraduate Enrollment	0.0509	Graduate Enrollment			0.572

Table 3 Total Engineering Faculty Requirements Regressions

Model No.	α Constant	Coefficient	Independent Variable	Coefficient	Independent Variable	Coefficient	Independent Variable	R^2
1	7.568	0.0991	Degrees					0.358
2	4.937	0.0842	BS Degrees	0.1450	Graduate Degrees			0.386
3	−12.765	0.8236	BS Degrees	−0.5918	MS Degrees	−4.2884	PhD Degrees	0.952
4	− 9.721	0.0664	Enrollments					0.699
5	−10.370	0.0711	Undergraduate Enrollment	0.0549	Graduate Enrollment			0.701

partment, the faculty required is indicated by the following relationship:

Faculty = 2.508 + 0.0394 × (Undergraduate Enrollment + 0.0509 X (Graduate Enrollment). Note that the R^2 are higher than for the departmental models, as given in Table 1. This model would be applied department by department. The total engineering faculty (for the 5 departments) would be obtained by summing up the model results for each department.

Another approach to obtaining a total engineering faculty requirement is to sum each of the nine annual observations across departments. The dependent variable would be total engineering degrees granted. This regression would have only 9 observations. The results of the 5 models are summarized in Table 3. The R^2 for these models are also higher than for the departmental models (Table 1).

Let us now consider the relationships between enrollments and degrees granted. Clearly, this is a function of many parameters including students withdrawing from the university, changing engineering departments, transferring into an engineering department at various levels, and, perhaps, even the state of the economy. Because the focus of the study is the impact outside the university, each of these issues has not been studied, but rather an overall approximation as a fraction was determined from historical records. At IIT, all freshman students in engineering and physical science take a common core. No attempt is made to separate the students by departments except for counselling on the second year program. Students do declare a departmental preference prior to admission. No data is available to determine the relationships of the number in sophomore class/number in the freshman class on a departmental basis. For the Engineering and Physical Science programs,

$$\frac{\text{No. in Sophomore class following year}}{\text{No. of Freshmen}} = 0.733$$

Data is available to estimate the number of B.S. degrees of the class/number in the sophomore class on a departmental basis. These results are presented in Table 4. The ratio of 1.570 for Industrial Engineering reflects the large number of transfers into that department after the beginning of the sophomore year. Assuming that the E & PS

Table 4 Number in the Class Receiving B.S. Degrees/Number in the Sophomore Class

Department	Ratio
Chemical Engineering	0.935
Civil Engineering	0.806
Industrial Engineering	1.570
Electrical Engineering	0.904
Mechanics, Mechanical & Aerospace Engineering	0.936

freshman ratio is typical for all departments, Table 5 is obtained. Table 5 can be used to relate undergraduate enrollment to the B.S. degrees granted.

Table 5 Number in the Class Receiving B.S. Degrees/Number Enrolled as Freshman

Department	Ratio
Chemical Engineering	0.685
Civil Engineering	0.591
Industrial Engineering	1.151
Electrical Engineering	0.662
Mechanics, Mechanical & Aerospace Engineering	0.686

At the present time, data is not available to determine similar factors for graduate enrollment and degrees granted. This is a subject for further analysis.

What should be the distribution by rank of the faculty? Linear regression models were very inconclusive. Table 6 presents the average distribution over the period 1965-1971.

Table 6 Percentage Distribution of Faculty by Rank

Department	Professors	Assoc. Professors	Assist. Professors	Total
Chemical Engineering	56.9	19.7	23.4	100.0
Civil Engineering	48.4	29.1	22.5	100.0
Industrial Engineering	25.4	42.2	32.4	100.0
Electrical Engineering	35.3	31.4	33.3	100.0
Mechanics, Mechanical & Aerospace Engrg.	42.9	31.9	25.2	100.0
Average	42.4	30.5	27.1	100.0

Externally funded research also affects the requirement for faculty, by providing release time to do the research. Data was available for only three years on the amount of external research funds, by department. Because of the small number of observations, the departments were assumed to be equivalent (i.e., α and all βi the same) giving $3 \times 5 = 15$ observations. The regression model yielded the result in Table 7.

Table 7 Research Funds

Faculty = −3.243 + 0.0320 (Undergraduate Enrollment) + 0.1129 (Graduate Enrollment) + 0.00003 (Research Funds)

$$R^2 = 0.916$$

This implies that 1/0.00003 = $33,000 of sponsored research is required for the equivalent of one faculty member. The externally funded research was funded to the extent of 95% by NSF, HEW, NIH and other federal agencies. The locally-oriented research was negligible.

FACULTY ACTIVITIES

A questionnaire was sent to 65 faculty members at Illinois Institute of Technology at the rank of Assistant Professor and above. There were 30 responses. A copy of the questionnaire is included in the Appendix. The questionnaire did not cover the entire range of faculty activities, but concentrated on those whose direct impact was outside the university.

Consulting has a potential impact on the economic health of private concerns and efficient utilization of the resources of governmental units. Because of this, the survey requested information based on the type of organization (private or public) and the location of the impact or client (local, state, or national). The results reported in days per year averaged over a two year period, 1969-1971, are presented in Table 8, and in percent of consulting in Table 9.

Table 8 Average Number of Days Consulting/Year

	Chicago	Illinois	National	Total
Type of Client				
Private	13.96	0.80	8.67	23.43
Government	0.10	0.93	1.30	2.33
Total	14.06	1.73	9.97	25.76

Table 9 Percent of Consulting/Year

	Chicago	Illinois	National	Total
Type of Client				
Private	54.24	3.11	33.60	90.95
Government	0.39	3.61	5.05	9.05
Total	54.63	6.72	38.65	100.00

Tables 8 and 9 show that a small amount of consulting is done for governmental units. This is particularly true for Chicago, which by contrast has a large amount of consulting for private clients. The relatively small amount of private Illinois consulting indicates that private consulting may be done for local production plants and for national headquarters of major firms, but relatively little at the plant level outside of the local community. Governmental consulting accounts for only 9.05% of the consulting at all levels, while private clients account for 90.95%. In spite of the proximity of Chicago, more governmental consulting is done at the national level. This might be because of contacts and mechanisms established at the National Science Foundation, Department of Health, Education and Welfare, and the National Institute of Health. It appears that similar mechanisms would be required at the state and local level to expand the utilization of the engineering faculty resources at these levels.

The average number of total consulting days by rank is given in Table 10.

Table 10 Average Number of Days Consulting/Year by Faculty Rank

Rank	Average Days
Professor	28.4
Associate Professor	29.5
Assistant Professor	15.4

Let us consider Professors and Associate Professors as one group (Tenured) and Assistant Professors as a second group (Non-Tenured). Strictly speaking, there could be non-tenured Associate Professors, but the data does not permit this distinction. This dichotomy was used to test statistically several hypotheses concerning the groups.

The tests all took as the null hypothesis that the two groups had the same characteristic and the alternate hypothesis that they were different. For an $\alpha = 0.05$, the null hypothesis that the variance of the number days spent consulting was the same was rejected. When the null hypothesis was that the means of the number of days spent consulting were the same (no assumption on equality of variances), it was rejected for $\alpha = 0.10$ and accepted for $\alpha = 0.05$. The conclusion is that there is a significant statistical difference between the distribution of the populations for tenured and non-tenured faculty members. A test (Bartlett test) of the null hypothesis that the variance of the number of days spent consulting was the same for the three groups, Professors, Associate Professors, and Assistant Professors, was rejected at $\alpha = 0.05$. No test was available for the hypothesis that the means were the same for the three groups. Differences between departments were studied next. No test was available for the equality of the three means with no assumption on the equality of the variances. However, representative departments could be tested two at a time for equality of means. This resulted in the null hypothesis of equal means between departments being rejected for $\alpha = 0.05$.

Another activity of engineering faculty is involvement in community service and participation on citizen boards. The percent of participation in these activities by faculty members by rank is given in Table 11.

Table 11 Percent Participation in Service Activities

Rank	% Participation
Professors	38.5
Assoc. Professors	55.5
Assist. Professors	57.2
Overall	46.7

Of those who participated, there was an average of 1.35 activities per faculty member. Statistical tests as to hypothesis testing were inconclusive for differences by rank and by department. A possible explanation of the lower percentage participation of Professors is that they are less likely to have school age children who are often a motivating factor for this type of involvement.

Another activity outside the university is lecturing to various groups. These groups were classified into the following categories: other universities, professional societies, local firms, national firms and miscellaneous. The average numbers of lectures per year to each of these categories are presented in Table 12 by faculty rank.

Table 12 Average Annual Number of Lectures

Rank	Other Universities	Professional Societies	Local Firms	National Firms	Miscellaneous	Total
Professor	1.38	1.77	0.00	0.23	0.46	3.84
Assoc. Prof.	0.11	0.78	0.33	0.33	1.00	2.22
Assist. Prof.	0.63	0.25	0.00	0.00	0.00	0.88

Statistical hypothesis tests were inconclusive with respect to differences between ranks. However, this is because of the shortcomings of the tests which have been developed, not the particular values of the data. It appears that the higher the rank, the more lectures are given per year. Observe that most of the lectures reported were to other uniiversities and professional societies. These activities are in large part interaction with other academic individuals. An increase in rank and academic expertise would lead one to expect more requests or opportunities to speak to individuals with similar academic interests. There was a complete absence of lectures to local firms. This might be a fruitful area to open by establishing mechanisms for the initial contacts. Surveys of recent developments in an engineering discipline may be included in professional society meetings, particularly at the level of local chapters.

Another activity is involvement in professional societies. All faculty members are members of one or more professional societies. The professional society activities tabulated were those other than membership. Examples include memberships on committees and offices held for local and national professional societies, and editorships of professional journals. The average number of such memberships are summarized in Table 13 by faculty rank.

Table 13 Professional Society: Offices and Committee Assignments and Editorships

Rank	Local Chapter	National Activities	Editor of Journal	Total
Professor	0.31	2.08	0.23	2.62
Associate Prof.	0.56	0.78	0.11	1.45
Assistant Prof.	0.38	0.25	0.00	0.63

The results are similar to those for lectures. The higher the faculty rank, the larger the amount of participation. Again, this seems likely as one develops professionally. This may also be the reason for the larger amount of consulting by tenured faculty relative to non-tenured faculty. This is likely to be related to professional development, rather than the tenure decision.

SUMMARY

The first section of this paper can be used as the basis for a model relating enrollment to degrees granted, by degree level and department and to faculty requirements, by rank and department. This gives a set of linear equations.

The "degrees conferred" can be used with data on the location of an initial job to determine the additions to the local, state and national pool of engineers. The "faculty requirements" are the input to another set of linear equations which give faculty activities as the output.

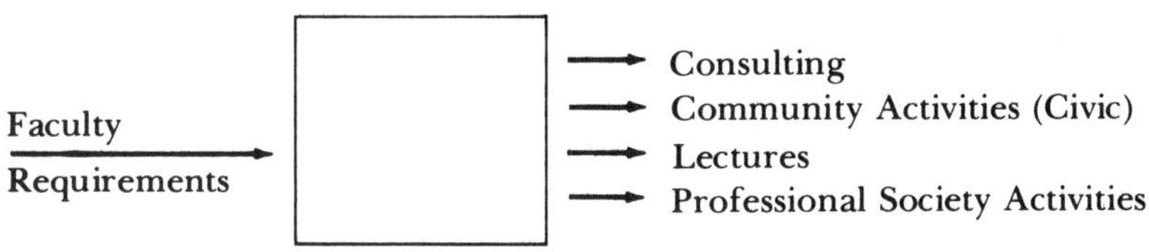

These models are sequential and static, but are useful for short-term estimation of the extent of activities which affect the community outside the university.

Appendix

Questionnaire on the Outside Activities of Engineering Faculty for the Project on the Regional Role of Engineering Colleges

A. Consulting

1. During the past two academic years I spent the following number of days consulting

	1969-70 Private Client	1969-70 Government Client	1970-71 Private Client	1970-71 Government Client
Chicago Area Client				
Illinois Client				
National Client				

2. About what percent of this *private* consulting indirectly affects the following organizations?

Local Government ________%
State Government ________%
Federal Government________%
Private Industry ________%
Other ________%

B. During the past two academic years I gave lectures for the following organizations outside the Engineering College: (Examples might include other colleges or universities, professional societies, civic organizations, etc.)

C. During the past two academic years, I had the following assignments in professional organizations: (Examples might include officer or committee assignments at local or national level, editor for professional journals, etc.)

D. During the past two academic years I participated in the following community and citizens' activities: (Examples might include school, village, or church boards or committees, Boy Scouts, Little League Baseball, etc.)

Department	Rank

Extra–University Activities of Engineering Faculties

Thomas W. Knowles

INTRODUCTION

The purpose of this study was to gather and analyze data on the nature of faculty activities. The results could be used as the data base for modeling efforts of the interface between the engineering college and the region. Obviously, faculty members provide services other than teaching and research. The other activities considered were consulting, professional societies, community services and other professional actitivies.

Data was obtained from all of the engineering schools in Illinois. This was obtained by a series of quesionnaires which were sent to faculty members. A pilot study was conducted at Illinois Institute of Technology. Upon analyzing the results, a revised questionnaire was sent to the faculty members of Bradley University, Northwestern University, Southern Illinois University at Carbondale and the University of Illinois at Chicago Circle. Data from the University of Illinois at Urbana-Champaign was already available from faculty activities reports. Bradley, Northwestern and IIT are private schools; UICC, UIUC and SUIC are state-supported schools.

The questionnaire forms are included in the Appendix. Unfortunately, not all of the data is directly comparable between schools.

Thomas W. Knowles is an Assistant Professor in the Department of Industrial and Systems Engineering at Illinois Institute of Technology, in Chicago, Illinois.

CONSULTING

Consulting has a potential impact on the economic health of private concerns and the efficient utilization of the resources of governmental units. Because of this, the survey requested information on the type of organization (private or government) and the location of the client. The results are comparable for all institutions except University of Illinois at Urbana-Champaign, which has a reporting procedure with different categories. The results reported in days per year averaged over a two-year period, 1969-1971, are presented in Table 1, for each institution.

Table 1 Average Days Consulting/Year/Faculty Member

School	Type	Location Chicago	Illinois	National	Total
Bradley	Private	0.89	4.11	0.22	5.22
	Government	0.00	0.39	0.83	1.22
	Total	0.89	4.50	1.05	6.44
IIT	Private	13.96	0.80	8.67	23.43
	Government	0.10	0.93	1.30	2.33
	Total	14.06	1.73	9.97	25.76
Northwestern	Private	5.84	1.74	1.01	8.59
	Government	1.72	2.69	2.03	6.44
	Total	7.56	4.43	3.04	15.03
SIUC	Private	0.00	0.92	0.38	1.30
	Government	0.15	0.15	2.70	3.00
	Total	0.15	1.07	3.08	4.30
UICC	Private	3.35	0.44	0.80	4.59
	Government	2.32	0.17	2.52	5.01
	Total	5.67	0.61	3.32	9.60
Composite	Private	5.68	1.46	2.11	9.25
	Government	1.30	1.38	1.94	4.62
	Total	6.98	2.84	4.05	13.87

The composite reveals that the average annual number of days consulting was 13.87. Of this, 9.25 days or 66.7% was for private concerns and 4.62 days or 33.3% was for government. More government consulting was done outside of Chicago (mostly for state government) than for Chicago, although three of the institutions were located in Chicago. This should be contrasted with the ratio for private consulting of 5.68 days in Chicago and 1.46 days in Illinois outside of Chicago. This indicates that the mechanisms for contacts between faculty members and private concerns are better established than those between faculty members and government. Similarly, because there are mechanisms such as HEW, NSF, NIH, etc., more governmental consulting is done at the Federal level than in either category within Illinois. The average number of days of consulting was higher in urban areas than nonurban areas, but there was no strong distinction between private and state universities.

The second questionnaire also surveyed the scope of the consulting. Thus, for example, although the consulting might be done in Chicago, the firm could be a national one and hence the scope would be national. The results are summarized in Table 2.

Table 2 Average Days Consulting/Year/Faculty Member

School	Type	Local	Illinois	National	Total
Bradley	Private	3.22	0.00	2.00	5.22
	Government	0.11	0.00	1.11	1.22
	Total	3.33	0.00	3.11	6.44
Northwestern	Private	1.30	2.26	5.03	8.59
	Government	1.00	1.13	4.31	6.44
	Total	2.30	3.39	9.34	15.03
SIUC	Private	0.46	0.08	0.76	1.30
	Government	0.08	0.62	2.30	3.00
		0.54	0.70	3.07	4.30
UICC	Private	0.13	2.50	1.96	4.59
	Government	1.84	0.30	2.87	5.01
	Total	1.97	2.80	4.83	9.60

The results for University of Illinois at Urbana-Champaign were categorized into governmental, industrial, professional (e.g. expert witness) and other. There was no differentiation with respect to location or scope. The results are summarized in Table 3.

Table 3 Average Days Consulting/Year/Faculty Member University of Illinois at Urbana-Champaign

Category	*Days*
Government	3.6
Industrial	7.2
Professional	0.6
Other	1.4
Total	12.4

The average consulting days is given by rank in Table 4.

Table 4 Average Total Consulting Days By Rank

University	Prof.	Assoc. Prof.	Ass't. Prof.	Unknown
Bradley	12.5	5.4	4.0	—
IIT	28.4	29.5	15.4	—
Northwestern	17.7	13.8	3.0	17.8
SIUC	2.8	8.3	1.3	—
UICC	15.8	9.7	6.0	12.0
Composite	18.7	13.3	5.8	17.1

In general, Professors and Associate Professors do more consulting than Assistant Professors. That dichotomy was used to test for significance of the difference between the groups. The tests showed significant difference in the means for $\alpha = 0.05$. No test was available for the equality of the means of the universities with no assumption on the variances. However, representative universities could be tested two at a time for equality. This resulted in the null hypothesis of equal means between universities being rejected for $a = 0.05$.

Another activity of engineering faculty is community service and participation on citizen boards. The percent of faculty members of a given rank participating and the average number of activities of the participants are given in Table 5.

Table 5 Percent Involvement in Community Service

Rank	Bradley	IIT	NU	SIUC	UICC	Composite
Professors	100.0	33.5	30.7	50.0	33.3	42.8
Assoc. Prof.	44.4	53.5	50.0	100.0	37.5	63.8
Ass't. Prof.	60.0	57.2	16.7	25.0	55.7	46.5
Unknown	—	—	44.0	—	60.0	46.6
Overall	50.0	46.7	35.6	58.3	50.0	49.6
Avg. No. of Activ.	1.80	1.35	2.26	2.43	3.23	2.30

The survey showed that Associate Professors were somewhat more likely to participate in community service activities than Assistant Professors and Professors. This is probably explained by the greater likelihood that Associate Professors would have primary school age children. It is interesting to note that of engineering faculty members who do participate in community activities (49.6%), most participated in more than one activity, (an average of 2.30).

Another faculty activity outside the university is lectures to various groups. These groups were classified into the following categories: other universities, professional societies, civic organizations and others. The average number of lectures per year to each of these categories is presented in Table 6.

Table 6 Average Annual Number Of Lectures

University	Other Univ.	Prof. Societies	Civic	Other	Total
Bradley	0.10	2.44	1.94	0.72	5.20
IIT	1.05	1.15	0.24	0.48	2.92
Northwestern	3.42	3.36	0.52	1.11	8.41
SIUC	1.00	0.69	0.61	0.54	2.84
UICC	2.31	1.84	0.91	0.39	5.45
Composite	2.42	2.53	1.22	0.81	6.98

There was no statistically significant (at $\alpha = 0.10$) difference between the means for private and state-supported universities. Engineering faculty give most lectures to groups related directly to their profession

— other universities and professional societies. With the exception of lectures to local chapters of professional societies, these have no direct local impact. There were essentially no instances of lectures to local firms. This might be a fruitful area to open by establishing mechanisms for the initial contacts.

The average number of lectures of all types is presented in Table 7 by rank.

Table 7 Average Annual Number of Lectures By Rank

University	Prof.	Assoc. Prof.	Asst. Prof.	Unknown	Overall
Bradley	4.50	5.22	5.40	—	5.20
IIT	3.84	2.88	0.88	—	2.92
NU	9.65	8.60	1.67	10.28	8.41
SIUC	3.75	0.40	3.25	—	2.84
UICC	5.83	4.75	3.63	3.40	5.45
Composite	7.52	6.00	2.70	9.13	6.16

Professors and Associate Professors gave significantly more lectures than Assistant Professors. However, this was not uniformly true at all universities. The probable reason for this is the greater recognition for research achievements of Professors and Associate Professors.

Another activity is involvement in professional societies. Almost all faculty members are members of one or more professional societies. The professional society activities surveyed were those other than membership. The average number of local committee, national committee and journal editorship assignments are presented in Table 8.

Table 8 Average Number Of Professional Society Assignments

University	Local Committee	National Committee	Editor of Journal	Total
Bradley	0.55	0.28	0.00	0.83
IIT	0.40	1.20	0.10	1.70
NU	0.67	1.62	0.16	2.65
SIUC	0.61	0.31	0.00	0.92
UICC	1.08	1.08	0.04	2.20
Composite	0.77	1.33	0.10	2.25

Analysis by rank shows that Professors and Associate Professors have more professional society assignments than Assistant Professors. Again, this is likely because of professional development.

*Short-Term Demand for Engineers In Illinois, 1972**

Hugh Folk, Bradley Parlin and Thomas Wicke

This study was undertaken in the summer and fall of 1971 to provide information on three questions: (1) What are the current employment opportunities for engineers displaced from defense industry in Illinois; (2) What is the current demand for new graduates in engineering from Illinois employers; and (3) What is the current specific demand for environmental engineers. A brief preliminary report based on interviews with ten employers was presented to a meeting of the Regional Role of Engineering Colleges staff in October, 1971 in which it was concluded that:

1. Government employment of environmental engineers will grow rapidly in 1971/72 and then level off.
2. Private employers did not expect to hire many environmental engineers.
3. Planned hires for 1971/72 for private employers were substantially below actual hires for 1970/71.

Hugh Folk is Professor of Economics, Center for Advanced Computation and Institute of Labor and Industrial Relations, University of Illinois at Urbana-Champaign.
Bradley Parlin is Assistant Professor of Sociology, University of Indiana at Fort Wayne.
Thomas Wicke is Graduate Research Assistant, Institute of Labor and Industrial Relations, University of Illinois at Urbana-Champaign.

*The research underlying this paper was supported by the Regional Role of Engineer-

This paper reports on a larger sample but the conclusions are the same.

The first section of the paper describes the survey, the second section reports on employer recruitment plans, sources, constraints, reactions to job-matching systems, and reaction to retraining and substitution. The third section discusses the coverage and consistency of the survey with respect to other studies. The fourth section considers the problems and desirability of a system of regular short-term surveys of recruitment plans for institutional planning.

I. THE SURVEY

The survey was directed to organizations expected to be large employers of engineers in the State of Illinois. For the results reported here, twenty-six organizations were interviewed (see Table 1). These included seven public and nineteen private Illinois employers. The private employers included public utilities, consulting firms, and light and heavy manufacturing. The public employers were balanced between Federal, State and local levels.

Information Sources. Interviews were designed to yield the following information:

1. Type and number of engineers employed (including job descriptions if available).
2. Character of labor market in past two years, present, and immediate future, including, where available, forecasts and areas of chronic shortages.
3. Major sources of recruitment.
4. Constraints on recruitment.
5. Experience and reaction to job matching systems (such as GRAD I and II, or Compujob).
6. Acceptability and willingness to participate in retraining programs, possibilities for substitution between degrees and specialties and related information.

ing Colleges study and from the Manpower Planning for the University of Illinois project at the Center for Advanced Computation. Mrs. Pauline Chapman, Placement Director of the College of Engineering, University of Illinois at Urbana-Champaign, helped us a great deal in planning the survey and we are grateful to her. The many anonymous respondents in the organizations contacted went to considerable trouble to assist us and we appreciate their time and trouble.

Table 1 RECRUITMENT—1971/72 Planned, 1970/71 and 1969/70 Actual, and 1971 Employment, by Organizations

Employer	1971 Illinois Engineer Employment	Recruitment Actual 1969/70	Recruitment Actual 1970/71	Recruitment Planned 1971/72	Change in Recruitment (per cent) 1971/72 / 1969/70	Change in Recruitment (per cent) 1971/72 / 1970/71	1971/72 Recruitment as per cent of Employment
PUBLIC							
				5			
				68			
	1.367			0			0
	97			95			98
	20	8		15	+ 88		75
	235	19	19	19	0	0	8
	205			25			12
PRIVATE MANUFACTURING							
		150		52	– 65		
		65	40	15-20	– 77	– 62[b]	
	280	46	35	35	– 24	0	13
	3.500	88	85	50-60	– 43	– 41[b]	1[b]
		170	85	50-60	– 71	– 41	
				15-18			
		29	–	9	– 69		
	1,500	520	39	26	– 95	– 33	2
	22	3	2	4-5	+ 33	+100 [b]	18[b]
	700	100	42	25	– 75	– 40	4
UTILITY							
				40	– 53		
	563	85	70	35	– 43	– 50	6
		61	100	60		– 40	
CONSTRUCTION and CONSULTING							
	300	6	10	10-12	+ 67	0[b]	3[b]
	60	10	8	8	– 20	0	13
	500	35	22	15	– 57	– 32	3
	600	122	175	150	+ 23	–15	25
	400	35	50	35	0	–30	9
	306	107	120	50	– 53	– 58	16
TOTAL	10,655	1,659	903	911[a]			
Organizations reporting employment	10,655	1,659		597[a]			6
Organizations with recruitment data 1969/70 and 1971/72		1,393		883[a]	– 47		
Organizations with recruitment data 1970/71 and 1971/72			903	587[a]		– 35	
Organizations with recruitment data 1969/70 and 1970/71			802				

a. Total based on lower value of range of values.
b. Percentages based on lower value of range of values.

II. SURVEY RESULTS

Types of Engineers Sought. Mechanical engineering was the most widely sought specialty and was specified by twenty-one employers, followed by civil engineering, specified by twelve employers. The third most sought specialty was electrical engineering (power), followed by industrial and safety engineers.

The public agencies were most interested in civil engineers (particularly sanitary and environmental options) and safety engineers. The private firms were most interested in mechanical engineers and industrial and electrical engineers. Much of the public recruitment effort is focused on testing and compliance activity while the private employers are directing their recruiting efforts toward staffing those areas which have been traditionally functionally specific. This would explain the differential emphasis by public and private employers on degree specialty.

The total number of engineers employed in Illinois by the sample employers is 10,655. This does not include nine employers who did not respond.

Outlook. The outlook for the recruiting year 1971/72 was mixed. Public agencies were increasing hiring while private employers had cut recruitment activity. The heaviest recruitment activity on the public side is in environmental and occupational safety agencies which anticipate volume recruitment activity.

Table 1 summarizes the recruiting activity of the sample companies for the 1969/70 and 1970/71 recruiting years as well as forecast data for 1971/72. Seven employers in our sample of twenty-six were unable to provide actual 1969/70 recruiting activity data and ten were unable to provide 1970/71 data. This of course precludes computing relative changes for those employers.

In all, twenty-six employers in our sample hope to hire 911 engineers in 1971/72. Those sixteen employers for whom 1970/71 recruiting activity data was available anticipate a gross decrease in recruiting activity of 35 percent in 1971/72.

As we noted above, the heaviest recruiting activity will be by public sector employers, a rate of recruitment of 12 per cent of employment, in contrast to a 6 percent rate for the entire sample. All three private industry sectors show percentage decreases in recruiting of 30 to 40 per cent below last year's hires.

Source. Seven of the twenty-six sample employers have either reduced college visits or frozen college hiring. There seems to be a tendency to recruit experienced applicants at the expense of college programs under the current market situation. A major difference in public and private employers derives from a differential emphasis on sources of engineering prospects. Public employers depend primarily on experienced engineers, while not avoiding new graduates. NIOSH is a specific exception which gets much of its field staff from conscientious objectors fulfilling alternative service requirements.

In sharp contrast private employers express a strong preference for, and sometimes restrict their hiring to, new graduates.

Constraints. Public agencies reported constraints in competitive areas, inability to pay travel, noncompetitive salaries for scarce environmental engineers, delays inherent in civil service. Private firms reported no serious constraints other than temporary freezes on new graduate hiring.

Reaction to Job Matching Systems. Only one public agency had any experience with computerized job matching systems. It had paid $500 and received four unsuitable resumes. Some interest was expressed in using such systems, but fee systems are generally unacceptable.

Private employer experience with computerized job placement systems was far more extensive but much less optimistic. Fifteen of the nineteen private employers sampled had experience with one or more of the common placement systems. All of the respondents found the College Placement Council's GRAD I to be almost useless from an employer standpoint. The major complaint seems to derive from the disproportionate number of marginal candidates who use the system as a last resort. The same negative attitudes characterize the employers' response to other systems such as Compujob. Employers were maintaining a wait and see attitude on CPC's GRAD II, but they were generally skeptical of the value of matching systems as a useful instrument in facilitating job placement.

In contrast to the negative response of private employers noted above, public agencies were generally receptive to any system that promised to assist them in meeting their needs at no significant cost. Recruitment is expensive and time consuming and is not adequately budgeted in most public agencies.

Acceptability of Retraining and Substitution. Public agencies in the environmental and safety fields demonstrated a high degree of flexibility with respect to field, stressing related experience rather than fields. They generally expressed a high degree of willingness to consider engineers individually in reference to their professional experience rather than rigid adherence to fields of training as evidenced by degrees.

In contrast, private employers seem much less willing to accept individuals outside of their formal degree areas. Eight of the nineteen private employers had experimented with limited substitutions across degree areas. More commonly, however, private employers suggested that lengthy and costly retraining (implicit is substitution) precluded serious interest in this alternative. In addition, current market conditions were so soft that it was not necessary for employers to experiment with substitution in order to meet their current engineering manpower demands. Private employers, unlike their public counterparts, were not generally expanding hiring in areas of environmental engineering. Where increased demand for environmental engineers was apparent among private employers, the tendency was to retrain and promote from within rather than increase outside recruitment. This contrasts with the response of public employers noted above.

Summary. There has obviously been a substantial increase in governmental employment of engineers in environmental services resulting from state and federal environmental protection and occupational safety and health legislation. This has resulted in the rapid growth of engineering employment in the Illinois EPA, OSHA, NIOSH, and will have effects of lesser degree but perhaps of greater magnitude in local agency employment. It cannot be expected that the rate of increase experienced during the past year in these agencies will be sustained for any period of time. Much of the hiring has been staffing up for new responsibilities and has drawn on experienced engineers, often with experience in environmental fields.

There is no evidence that the private industry in our sample has yet made or plans to make any substantial increase in engineering staff devoted to environmental or other problems. Just the reverse, the number of new hires expected for 1971/72 will be reduced to almost half the levels of 1969/70. The findings of this survey are entirely consistent with our preliminary survey findings as well as surveys on engineering demand on a national level. The employment situation for new graduates in 1971/72 and for experienced engineers in the coming year will be very serious indeed.

III. HOW REPRESENTATIVE ARE THESE RESULTS?

While no attempt was made to design a representative sample, we believe we have covered all major types of engineering employers and have included many of the major employers. There is no good estimate of the number of engineers employed in Illinois and will not be until Census data are available for 1970. In 1960, Illinois employed about 6.6 per cent of the engineers in the United States. Engineering employment in the United States probably grew somewhat faster than did Illinois employment. Illinois might be expected to have about 72,000 engineers in 1971. For those employers reporting employment in Illinois, the number of new hires represented about 6 per cent of employment. If the same ratio held for the entire sample, then the sample firms would account for about 15,800 engineers, or roughly 22 per cent of total engineering employment in the State.

If those estimates are worth anything (and we do not argue that they are) and if the same ratio of new hires in 1971/72 to employment holds, then the 911 new hires represents a total new hire population for Illinois of about 4,100 engineers.

If it were possible to compare this figure to the number of new entrants, it would suggest that there are ample new jobs for new graduates, who will number about 1,477 in the State. Obviously, this cannot be done. Many of the new hires represent the filling of jobs for engineers who have quit and are moving to other jobs. In any year, about 30 per cent of engineering openings have been filled by new graduates, and this includes graduate degrees and the B.S. graduates who went to work. This suggests that about 1,500 engineering graduates in Illinois will be seeking jobs and about 1,200 jobs will be available, if the ratio between new graduates and experienced engineers holds. This suggests about a 20 per cent deficit in jobs. It does not mean that 20 per cent of the engineering graduates in Illinois will be employed or underemployed. It may mean some Illinois graduates must go far afield, and the immigration of graduates will be smaller.

It must be pointed out that the comparison is quite approximate, but if the 1970/71 recruiting year was one of approximate balance with a slight bias toward oversupply (as indicated by underemployment and unemployment of new graduates), and if the 1971/72 recruiting year is 35 per cent below 1970/71, then the 20 per cent deficit of jobs which we have indicated in the preceding paragraph may be approximately correct.

Other studies of recruiting this year show somewhat different results. The College Placement Council's survey which we have read about but not seen suggests some improvement in recruiting over last year. This is not necessarily inconsistent with our findings since last year was probably poorer nationally than it was in Illinois.* Just because the Illinois market was better than the national market last year does not mean that it will be better this year. The poor national picture last year may in fact account for the poor Illinois picture this year, since the relatively small number of vacancies may come from a decline in the number of voluntary job leavers who in the past have left jobs to head West. With fewer openings in California, there were fewer replacements needed in Illinois and, consequently, fewer job openings in Illinois.

IV. SHORT-TERM DEMAND SURVEYS

The results of the survey reported in this paper suffer from the problems common to short-term labor demand surveys. Coverage is never complete, comparability to previous year's results is lacking, plans are subject to revision and are therefore subject to internal non-comparability if contact time differs for the several respondents. While it is impossible to say whether the plans expressed by the respondents to this study will be carried out, there are some reasons to believe they will be. Engineering recruiters have the organizational responsibility to fill openings, and are usually judged by their performance of this task.[1] Openings are budgeted and unless the economic situation of the firm changes considerably in the course of the recruitment year, revisions are relatively infrequent.

The situation in short-term demand forecasting for workers such as engineers differs considerably from the situation for employees in general, since employers understand that engineering recruitment is a

*The College Placement Council reported a decrease in the number of job offers of 56 per cent for technical graduates from 1969/70 to 1970/71. Our survey showed a decrease in hires of 43 per cent for these two recruiting years.
See *A Study of 1970/71 Beginning Offers by Business and Industry,* Bethlehem, Pa.: The College Placement Council, 1971, p. 1.

lengthy and time consuming task involving considerable expense and not easy to turn around.

It is planned to follow up these interviews in the 1972/73 recruitment year in order to see how well plans predicted outcomes. When adequate data on the employment and distribution of engineers in Illinois is available from the 1970 Census of Population it will be possible to project the total population demand for engineers from a fairly small sample of firms and agencies.

We believe that an effective and inexpensive short-term forecasting system could be developed from organizational interviews and questionnaires, but we doubt that such a system would have much value for the kind of planning that educational institutions, public agencies, and students must make. Year-to-year fluctuations in demand are substantial, and "next year" is not a very useful guide to action when dealing with an educational system which has a four or five year period.

This study was not intended to provide reliable estimates of total demand for engineers, but to indicate how the current situation compared to earlier years. Properly controlled, even a simple survey can accomplish this, and we believe this survey does. We found that government agencies had substantially increased demands for engineers arising from environmental concerns, but that the plans amounted to a once and for all step increase which cannot be expected to continue past this recruiting year. We found that most private employers had no immediate plans to increase hiring of environmental engineers. We also found that new hire plans by private employers were substantially below last year's plans and that this suggests substantial employment problems for new graduates and unemployed engineers in the present recruiting year.

REFERENCE

1. Folk, H., *The Shortage of Scientists and Engineers*, Lexington, Mass.: D.C. Heath, 1970, Chapter IV.

Supply and Demand for Engineers In Illinois: 1975 and 1980

Michael Babcock, Keith Erickson, Hugh Folk, and Walter Nidzieko

Educational decision-makers require reliable forecasts of occupational supply and demand. Recognizing this need, the University of Illinois undertook the development of a manpower and educational forecasting system known as STEP I at the Center for Advanced Computation (CAC). The STEP I system includes population, enrollment, mobility and demand models for the United States and Illinois. The STEP I system is comprehensive, but it is necessary to make numerous detailed assumptions and specifications. Forecasting is an inherently hazardous activity. Changes in underlying economic and social conditions can make a forecast which appears reasonable at the time it is made miss the actual outcome widely. The widely expected and forecast "engineer shortage" for 1970, for instance, did not occur because of the unexpected combination of recession and cutbacks in military and R and D spending.

Given the sudden changes in underlying conditions which occur from time to time, there is little reason to rely on a system which merely projects recent trends. The STEP system is free of this shortcoming. The results presented here suggest that the engineering short-

The authors are members of the Center for Advanced Computation, University of Illinois, Urbana-Champaign. Professors Roger Bezdek and Robert Dauffenbach provided helpful advice and assistance and we are grateful to them. Research for the paper was supported by the Office of the Provost of the University of Illinois.

ages experienced in the 1950's and 1960's will return in the mid-1970's. Thus, the recent unemployment or "surplus of engineers" appears to us to be a temporary situation.

Despite the attempt to take account of conditions which might influence the supply of and demand for engineers our results necessarily may turn out to be wrong. Readers are cautioned against placing excessive trust in what must necessarily be an exercise in forecasting.

I. THE FORECASTING MODEL

The STEP system includes the following components:

1. a set of alternative projected patterns of GNP for future years
2. an activity-industry matrix which converts GNP by program of expenditure into final demands by industry for the United States
3. an employment inverse matrix which generates total employment generated in each industry by unit expenditure in every other industry for the United States*
4. a set of regression equations which estimate Illinois employment by industry from U.S. gross exports, U.S. total employment, Illinois total employment, and U.S. total employment by industry**
5. an occupation-industry matrix which distributes the total Illinois industry employment among the various occupations.***

The present paper presents results of STEP I forecasts of Illinois demand for engineers in 1975 and 1980 using the national GNP forecasts developed by the Bureau of Labor Statistics of the U.S. Department of Labor.†

*For a discussion of the national model, see Roger H. Bezdek, *Manpower Implications of Alternate Patterns of Demand for Goods and Services* (Ph.D. thesis, University of Illinois, Urbana-Champaign, 1971).

**For a discussion of the state model, see Michael W. Bobcock, *Employment Implications of Alternate Federal Spending Priorities for the Illinois Economy* (Ph.D. thesis, University of Illinois, Urbana-Champaign, 1973).

***One important shortcoming of this study is the use of an estimated occupation-industry matrix for Illinois rather than one based on the 1970 census for which data was not yet available. The estimate is discussed in Babcock, *op. cit.*

†The occupational demands implied by three other patterns of national GNP is presented in Babcock, *op. cit.* While total Illinois employment varies considerably under the four alternative GNP patterns, the variation in engineering demands are within five percent of those presented here. These variations are too small in view of the inherent uncertainties in the forecasts to warrant separate discussion here.

The actual forecasts were made by first projecting GNP by industry of final demand. The pattern of final demand projected implied necessary levels of total demand by industry, and total output, in turn, indicated total labor requirements. The final demand vector was converted to total employment by use of an employment inverse matrix based on input-output studies prepared by the Bureau of Labor Statistics. The employment forecasts for private wage and salary employment were input to the State employment model, along with forecasts of total Illinois employment.

II. THE STATE EMPLOYMENT MODEL

The Illinois employment model consists of 60 equations which relate Illinois employment in each industry to the following variables: (1) U.S. total employment (2) U.S. employment in the industry (3) Illinois total employment (4) U.S. exports and (5) linear trend. Only variables for which forecast values will be available in the system are included in the forecasting equations. Despite this limitation, the proportion of variance in industry employment explained by the equations (R^2) is quite high, suggesting that this set of independent variables is sufficient to explain employment changes in most Illinois industries over time (see Table 1). For 1975 and 1980 the following general assumptions were made: (1) economic, social, technological, and scientific trends were continuing about the same as in the recent past; (2) the institutional framework of the economy will not change radically; (3) values placed on work, income, and leisure will not change drastically; (4) fiscal and monetary policies will be generally successful in achieving a satisfactory balance between unemployment and inflation without reducing the long-term economic growth rate; (5) the Vietnam War will have ended by 1975 and the U.S. will not become engaged in any other wars in this decade; (6) fertility rates will be lower than they have been in recent years.

The forecast value of Illinois total employment was derived from Illinois population forecasts of the U.S. Bureau of the Census and from BLS forecasts of labor force 16 years old and over. We assumed an unemployment rate of four percent and Illinois total employment at 4,456,000 in 1975 and 5,190,000 in 1980.

The estimate of total exports was obtained by linear extrapolation of the 1958 to 1970 trend in the series of U.S. merchandise exports prepared by the U.S. Department of Commerce.

Table 1 Illinois Industry Total Employment Forecast, 1970, 1975, and 1980 (Thousands of Employees)

Industry	R^2 of the equation	Private wage and salary 1970 Actual	Private wage and salary 1970 Foreast	All employees Forecast 1975	All employees Forecast 1980
Agriculture	.99+	160.0	163.7	148.7	132.6
Bituminous coal mining	.96	9.9	9.1	8.5	8.6
Oil and gas extraction	.99+	6.1	6.7	5.1	2.8
Nonmetallic mining, except fuels	.97	7.0	6.8	11.0	8.2
Food and kindred products	.92	132.7	131.7	141.9	131.3
Apparel and other textile products	.97	36.0	38.1	30.9	32.1
Construction	.85	184.4	174.7	263.2	280.8
Lumber and wood products	.94	11.8	12.2	14.5	14.4
Furniture and fixtures	.98	26.7	25.2	36.6	38.2
Paper and allied products	.99	39.9	40.7	48.4	53.2
Printing and publishing	.99+	117.6	117.3	119.9	134.0
Chemicals and allied products	.99+	65.9	64.2	71.6	68.0
Petroleum and coal products	.97	14.5	16.8	13.0	13.9
Rubber and plastics	.99+	43.6	43.9	52.4	65.5
Leather and leather products	.99+	13.8	13.9	11.8	9.5
Stone, clay, and glass products	.78	37.9	36.6	35.2	39.8
Primary metals manufacturing	.95	109.3	106.0	111.8	117.7
Fabricated metal products	.99+	143.3	138.6	153.9	163.5
Machinery, except electrical	.94	221.3	231.7	241.9	279.6
Electrical equipment and supplies	.95	198.4	197.7	195.1	217.9
Transportation equipment	.92	46.4	44.4	43.4	51.2
Instruments and related products	.99	43.6	45.3	50.6	53.1
Miscellaneous manufacturing	.99	33.3	33.9	48.5	43.7
Railroad transportation	.99+	59.2	62.4	75.7	45.6
Local and interurban passenger transit	.90	17.8	16.7	20.5	31.3
Trucking and warehousing	.99+	71.2	77.3	89.6	93.7
Air transportation	.99+	25.2	24.8	27.4	38.0
Pipeline transportation	.70	.9	1.3	1.1	1.1
Transportation services	.94	7.5	6.6	14.5	11.5
Communications	.93	61.8	61.5	49.6	52.0
Electric, gas, and sanitary services	.95	36.8	37.7	40.2	45.3
Wholesale trade	.99+	278.3	273.7	365.0	392.5
Building materials and farm equipment	.87	25.9	27.6	38.3	37.1
Retail general merchandise	.99+	165.3	168.4	214.8	234.9
Food stores	.99+	107.9	108.7	152.4	165.5
Auto dealers and gas stations	.99+	77.3	78.9	103.3	119.3
Apparel and accessories	.99	44.1	44.6	52.3	53.5
Furniture and home furnishings	.99+	24.4	25.4	32.2	35.9
Eating and drinking places	.92	147.3	139.2	198.8	221.3
Miscellaneous retail stores	.97	68.9	75.0	93.9	91.8
Banking	.99+	61.3	61.8	68.9	75.4
Credit agencies other than banks	.99+	22.7	23.9	28.3	31.8
Security and commodity brokers and services	.99	11.4	11.4	11.1	12.5
Insurance carriers	.99+	71.5	72.0	80.6	97.1
Insurance agents, brokers, and service	.99+	17.6	17.6	21.3	25.1
Real estate	.76	47.3	55.3	59.3	65.4
Hotels and other lodging places	.97	37.0	36.9	45.5	43.6
Personal services	.99	60.5	60.3	108.9	121.7
Miscellaenous business services	.99+	102.9	101.4	138.6	188.6
Auto repair services, garages	.99+	19.4	18.6	34.2	37.8
Miscellaenous repair services	.99+	19.4	18.6	20.0	23.2
Motion pictures	.95	10.3	10.3	9.2	10.4
Amusement, recreation, exc. motion pictures	.98	9.2	9.2	25.3	30.1
Medical and other health services[b]	.98	25.0	25.7	317.8	402.1
Legal services	.99+	18.0	18.6	31.3	39.5
Education services—private[b]	.99+	14.4	14.7	110.9	148.2
Education services—public		204.6	252.4	288.5	385.3
Nonprofit membership organizations	.98	73.7	76.5	121.1	122.4
Miscellaneous services	.96	39.3	39.4	45.3	65.6
Federal public administration				88.4	91.0
Post Office				58.7	60.4
Other federal public administration				29.7	30.6
State public administration				34.3	40.0
Local public administration				86.4	100.0
Total		3,669.5	3,692.2	5,201.4	5,782.9

[a] Total employment all years.
[b] 1970 coverage incomplete.

III. FORECASTS OF INDUSTRY DEMAND FOR 1975 AND 1980

The forecasts for private wage and salary employment in Illinois for 1970 and the actual values estimated by the Illinois Department of Employment are given in Table 1. There is fairly close correspondence between the forecast and actual values. The forecast values for total employment by industry for 1975 and 1980 are also given in Table 1. These are not directly comparable to the private wages and salary employment figures for 1970. The pattern of change suggested by the forecasts are consistent with those expected by most observers. Agriculture and the extractive industries decline, manufacturing grows modestly, and services grow rather rapidly.

It is important to note that the industry forecasts implied by the BLS national model lead to industry employment far in excess of the total employment derived from BLS labor force forecasts. This could mean that the regression equations of the model inherently exaggerate the amount of employment growth expected. We do not think this is the case, since application of the regression model to the alternative national patterns of GNP developed at CAC are much closer to the total employment figures derived from labor force forecasts. Even in these forecasts, there is a tendency for the sum of industry forecasts to outstrip the labor force. We think this means that the structure of Illinois industry corresponds favorably with the patterns of demand expected for the United States economy. This contrasts somewhat with the situation of the 1950's and 1960's. The user must note this important inconsistency and tendency of the model to exaggerate total industry employment.

IV. FORECASTS OF ENGINEERING DEMAND IN 1975 AND 1980

The employment by industry forecasts presented in the previous section when distributed according to the estimated percentages of workers in the various occupations in each of the industries provide forecasts of number of engineers by specialty. Forecasts derived by this method for 1975 and 1980 are presented for Illinois in Table 2. In addition to the forecasts prepared by CAC, a forecast developed by the Illinois Department of Labor by a similar method is presented for the sake of comparison.

The pattern of occupational demand forecast for 1975 and 1980 is

Table 2 Projected Illinois Employment of Engineers

Occupational Title	Actual 1960 (a)	Actual 1970 (b)	Forecasts CAC 1975	Forecasts CAC 1980	Illinois Dept of Labor (c) 1975
Aeronautical	300	300	428	499	600
Chemical	2,500	3,200	3,370	3,783	3,800
Civil	9,700	11,200	14,133	16,379	13,500
Electrical	10,700	11,000	14,923	18,069	18,000
Industrial	8,100	11,000	13,358	16,416	15,600
Mechanical	12,700	14,000	17,408	20,731	20,300
Metallurgical	1,400	1,600	2,078	2,427	2,400
Mining	300	300	315	308	200
Other Technical	11,100	13,000	19,833	25,221	22,100
Total	56,900	67,000	85,840	103,816	96,600

(a) U.S. Bureau of the Census
(b) Estimated by the authors
(c) Source: Bureau of Employment Security, *Occupational Manpower Requirements: Projections for Illinois 1960-1975,* May 1970

similar to that of 1970, and each of the occupations is expected to grow. The CAC forecast for 1975 is somewhat lower than the Illinois Department of Labor forecast, suggesting that the demand presented here is rather conservative. Forecasts based on three alternative national GNP patterns were also prepared and fell fairly close to the figures presented here, although a forecast based on GNP which emphasized spending on health, education, and welfare was about 3,500 engineers lower in 1975 and about 10,000 lower in 1980 than the BLS GNP pattern.

The method of estimation used assumed that the proportion of an industry's workers who would be engineers in 1975 and 1980 was no higher than the estimated 1970 proportion. This method was also conservative, since there has been a general tendency in manufacturing for the proportion of engineers in the work force to increase.

These forecasts are forecasts of demand for employment on the assumption that sufficient numbers of workers are available and that each industry's propensity to hire engineers remains unchanged over time. It is possible that demand for engineers will be either greater or less than the figures given here, since demand for an occupation depends on technology, factor prices, the availability of substitutes and many other factors.

These estimates of demand can be used to derive the number of additional or new engineers required by employers. Total requirements equals the sum of replacements for engineers leaving work from death, retirement, or other reasons and growth. To estimate the number required for replacement, we assume that employment grows linearly from 1970 to 1975 and then linearly from 1975 to 1980. Using an annual attrition rate of 2.185 percent, estimated from 1950 to 1960 experience, the aggregate attrition is 8,143 for the period 1970 to 1975 and 10,164 for the period 1975 to 1980, or 18,307 in total.* In addition to the replacement requirements, 18,840 new engineers would be needed to increase employment from the 1970 estimated to the 1975 forecast and 17,976 to increase employment from the forecast 1975 to the forecast 1980 level. Combining the growth and replacement requirements leads to 26,983 new engineers required from 1970 to 1975, and 28,130 required from 1975 to 1980, or a total of 55,113 required for the decade 1970 to 1980.

V. ENGINEERING GRADUATES

The number of graduates from engineering schools in Illinois is given in Table 3, for the period 1962-1971. The University of Illinois has graduated more than half of the engineers in each of these years. The proportionate share of the total has declined for the Urbana-Champaign Campus and grown for the Chicago Circle Campus such that they now rank first and second in the State if they are considered as separate institutions (Table 4.) Total engineering enrollment by institution moves in a slightly different way from degrees (Table 5). The movement is particularly clear in sophomore enrollments which clearly lead total enrollment (Table 6). Sophomore enrollment peaked in 1968, but total enrollment did not peak until 1970, and bachelor's degrees peaked in 1971. Thus engineering students seem to have responded very rapidly to the recent downturn in engineering employment.

*Hugh Folk, *The Shortage of Scientists and Engineers,* Lexington, Mass., D. C. Heath and Company, 1970, p. 279.

Table 3 Illinois Engineering Bachelors Degrees, 1962-71

INSTITUTION	1962	1963	1964	1965	1966	1967	1968	1969	1970	1971
Bradley	60	61	59	65	72	57	66	84	119	113
Northwestern	111	105	112	133	134	127	125	114	130	137
U of I (Urbana)	635	711	743	700	692	694	674	868	683	680
U of I (Chicago)	—	—	—	—	—	—	64	196	294	330
IIT	335	340	360	321	324	337	305	366	290	289
SIU	—	—	—	—	16	27	24	34	36	27
TOTAL	1141	1217	1274	1219	1238	1242	1258	1480	1552	1576

Table 4 Distribution of Illinois Engineering Bachelors Degrees by Institution, 1962-71

INSTITUTION	1962	1963	1964	1965	1966	1967	1968	1969	1970	1971
Bradley	5.3	5.0	4.6	5.3	5.8	4.6	5.3	5.7	7.7	7.2
Northwestern	9.7	8.6	8.8	10.9	10.8	10.2	9.9	7.7	8.4	8.7
U of I (Urbana)	55.6	58.5	58.3	57.5	55.9	55.9	53.6	46.4	44.0	43.2
U of I (Chicago)	—	—	—	—	—	—	5.1	13.2	18.9	20.9
IIT	29.4	27.9	28.3	26.3	26.2	27.1	24.2	24.7	18.7	18.3
SIU	—		—		1.3	2.2	1.9	2.3	2.3	1.7
TOTAL	100	100	100	100	100	100	100	100	100	100

Table 5 Total Engineering Undergraduate Enrollment in Illinois Engineering Colleges

INSTITUTION	1962	1963	1964	1965	1966	1967	1968	1969	1970	1971
Bradley	625	614	632	641	631	652	667	763	739	659
Northwestern	844	846	850	784	744	772	784	832	792	781
U of I (Urbana)	3646	3677	3573	3572	3674	3703	3666	3502	3599	3151
U of I (Chicago)	—	—	—	—	—	1545	1890	2051	2139	2122
IIT	819	705	745	779	837	794	765	744	730	698
SIU	—	—	134	346	341	432	514	468	501	466
TOTAL	5939	5922	5934	6122	6277	7898	8286	8369	8500	7877
% of 1962	100	99.7	99.9	103.1	105.7	133.0	139.5	140.8	143.1	132.6

Table 6 Sophomore Engineering Enrollment in Illinois Engineering Colleges

INSTITUTION	1962	1963	1964	1965	1966	1967	1968	1969	1970	1971
Bradley	218	115	112	92	117	157	164	173	154	144
Northwestern	283	251	251	211	205	243	221	235	208	217
U of I (Urbana)	773	753	753	843	922	902	942	764	788	759
U of I (Chicago)	—	—	—	—	—	376	416	381	377	405
IIT	236	246	235	302	309	266	261	270	233	243
SIU	—	—	35	90	102	112	108	147	123	127
TOTAL	1420	1365	1386	1538	1655	2056	2112	1970	1883	1895
% of 1962	100	96.1	97.6	108.3	116.5	144.8	148.7	138.7	32.6	133.5
% of 1967	—	—	—	—	—	100.0	102.7	95.8	89.2	92.2

Engineering degrees have been projected by applying a survival rate for sophomore enrollments for the period 1972 through 1974. For the period 1975 through 1985, projections were based on the share of engineering in total Illinois degrees. The basic Illinois first degree projections are those of the University of Illinois, *Supportive Study II,* which are probably somewhat too high. Four series of engineering first degree projections are made:

Series A assumes a linearly increasing proportion of all first degrees will be engineering degrees, growing from 2.71 percent in 1973 to 7.0 in 1985. This series is excessively high, but the 1985 proportion is substantially below the level of 9.3 percent reached in 1958.
Series B assumes that the proportion of engineering degrees in the total will grow to 3.9 percent of all first degrees, the proportion projected by *Supportive Study II* for 1980. This will serve as the high alternative.
Series C assumes a rise in the proportion of engineering degrees to 3.4 percent of the total.
Series D assumes that engineering degrees will remain at the level of 2.75 percent of all degrees to be reached in 1974.

All four series are tabulated in Table 7.

For the period 1970-1974, the forecasts suggest a total number of engineering graduates of 7,438, and for the period 1975-1979, numbers ranging from a low of 7,439 for Series D, 9,326 for Series C, 9,928 for Series B, and 10,537 for Series A.

The number of degrees for the period 1970 to 1980 ranges from a low of 14,777 to a high estimate of 17,975. It must be recognized that not all persons graduating with engineering degrees will work as engineers. Some will leave the profession entirely and others will attend graduate school, thus delaying their entry into work. The number entering graduate school, of course, will be approximately offset by the entry of engineering graduates of periods before 1970 who obtain terminal degrees in the 1970's and enter engineering work. In addition to the normal non-entry of engineering graduates, some will leave the State to work as they have in the past, and other engineering graduates from out of State will enter Illinois to work. In the past, Illinois was a net exporter of engineers, but it may possibly become a net importer during the next decade.

Table 7 Illinois Engineering Bachelors Degree Projections

YEAR	SERIES A	SERIES B	SERIES C	SERIES D
1972	1477	1477	1477	1477
1973	1412	1412	1412	1412
1974	1421	1421	1421	1421
1975	1624	1630	1786	1441
1976	1850	1855	1807	1457
1977	2095	2106	1941	1483
1978	2344	2141	1872	1510
1979	2624	2196	1920	1548
1980	2929	2762	1978	1593
1981	3231	2317	2025	1634
1982	3469	2321	2029	1636
1983	3657	2293	2005	1617
1984	3814	2250	1968	1587
1985	3906	2176	1903	1534

In addition to engineering graduates there are other sources of supply for engineering jobs, and these include graduates of non-engineering curricula, upgraded technicians, foreign immigrants. In the past, fewer than one-half of the persons counted as engineers in the census have lacked college degrees.

VI. CONCLUSION

Even if we assume that the maximum degree forecast occurs, it is clear that the number of engineers required for growth and replacement under our demand forecast will far exceed the numbers of new graduates. During the 1970's there will be approximately four jobs for each new graduate. The shortfall of engineering degrees under requirements will grow increasingly as the decade passes, and it seems likely that the phenomenon of the shortage of engineers which marked the 1950's and 1960's will return during the middle and late 1970's.

There is little that engineering colleges can do to change the shape or timing of the shortage. It is important, however, to point out to prospective engineering students that there is every likelihood that engineering will continue to be a highly rewarded and widely demanded occupation. While all forecasts are subject to uncertainty, the wide discrepancy between demand and supply demonstrated in this paper suggests that a remarkable reduction in engineering demand would have to occur for the shortage to disappear.

While the forecasts presented here suggest that the economic future of engineering is quite bright, the impending shortage of engineers may not be a cause of major national alarm. During the 1950's and 1960's American industry showed remarkable ability to substitute other kinds of workers for engineering graduates in jobs which had traditionally been held by engineering graduates. The same kind of process can continue in the 1970's, especially since there is likely to be an ample supply of college graduates in non-engineering fields who will have difficulty in finding work in the vocational areas related to their training.

Influence of Engineering Colleges On State Policy

Robert G. Granda

This paper will attempt to analyze the current extent of interaction between engineering colleges in Illinois and Illinois State government, especially the General Assembly. It will discuss those steps which need to be taken and will explain and analyze a proposed comprehensive solution to the problem of lack of interaction.

PRESENT SITUATION

Given the size and complexity of State government, engineering colleges have almost no policy-making function with regard to the passage or implementation of legislation. The term "policy-making function," apart from its obvious meaning, includes the furnishing of technical information on which significant legislative and administrative decisions are based. The preceding statements do not take into account those situations where faculty or graduates of engineering schools are paid — either as permanent employees or on a contractual basis — for solving specific, purely technical problems, such as designing a bridge or highway. These functions, although necessary and valuable, are not of the same nature as is assistance in planning a future course of action for the State in decisions involving technology.

Robert G. Granda, a lawyer, is a member of the staff of the Illinois Legislative Council.

The main reason for this lack of policy-making import seems to be lack of communication between the colleges, the General Assembly, and the State departments and agencies. This state of affairs indicates the need for an effective system of liaison or a campaign aimed at encouraging individuals in colleges and State government to volunteer assistance in specific areas on one side, or to ask for it on the other.

It appears that the engineering schools in Illinois and their faculty members are prepared, if asked, to assist members of the Illinois General Assembly by giving advice and information on technological matters but that they are called on to do so relatively seldom. The conclusion is based on interviews with top staff personnel or leaders of the Illinois Commission on Atomic Energy, the Science Advisory Council, the Commission on Technological Progress, the Commission for Economic Development, the Director of the Illinois State Museum (who acts as a formal liaison between the Illinois Academy of Sciences and the Illinois General Assembly), the assistant to the President of the University of Illinois, and with officials of other offices. Written reports of these and other sources such as the Intergovernmental Cooperation Commission, and the Midwestern Regional Conference on Science and Technology and State Government were also examined.

The general conclusion is not meant to minimize the contributions of members of engineering college faculties who assisted the General Assembly through such activities as participation in conference, etc., but it is meant to point out the almost lack of contact between engineering schools and individual legislators on the hundreds of bills and issues on which such contact could be helpful.

Some specific comments from persons contacted may be helpful. William Rose, of the Commission for Economic Development, said that he feels there is a need for a formal liaison officer for each or all universities. This liaison officer should be available to any member of the General Assembly for the purpose of putting him in touch with one or more university people who could advise him on questions of technology related to legislation. He also suggested that the informational contacts available through this liaison be interdisciplinary, so that the ramifications of technological discussions could be explored.

The usefulness of such a liaison was agreed to by Clinton Stone, a member of the Commission on Technological Progress, who added that he feels there is a real need for advice on specific pieces of legislation. He felt that there would need to be some type of control, however, so that one expert in a field would not find himself contacted by 40 or 50 legislators with the same question on the same bill.

The awareness of the problem recognized in the immediately preceding statements is unusual. While there is a great *need* for cooperation between the engineering colleges and the State, there seems to be little *demand* for it in the form of requests for interaction. This may be the result of a lack of any organized State science policy. The need exists; the demand may have to be created.

The recent (and continuing) legislative session does not hold out any startling prospects of significant change. It is obviously erroneous to preclude the possibility of legislator A and professor X being brought together on a specific problem. There are no barriers whatsoever to an administrator or legislator asking for help from a member of the university community and receiving that help, possibly on a continuing basis. Nor is there any reason why some or all of the faculty or students of one or all the engineering colleges could not offer, on their own initiative, to assist the State decision-makers in attacking various problems. It is equally obvious that this is unlikely to occur spontaneously, or even that such an occurrence, or cluster of occurrences, would be desirable on such an unorganized basis.

There seems to be a need for an organized approach to the problem of unexploited expertise co-existing with imperfect State knowledge in that area of unexploited expertise. The benefits of such a bringing together leap out at the investigator. Legislators — who, in Illinois, are supposed to be "citizen-legislators", and therefore able to devote less than all of their time to legislating — are forced to deal with problems requiring more technical knowledge than they possess. Members of the engineering college community, possessing varying degrees of expertise about one or more technical areas, find themselves reading about legislative decisions in those areas — decisions which they feel should be improved upon. These same engineering people bemoan lost opportunities for knowledgeable, perhaps innovative, solutions to the problems of the State. They see necessary projects, projects that might brighten the employment picture for their cohorts, unnecessarily rejected or reduced in scope, because of a lack of understanding by legislators or administrators. Finally, these same people, along with all the other types dependent to one extent or another on legislative appropriations, suffer through a period of uncertainty while the General Assembly debates about how to slice up the fiscal pie.

Picture the optimum situation in which Senator Smith or Director Jones wrestles with a problem, goes to his phone and contacts Professor Brown or intermediary Johnson, knowing that he will be able to obtain accurate, expert, unbiased information on the problem. If the contact produces the expected results everyone is happier, one for having acted with benefit of good information, the other for having helped the State and for having shown the benefits to be obtained by the continuation, or enhancement, of his work. Apart from personal satisfaction and gratification of altruistic desires, the academic has made himself and his cohorts, necessary and desirable to the legislator or administrator. The world runs on self-interest and mutual back-scratching. There is no virtue in ignoring that fact. If the engineering colleges make themselves indispensable to the smooth working of State government, they will be given more enthusiastic economic and legal support. That statement may seem more offensive than a softer statement about scholarly pursuit and high statesmanship but it is accurate.

At the outset, it is necessary to recognize the fact that there is no such thing as a "pure" engineering problem, at least in the context of this particular study. This study section relates to the application of engineering expertise to the problems of the State. Short of magic, any suggested solution adopted for State problems will require the expenditure of the economic resources of the State. The State, as I note elsewhere, is not an abstract concept — it is 11 million people. The money spent is their money. The people who spend that money — the legislators and State executives — are, besides being taxpayers themselves, individuals who feel that they can make a significant contribution to the welfare of the State and who want to remain in office so they can carry out their long-term projects. In less abstruse terms, legislators and administrators want to keep their jobs. One of the ways to do this is to avoid squandering the people's money. In addition, they must protect the people they represent. At the same time, they must avoid unnecessarily antagonizing those people or unnecessarily disrupting their lives.

Returning to the idea that there is no such thing as a pure engineering problem, with the preceding statements in mind, consider that almost every State action has economic, social, and political repercussions. The various factors must be balanced in an attempt to provide a solution which is technically feasible, economically acceptable, and causes as little disruption as possible.

The point of this stage-setting is that an interdisciplinary approach is necessary to technical problems if the solutions are to be acceptable. This is not to imply that engineers are ignorant of the real world. It is meant to suggest that, just as legislators and administrators lack technical expertise, so may engineers lack an appreciation of the socio-economic and political framework into which their proposed solutions fit. The answer seems to be the provision of a means by which various participants in a joint project can be brought together and can be helped to communicate, even though they may be accustomed to speaking different languages.

The total picture may not be as bleak as it looks. There are some efforts now to alleviate the problem. The Board of Higher Education has issued a policy statement to the effect that the universities have a mandate for public service work. At least one school, the University of Illinois, already offers assistance to legislators, without limiting that assistance to engineering questions. William Rice, assistant to the President of the University of Illinois, said that if a legislator asks for information, he will refer him to a faculty member. He gets 12 or 15 such requests annually, but he cannot recall if any of them dealt with engineering. Few individual contacts between members of the General Assembly and college engineering personnel were reported by the other persons interviewed, although a small number of faculty people testified at legislative commission hearings. The Conference on Science and Technology, which has already been mentioned, is not stressed in this memorandum because it was not aimed at advice on individual pieces of legislation, but rather with the development of guidelines for the General Assembly and the Governor.

These contacts, helpful as they are, are relatively insignificant when compared to the needs of the State as a whole. As will be discussed shortly, there are at least two more examples of attempts at the desired interaction. Before discussing them, however, it may be helpful to absorb some background on the workings of State government.

State laws are passed by the General Assembly and signed by the Governor. Many of them are administered by agencies, such as the big State departments, controlled by the Governor. The Governor and some of the legislators have staff aides. The most recent annual budget for the State amounted to six billion dollars. The State population is approximately 11 million. The General Assembly, comprised of 235 members when at full strength, has considered approximately 6,000 bills and resolutions during its current session. Most of these bills and resolutions are considered by one or more of 35 standing legislative committees. An informal check indicates that at least 250 of these bills are clearly science-related. The actual total in which some technological expertise would have been helpful is probably double or triple that number. A survey of the General Assembly shows that only two legislators are engineers, in contrast to the 113 who are lawyers, insurance men, or real estate men. These three occupations were singled out to show the apparent tendency of legislators to engage in occupations where they deal with people, rather than technology. The General Assembly is assisted by several groups dealing with technical questions. These are described in the attached Appendix.

Some of these laws require State agencies to set technical standards within a range established by the General Assembly, as where the legislation prohibits smoke emissions in an amount exceeding standards set by an administrative agency. Besides showing the interrelationship of the executive and legislative branches, this should show the need the General Assembly has for accurate technological information to enable it to give the administrative agencies workable statutes.

These statutes, especially those involving the police power of the State, greatly influence the lives of each citizen. They may deal with air or water pollution, pesticides, vehicle safety and inspection, industrial safety, noise pollution, fabric flammability, or lead poisoning. They deal with important problems and the legislators who make decisions on them require a great deal of information.

One attempt to provide that information, which is currently underway, is the action of Dr. Milton Thompson, Director of the Illinois State Museum, in transmitting research requests from legislators to members of the Illinois Academy of Science. In the two years in which he has offered this service, only four or five legislators have requested assistance. This is probably due to lack of publicity (one letter announcing the service was sent to each legislator).

This may be contrasted with the activities of the Illinois Society of Professional Engineers which, in approximately the same time period, supplied testimony at committee hearings on 15 of the 101 bills on which the group took a position. Some of this testimony was by staff people, aided by engineers' reports, and some by consulting engineers or engineering faculty. This, however, was expert testimony on specific issues as opposed to the furnishing of information on broad policy questions.

Possibly the most successful venture into the field of voluntary cooperation between engineering faculty and the State is the program now in existence at the Institute for Environmental Quality. The Institute director, Michael Schneiderman, reports that he has had very significant voluntary help from engineering faculty members, as well as faculty people in other disciplines. This has included several appearances before legislative committees. He reports that he receives a great deal of informal help, some in the form of informational assistance by telephone. The Institute's Solid Waste Management Task Force is an unpaid group of volunteers. Mr. Schneiderman says that the entire State noise control program is being designed by a task force from the University of Illinois College of Engineering. Paid assistance is used, some of it consisting of part time employees from colleges of engineering. It is felt that these people, paid and unpaid, are "essential" to the Institute.

Mr. Schneiderman said that, in the beginning, there was a need for mutual education. He recalled bringing the entire House Environment Committee to the University of Illinois to encourage State/University interaction. The Institute has also made a microbiologist available to that committee. This is in addition to the creation of an Environmental Resource Center, headed, without pay, by Dr. Bertram Carnow, a distinguished physician at the Univeristy of Illinois. Dr. Carnow has worked with the House Environment Committee and testified at several of its subcommittee hearings.

It should be noted that the legislation creating the Institute for Environmental Quality (Ill. Rev. Stat., 1971, ch. 111½, sec. 1006) says that:

> Maximum use shall be made of existing federal and state agencies, facilities and personnel in collecting research under this title.

NEEDED IMPROVEMENTS

Hopefully, the foregoing has shown the need for competent, unbiased, policy research assistance to the State on a comprehensive, well-organized basis. This need is magnified in the area of technical problems, where the daily experience of legislators may not supply them with needed knowledge. This is the gap the engineering schools can fill.

If their assistance is forthcoming, it must be objective. Each of us has personal feelings on many issues but these feelings cannot be unleashed to the extent that they would influence the researcher in his relationship with the beneficiaries of his knowledge. The expression of opinion, *clearly labeled as opinion,* is one thing. The injection of personal attitudes into a supposedly neutral work product is another, completely different, thing.

There is a need for easy lines of communication between the researcher and the legislator or administrator. Both types of people involved are busy and would greatly resent having to cut through red tape to do something meaningful. Simple as it may seem, there would be much value in the provision of a central office and a central phone number known to, and easily available to, participants from the State and from the universities.

Another needed element of a solution is the assurance that those persons who offer their expertise in testimony before a legislative committee will not be harassed. Deliberate rude treatment does not often occur but, apparently, one or more engineers have resented the treatment they received at the hands of a legislative committee at some time in the past. In order to overcome the reluctance of those who hear rumors, perhaps exaggerated, about such past incidents, deliberate efforts should be made to avoid a recurrence.

On the other hand, some legislators and administrators harbor prejudice against what they feel will be "ivory tower" suggestions from faculty people. The best way to dispel this attitude is probably by giving publicity to examples of beneficial results obtained with the help of faculty people, as in the case of the Institute for Environmental Quality.

Whatever system is adopted must be the subject of sufficient publicity so that potential participants are made sufficiently aware of it. There should also be emphasis on the fact that contact can be initiated from either side.

It would be helpful to have a liaison officer at each school so that State people would know whom to contact. Depending on the extent of contacts, this could be one person with other duties who could be available for sporadic comments or it could be a full-time staff person.

SUGGESTED SOLUTION

Assuming the accuracy of the preceding statements of what is needed, the question arises "What is the best way to bring this situation about?" An answer, perhaps the best answer, lies in the promulgation of information about the existence of a virtually untapped pool of knowledge and the creation of a conduit to facilitate the transfer of that knowledge from its source to the places where it is needed. One approach to this is the type of facility proposed by the Legislative Council to the National Science Foundation — the creation of the position of "Staff Scientist", to be filled by a person who will, in cooperation with the Illinois Academy of Sciences, act as a liaison between technological experts and the members of the Illinois General Assembly. If such a position is created, it is reasonable to assume that some degree of cooperation with the Governor's Science Advisor and administrators will arise, thus tying together the engineering schools and the legislative and executive branches of State government. The remainder of this paper will be devoted to an explanation of that proposal and an explanation of the various elements of the proposal. It is assumed, and hoped, that the participants in this project, and the later readers of these words, will examine the suggestions with a critical eye and will make constructive suggestions for the improvement of the proposal. I ask only that these people not discount what follows as mere chauvinsim, but that they examine it as an explanation of what seems the most feasible proposal offered to date.

The planned project, as described in the grant application[1] submitted to the National Science Foundation, may be summarized as follows:

The proposed program would be a 2-year cooperative project establishing a linkage between the Illinois Legislative Council and the Illinois State Academy of Science and other persons with scientific expertise to enable Illinois legislators and legislative committees to take advantage of this knowledge. A new position on the Council's research staff, Legislative Staff Scientist, would be created to work with the Executive Secretary of the Academy to bring legislators and scientists together in individual conferences and symposia. If the program is successful, it is anticipated that the State would fund it permanently. The grant request is for $77,850, with anticipated contributions of $9,218 by the Legislative Council and $7,150 by the Academy. It is proposed that a steering committee of legislators and Academy representatives be created. It is assumed that the Legislative Council staff would assist the Staff Scientist.

As to the exact duties of the Staff Scientist, his relationship with the Academy, and the mutual question of making recommendations, the following excerpt from the grant application[2] will be instructive:

Linkage Between Academy and Council

Both the State Academy and the Legislative Council must avoid policy recommendations—the Academy, under its constitution, in order to maintain its tax exempt status as a scientific-educational organization, and the Council because its enabling legislation does not permit recommendations. Any plan for linking these two organizations together to make information about scientific findings available to legislators must keep these restrictions in mind.

It is believed, however, that scientific and technological information pertinent to public policy formation can be conveyed to legislators without constituting recommendations by the following procedures:

1. Problem areas where it is considered that science and technology can be immediately useful will be identified and symposia of scientists and legislators will be arranged.
2. Problems requiring further research to which the scientific community might relate its interests will be identified and described for the members of the Academy and others.
3. Specific inquiries from legislators will be invited on an individual basis and will be individually answered, either by a written response or by arranging a consultation with an appropriate scientist or scientists.
4. Where indicated, scientists will be asked to present information to legislative committees and commissions.
5. Informed assistance can be rendered in other ways in organizing specific scientific efforts to serve the legislature.
6. While this proposal is stated in terms of a liaison between the Legislative Council and the State Academy, opportunities for involving other legislative and scientific groups will be explored and developed.

General guidance will be given by a steering committee consisting of legislators and Academy representatives. The steering committee will assist in recruiting a Legislative Staff Scientist for the staff of the Council. The Staff Scientist, working with the Council's research director and the Executive Secretary of the Academy, will coordinate legislative needs for scientific information with Academy resources.

The Legislative Staff Scientist will be primarily responsible for identifying the problem areas where symposia seem appropriate (item 1 above). He will do this by analyzing legislative bills in previous sessions in the Illinois General Assembly and other legislatures, utilizing the staff and facilities of the Council for this purpose, and by inquiries directed to legislators, especially those who have evidenced interest in this kind of assistance. He will also appear before legislative committees and commissions from time to time, to explain the liaison program.

The Executive Secretary of the Academy will be primarily responsible for developing procedures for identifying and locating research studies that have been or presently are being carried out in Illinois which can serve as sources of information for legislators and other interested groups (items 1, 3, and 4). He will also survey the members of the Academy for special knowledges and skills that are of potential interest in meeting legislative inquiries and will prepare a roster of possible consultants on particular topics.

Where inquiries can be readily answered from standard reference sources, articles in the literature, etc., it will be the responsibility of the Legislative Staff Scientist to prepare responses (item 3). Where consultation is desired, he will either arrange a meeting between the legislator and a scientist or scientists, or he will call on the Executive Secretary of the Academy for assistance (item 3).

Symposia will be jointly arranged in consultation with interested legislators and scientists, the scope of these being dependent on the interests of both groups. It is the intent to suggest by "symposium" an emphasis on small group contacts where participants can feel free to exchange information and views without the necessity of preparing formal papers. If further research and documentation is necessary, it can come later (item 2). The vehicle of communication leading to further research will be the small gathering of legislators who are concerned with a specific problem and scientists who have indicated an interest in this problem area.

Legislative identification and study of problem areas often surfaces in proposals (resolutions or bills) to create temporary study committees or commissions. Such proposals can be reviewed in their early stages by the Legislative Staff Scientist to ascertain the extent to which they may involve scientific problems, so that the appropriate mechanisms can be suggested for obtaining the necessary scientific assistance (item 5). This will lead to involvement with legislative groups beyond the Council and with scientific interest in addition to the membership of the Academy (item 6).

Budget estimates allow for four symposia annually and for enough money (at $100 per diem) to underwrite the services of scientific consultants and to carry out research on problems assigned by the Staff scientist. The amount would be sufficient to cover 50 days the first year and 100 days the second year.

The need for scientific expertise has already been demonstrated. The logic of involving the Illinois Legislative Council as an intermediary and participant in the exchange of knowledge may need to be explained. It is felt that the Legislative Council would have the machinery to bring to bear diverse elements on a single problem and could ease the communication problem. The Legislative Council has attempted to operate for 35 years on the basis of objective, nonpartisan, scholarship, diplomacy and practicability. Its staff members tend to acquire a general knowledge. Its work requires the establishment of contacts in a variety of disciplines, and a variety of locations—public, private, and academic. The sheer variety of questions researched tends to build up a feel for the overall priorities and problems of the State. The fact that the beneficiaries of Legislative Council research are elected State legislators requires an understanding of the various elements which enter into legislative decision-making. It also requires an ability to anticipate those issues likely to become ripe for decision-making at each session. Further, it requires the ability to suppress personal feelings and keep the work product objective and non-partisan.

In considering the merits of this proposal as a solution to the problem of lack of State/University interaction, the reader should keep in mind the potential benefits to the State, and to the individuals who are the State, if the proposal is implemented successfully. Those who appreciate the pervasive influence of technology in modern life can understand the need for wise policy decisions on the use and control of that technology.

REFERENCES

1. "Science Research Resources for State Legislators," Grant Application to the National Science Foundation by the Illinois Legislative Council, 1971.
2. *Idem*, pp. 16-19.

APPENDIX*

Existing S/T Groups. The Legislature has created a number of commissions dealing with science and technology (S/T) generally or with particular aspects of science and technology. These include:

Illinois Commission on Atomic Energy, consisting of four members of the Legislature, eight public members (who are appointed by the Governor) and seven State officials as ex officio members. The Commission's principal responsibility is to "investigate, support, encourage, and promote the economic, social, health, and technological impact of atomic energy on the citizens of this State. . . ."

Data Information Systems Commission, consisting of six members

*"Science Research Resources for State Legislators," Grant Application to the National Science Foundation by the Illinois Legislative Council, 1971, pp. 4,5.

of the Legislature and four public members (who are appointed by the legislative leaders). The Commission is directed to "establish a master plan for management information and data" for the State and its local governments and to monitor electronic data processing installations, budgets, and applications.

Joint Committee on Legislative Information Systems, consisting of fourteen legislators (chairmen and ranking minority members of major committees and service agencies, including the Legislative Council), the chief clerks of the two Houses, and five staff directors of service agencies. The Committee is directed to "review those aspects of legislative procedure which involve the use of information and .... make recommendations .... that will enable (the General Assembly) to function more effectively" through the use of electronic data processing, etc.

Science Advisory Council, consisting of four members of the Legislature, three members of the Commission on Technological Progress (appointed by the chairman thereof), seventeen public members (of whom six are appointed by the presiding officers of the two legislative Houses and eleven by the Governor), the director of the Department of Business and Economic Development, ex officio, the chiefs of the three State scientific Surveys (Geological, Natural History, and Water), also ex officio. The assignment of this Commission is seeking ways to connect the present and potential capabilities of Illinois with emerging sciences and technologies having important economic potential.

Commission on Technological Progress*, consisting of ten members of the Legislature and four public members (who are appointed by the presiding officers of the two Houses). The Commission's primary assignment is "to study, investigate, analyze and assess problems, growth, effect and future tendency of scientific and technological changes" on the State, future manpower needs, educational requirements, etc.

Some idea of the activity of the foregoing commissions can be gained from the appropriations made to them in fiscal 1971: Atomic Energy Commission, $15,000; Data Information Systems Commission, $12,500; Science Advisory Commission, $10,000; and Technological Progress Commission, $15,000. The Joint Committee on Legislative Information Systems, essentially an inter-agency body for legislative agencies, was funded out of an $18,000 item in the Legislative Coun-

*Now defunct.

cil appropriation for improving bill and report printing and status reporting, etc. The Atomic Energy Commission has been seeking to encourage other states to participate in the Midwest Nuclear Compact (which Illinois ratified when it was first proposed in 1967).

Consumption of University Technical Expertise by Governments

Clark H. P. Schneider

A question related to the extent to which engineering colleges supply technical expertise to local and municipal governments is how do universities stand among sources of technical expertise for local governments? Underwood and Fritschler (1970) in their study entitled "Science-Technology Advice in Local Governments" report upon a questionnaire survey of chief governmental officials of all cities in the United States whose population exceeded 25,000. Two hundred ninety-five responses were returned—a response rate of 34%. A more complete picture than that presented in this summary can be achieved by referring to the Underwood and Fritschler report.

The results of this survey with regard to the role of universities in supplying technical advice are not clear cut. On one hand respondents reported that universities are the third most heavily used source, ranking after city departmental personnel and private consulting firms, with 60% of the responding cities reporting that they utilize university personnel for technical advice.

Opposed to this vote of confidence in universities is the response to a question asking which sources of advice are heavily relied upon in eleven functional areas. Table 1 is derived from materials in Tables 5

Clarke H. P. Schneider is an Assistant Professor in the Center for Urban Studies at the University of Illinois at Chicago Circle.

and 6 of the survey report. This Table presents the rank of university personnel among twelve possible sources for each of the functional areas. The question was stated to allow multiple responses for each functional area. The eleven other possible information sources considered were: local government science unit, city departmental personnel, state government science unit, other state offices, Federal Government, Regional governmental units, industry and/or business, private consulting firm, local citizen advisory groups, national associations, and state leagues. On this question, universities do not rank nearly so high among sources. Over all functional areas, the average rank of university personnel as a source is 7.4. It seems that university expertise may not be utilized as heavily as the initial response reported would seem to indicate. Note too that functional areas which are more likely to involve technical questions—water supply, sanitation, sewage, environment, transportation, and housing—are those in which University ranks lowest. For these areas, the average rank of university personnel as an information source is 9.5.

Table 1 Rank of University Personnel as Sources of Advice in Eleven Functional Areas

Functional Area	University Personnel Rank	Functional Area	University Personnel Rank
Public Safety	6	Environment	7
Water Supply	9	Transportation	10
Sanitation	10	Health	7
Sewage	9	Housing	11
Welfare	6	General Management	5
		Education	2

Further insight may be gained by inspecting the pattern of replies to a question dealing with the usefulness of sources to the chief administrative officer of the city. In percent responding "very useful," Universities rank fourth. However, as Table 2 reveals, this does not represent a particularly high standing. From this Table, it is clear that there are at least two distinct response groups: a high group of sources labeled "very useful" by 65% or more of those responding, and a lower group with less than 40% regarding these sources as "very useful". Universities are at the top of the low group, but there is no way of escaping the fact that they are in the low group and do not score much better than any other member of this group. Only 10 percentage points separate the high member (Universities) from the low member (Other State Offices) of this group whereas 29 points separate the high member of this group (University Personnel) from the low member (City Department Personnel) of the higher group. Clearly, from this survey we must infer that universities are deemed less than outstanding as sources of technical advice to local area governments by governmental administrators.

Table 2 Sources by Usefulness to Chief Administrative Officer

Source	Number Rating It	% Very Useful	Source	Number Rating It	% Very Useful
Sci. Tech. Unit	7	86	National Assn.	178	35
Pvt. Consultants	243	72	Regional Govnmt.	120	32
City Personnel	235	68	Industry	105	32
Univ. Personnel	164	39	Sci. Tech. Task Forces	34	32
State Leagues	164	37	Per. Sci. Tech. Agency	17	29
Federal Govnmt.	98	37	Other State Offices	6	29
State Govnmt. Sci. Agency	102	36			

Unfortunately this survey provides no real data upon the extent to which local governments utilize technical advice *per se*. It seems likley that the selection bias, which would operate here in terms of responding to the questionnaire utilized, would favor response by these governments which do in fact utilize or at least solicit technical advice regularly. That is, governments which do not extensively utilize such advice are probably more likely to have discarded the questionnaire. Nevertheless, among cities which do utilize technical advice, this survey certainly supplies insight into the role of universities as sources of such information. Briefly, we have found that they do not seem to be perceived as particularly useful (compared to other sources) nor are they extensively used in the functional areas considered. Interestingly, however, they are frequently cited as sources of information. It seems likely that this may be a case where a general perception has been elicited and found upon closer examination to be contrary to other facts or perceptions.

Yet another approach to ascertaining the extent to which governments employ universities as technical information sources is to inspect in general the clients of academic research—namely what groups purchase technical expertise through grants for research or through contracts awarded to engineering colleges and their research affiliates in Illinois.

Two sources of data were used. First, the annual directory published by the American Society for Engineering Education entitled *Engineering College Research and Graduate Study,* which reports on engineering activities nationwide, provides some data on magnitude and sources of funds utilized for research in reporting institutions. In an attempt to elaborate and extend the figures published there, each of the six universities were queried concerning sources of research support.

Several institutions were unable to provide the breakdown requested, but even these supplied their feelings upon the disposition. At Northwestern, for example, it was felt that over 90% of research expenditures came from the Federal government.

A further difficulty with the figures used involve accounting differences between institutions. Thus variations in the inclusion of cost-sharing and the recovery of indirect costs exist. This however is between institutions, and the purpose here is not aimed at interinstitutional comparisons, but rather at ascertaining the relative importance of clients for research. Since accounting procedures are presumably uniform within each institution, these figures may be used to establish the relative importance of funding sources.

Table 3 presents the data found by the above means. A wild guess, similar to that made for Northwestern, gives IIT Research Institute's research funding on the order of 80% government, 15% industry, and 5% IITRI. Of the total research monies provided Illinois graduate engineering institutions and affiliates—nearly 50 million dollars—IITRI receives nearly one half (bearing in mind the caveat above about interinstitutional comparisons), so this guess must receive some consideration.

Table 3 Sources of Funds for Research in the Six Graduate Degree Granting Engineering Colleges of Illinois for Fiscal Year 1970-71.

	Governmental Sources				Other Organizational Sources				
	Federal	State of Illinois	Other States	Local	Private Not For Profit	Business and Industry in Illinois	Business & Indus. Not in Ill.	Other and Internal	Total
UICC	324,000	0	0	0	1,000	9,000	20,000	37,000	391.000
SIU	26,000	37,000	0	0	0	9,000	0	0	72,000
IIT	720,000	0	0	0	24,000	20,000	25,000	0	789,000
URBANA	9,595,000	293,000	0	139,000	192,000	334,000		2,851,000	13,404,000
NU	4,290,210	?			334,953		?	469,180	5,104,343
Bradley	4,000	0	0	0	0	0	0	4,000	8,000
Institute of Gas Technology	3,433,000	?			1,085,000	1,123,000		81,000	5,722,000
IIT Research Institute	?								24,280,000
	18,392,210 77.3%	430,000 1.8%	0 0	139,000 0.6%	1,302,000 5.5%	38,000 0.1%	45,000 0.2%	3,442,180 23,788,390 14.5%	49,770,343

The most obvious conclusion to be drawn from scrutiny of this table is that almost no research is purchased by state and local governments from Illinois engineering schools. Note by the way that the only distinct entry in the column Local Government, which is for the University of Illinois at Urbana, represents a purchase by a local government which is not in Illinois; and thus there are no clear instances at this level of local Illinois government purchasing technical research from the universities.

In conclusion then, the reasonable interpretation of the results of these two surveys presented here is that universities are not perceived by administrations of non-Federal governmental units as either being sources of information or technical expertise, nor are universities used in this role by local units.

REFERENCE

Underwood, J. E. and A. L. Fritschler, "Science Technology Advice in Local Governments," *Urban Data Service*, Washington, D.C.: International City Management Association, Vol. 2, No. 11, November, 1970.

Engineering Colleges and Their Impact On the City of Chicago

Judy Millaney

The City of Chicago was pleased to have been invited to be a participant in the "Regional Role of Engineering Colleges" study. Previous efforts toward increasing cooperation and interaction between local governments and universities have at best been perfunctory. We have seen this study as a practical means of identifying barriers to interaction and implementing measures to overcome these obstacles. We look forward to increased communication between the engineering colleges and the City in order that we may benefit from each other's knowledge and experience.

The position of the City of Chicago is that City officials are responsible for projects that are undertaken by, and for, the City. Engineering colleges are encouraged to offer suggestions, design programs (which are taken under advisement by the proper department heads), and review present City policies; but the colleges are considered to be advisory bodies and have no real decision-making powers.

City leadership is accountable to the electorate and this alone places it in a unique position vis-à-vis the universities. Since municipal government is responsible for results of particular projects, it is proper that they dictate final policy. With engineering colleges acting in an

Judy Millaney is a Research Assistant in the Department of Public Works of the City of Chicago.

advisory role, City government can rely on expertise in the fields of program planning and program execution.

There are various positions taken by top level personnel with regard to what types of engineers are needed for Chicago's future. The most prominent of these is what is commonly called "the whole man approach."

This new approach to education entails broad basic courses in the sciences and the humanities, courses which should be designed to give an overview of not only traditional engineering functions but also areas of management and problem solving. In addition, it was suggested that the management field include study in finance (cost control, cost engineering) and real estate or property value. Moreover, it was noted that engineers entering public service should be aware of how and why decisions are made. It is imperative that engineers understand how city government functions and the sources of funds. A realistic appraisal of a proposed project will result if the engineer is aware of the decision-making process and especially the source of funds available. To round out the education of the engineer, courses in Systems Analysis, Project Management, Linear Programming and Public Administration were considered most valuable.

The City feels that this approach will make the engineer more conscious of events around him which will enable him to deal effectively with problems resulting from projects which create social interaction. He will be able to see a project in its entirety rather than as parts of the whole.

Other areas dealing with the training of engineers that can be explored are: (1) Planning and research projects where universities have the necessary expertise and practical approach, (2) Summer programs where students get firsthand knowledge of projects and participate in them (i.e., cordon counts), (3) On-going co-op programs with students, and (4) Environmental studies.

In summary, the City of Chicago needs engineers who are well rounded individuals, who can deal effectively with social and environmental problems, who are aware of how City government operates, and especially those who are concerned with the future of the City.

In order to obtain a more specific view of interaction between universities and government, we examined the types of communication which have been prevalent. We approached various City Council committees and corresponding City departments which have dealt with engineering colleges to learn what their past interaction has been and what possibilities might exist for the future.

1. Aviation

a. *Committee on Aviation.* Few research reports of universities are forwarded to the Committee. However, those received are sent to the Department of Aviation for review. Any proposals resulting would, of course, have to be approved first by the Committee on Aviation and second by the Committee on Finance. Final approval rests with the Chicago City Council.

b. *Aviation Department.* Relatively few reports are received by the Department. Most contact with engineering schools results from inquiries made by students in regard to thesis papers. Also noted by the Department was the fact that almost all construction (i.e., runways, lighting, etc.) must meet FAA regulations. Many engineering students were unaware of these stringent regulations and/or the operation of the FAA/Department of Transportation.

2. Buildings and Zoning

a. *Committee on Building and Zoning.* There has been no communication with engineering colleges. However, the Committee may receive information from the Department of Development and Planning or the Building Department concerning changes in the Municipal Code which would be of interest to engineering Colleges. Perhaps universities could work through these two agencies.

b. *Building Department.* This may be an area where engineers could exert a positive influence. The Department receives reports of research programs on architectural and structural matters relating to building and zoning. In addition, the Department maintains close contact with the Mayor's Advisory Committee on Building Code Amendments which is made up of department heads of universities, select faculty and members of major engineering and architectural firms.

3. Public Works

a. *Committee on Local Transportation.* Reports of transit projects or similar material are reviewed by the Committee which employs a transit engineer. After a reply is sent to the submitting group, the report is forwarded to the Department of Public Works or other agencies that are involved. Final decision on a proposal is usually based on the recommendation of the Department head.

b. *Department of Public Works.* We have received various reports but not the kind used in decision-making. Usually, the reports are senior thesis papers which recount the latest developments in the field of transportation. One recent acquisition (1969) dealt with High Speed Ground Transportation (HSGT) between Midway and O'Hare Airports. The mode selected for study was a Gravity-Vacuum Tube. It was an excellent presentation, but no new proposals were suggested which could be useful.

Our survey showed in summary that engineering reports and proposals directed to the City are usually reviewed by both the department involved and the corresponding Council committee. Thus far, most of the studies have come from students rather than the colleges and have shown little practical application. However, members of the departments and the committees were enthusiastic about receiving more studies conducted by engineering students, especially studies related to "actual" problems, whether in the area of transportation or related areas.

A No-Fee Contract was entered into by the City and a major university for a housing study. Only a portion of the students involved were engineers. They explored various ideas—day care centers, legal centers and land usage. Concepts expressed by the students were sound; however, they were unclear as to the source of funds available for completion of their proposed projects. All students, not just engineering students, who will have contact with municipalities should be informed as to the meaning and status of funds available for financing capital improvement projects. These sources include motor fuel tax funds, general obligation bonds, revenue bonds, local funds, income funds, State and Federal aid, and the process by which these funds are obtained must be understood.

As a final check on what the City believes is necessary to prepare engineering students for public service, we interviewed four top level personnel in the Department of Public Works which is the largest employer of engineers in the City. Their suggestions are outlined briefly below.

Position I:

1. Establish co-op program for students.
2. Consult on planning and research projects when universities have necessary expertise and practical approach.

3. Have summer programs or part time employment for projects such as cordon counts.
4. Establish Technician Training Program for high school students.

Position II:

1. Whole man approach necessary.
2. Desirable curriculum.
   a. Broad basic courses in science and humanities.
   b. Management courses.
      1) Finance.
      2) Real estate.
   c. Policy courses.
      1) Sources of public funds.
      2) City functioning structure.
      3) Industry functioning structure.
3. Interaction between universities and local government.

Position III:

1. Courses in "management engineering."
2. Courses less technically oriented—minus traditional engineering functions.
3. Alternate concepts to perform a job.
4. Relationship between the "social" engineer and the "technical" engineer.
   a. Social engineer describes the problem.
   b. Technical engineer solves the problem.
5. Curriculum suggestions.
   a. Systems analysis.
   b. Project management.
   c. Linear programming.
   d. Decision-making—public administration.

Position IV:

1. Broader base of training to include social and environmental aspects.
2. Vocational training.
3. Awareness of government operation.

Information was obtained regarding faculty employed by the

City—faculty who act as consultants or who are employed by the City on a part time basis. Only a small percentage of these are engineers. Departments such as Development and Planning, and Urban Renewal employ many faculty who have been trained in the social sciences. One of the ways in which faculty can influence policy is in testifying at Committee hearings. In these cases, material can be presented directly to the Committee chairman and the issue discussed immediately. This may be an especially good approach when environmental issues are involved.

It is clear that the interaction between the City and the engineering colleges has been limited in the past. The colleges have not exerted direct influence on policy decisions; rather, their faculty have in some instances been engaged as consultants for specific problems, and small numbers of their students have been involved in reports for courses and in senior theses, which have not been directed to practical applications.

There has, however, been a significant indirect impact by the Illinois colleges through the considerable number of their graduate engineers that have been employed by the City as shown in Table 1. This impact can be increased if, in developing their educational programs, the colleges consider the needs of the City as set forth in this paper.

Table 1 Percentage of Illinois Engineering College Graduates Employed by the City of Chicago

	Total	*Ill. Grads.*	*Percent*
Department of Public Works	170	110	65
Department of Streets & Sanitation	115	29	25
Department of Water & Sewers	100	28	28
Chicago Transit Authority	43	35	81

Hopefully, as a result of the study of the "Regional Role of Engineering Colleges," wider avenues of communication and cooperation will be developed, and mechanisms for interaction will be established to increase the opportunities for the colleges and universities to participate with the City in meeting some of the complex problems which are faced.

The Engineering Experiment Station of the University of Illinois at Urbana

John J. Desmond

Before one can predict adequately what the role of the Engineering Experiment Station at Urbana-Champaign can or should be in the future, it seems necessary and appropriate to describe what its role in the past has been. Its role has changed in quite perceptible ways in its 70 year history. Some fundamental changes are beginning to appear now which perhaps suggests characteristics of its future role. First some history.

In 1945 Professors Baker and King wrote a two-volume work entitled *The History of the College of Engineering of the University of Illinois 1868-1945*. The volumes comprise over a thousand pages. Within this lengthy account the history of the Engineering Experiment Station is also contained. A very broad description of the nature and role of the station could be divided into two periods of history. The first is discussed thoroughly and documented by Baker and King covering the period from the Station's establishment in 1903 until 1945. I shall attempt to add a second chapter dealing with the post-war years to the present. Then I shall attempt to forecast its future roles.

John J. Desmond is Associate Director of the Engineering Experiment Station of the University of Illinois at Urbana-Champaign.

When Baker and King looked back upon 32 years of history, they concluded rather dramatically that "While the station has never had such financial support from federal funds as that allotted to the agricultural experiment stations connected with some of the larger land-grant colleges, its work has gone forward with steady but substantial progress under the allowances and contributions provided from State and proprietary sources; and there can be little question but that the reputation the station enjoys in the estimation of both the engineering profession and the general public as to the authoritative quality of the results attained and the unprejudiced character of the publications, has gone far in establishing the prestige which the University maintains among the educational institutions in the United States and the world at large."

What Baker and King were looking back upon was a history of great productivity in engineering research, conducted by a group of professors who were giants in their respective fields. At that time, the Station bulletin series, as opposed to scholarly journals, was the prime means of communicating research results to the rest of the world. Three hundred fifty-eight such bulletins and forty-nine of the circulars had been published by 1945. Authorship is attributed to such men as Talbot, Parmelee, Johnstone, Parr, Swann, Richart, Cross, Moore, Seeley, Wilson, Enger, Babbitt, Tykociner, Kunz, Willard, Kratz, and Goodenough. Late in this early period there were names of young men such as Newmark and Dolan beginning to appear in the record.

The conceptualization of the Station is attributed to Professor L. P. Breckenridge, the then head of the Department of Mechanical Engineering. Breckenridge had proposed that the federal government create in every appropriate land grant college Engineering Experiment Stations which would parallel the federally created Agricultural Experiment Stations which had existed for several decades. He was unsuccessful in this attempt, but it is interesting to note, though equally unsuccessful, very similar concepts were being discussed by the United States Department of Commerce in the 1960's, during the period of the State Technical Services Act.

Breckenridge then proposed that the University and State of Illinois take the leadership in creating such a Station, and by an act of the University Board of Trustees it was established in 1903. In that same year, the General Assembly appropriated what was then a generous figure of $150,000 to initiate the Station.

Breckenridge's concept of the work of the Station is best characterized by a quote from a 1906 Bulletin describing the Station. "In determining the character of the work which the station shall undertake, the most careful consideration will be given to the needs and the interests of the state. At the same time we shall not forget the debt which Illinois owes to her older sister states or to foreign nations for furnishing freely the results of scientific investigations or experimental determinations, making it possible for Illinois more cheaply to construct its railroads, mine its coal, generate its power, harvest its crops, communicate with its neighbors, and build its factories, its public buildings and its homes.

"The work of the station will also be largely determined by the funds and facilities which are available for its work. It will seek the cooperation of all the industrial enterprises of the state, both great and small, and it will give help along those lines that promise to aid the greatest number of its people.

"The work of the station should also extend into some fresh fields, seeking to discover new ways and means for economizing energy and materials, for the prevention of waste, for the perfection of labor-saving machinery, for safer methods of travel, and for surer sanitary methods of water supply and sewage disposal."

The concept clearly had its roots in the land grant tradition of public service to the State. Following the model of the Agricultural Experiment Station as it did, its intent was to foster a research enterprise with the specific charge of serving the economic interests of the State. It was acknowledged that such an enterprise could also provide an educational and research experience for students, but this was a secondary role.

During the next several decades and particularly during the 20's, many state universities followed the Illinois pattern. By 1930 there were of the order of 30-35 Engineering Experiment Stations created in the United States—all attempting in some way to be responsive to the economic and technical needs of their individual states.

While, of course, the research results growing out of the efforts of Illinois were not limited in their application to Illinois industry, the investigations had a decided Illinois flavor. Research and testing projects during many of the early years revolved around the mining, railroad, power construction and manufacturing interests of the state. The Appendix provides a selective grouping of project titles. This grouping highlights the Station's responsiveness to the State's economic interests.

During the first 20 years, support from industry tended to be other than direct financial support of the research. Several industrial committes were formed to aid in the selection of problems to be worked upon. The Industrial Conference on Fuel Testing and The Industrial Conference on Heating and Ventilating Research are examples. Companies during the early years were also prime suppliers of test materials ranging from coal samples to railroad cars. The direct financial support of the research was supplied by the State's appropriation to the Engineering Experiment Station.

Starting with the early 20's, the Station began to put its financial support on a footing which asked of industry greater participation. Dean Charles R. Richards prepared a Station circular in 1921 which can be contrasted to Breckenridge's earlier philosophical bulletin dealing with the functions of the Station. "In establishing relations for the conduct of a cooperative investigation, the Engineering Experiment Station will contribute the use of all of its facilities including its laboratories and equipment, together with heat, light, power, water, etc. It will assume the general direction and administration of the investigation, and it will publish all of the results secured and conclusions formulated as a result of the investigation. The cooperating agency is required to provide such funds as are necessary to pay the salaries of special investigators employed exclusively in the conduct of the particular investigation for the purchase of materials and special apparatus needed for the work and not already available in the Station laboratories, and for other necessary expenses of the investigation. In all contracts for cooperative research work which have been entered into by the University of Illinois through its Engineering Experiment Station, ownership of data secured, together with the right to publish these data and conclusions resulting therefrom, is reserved by the Station. The public is thus protected against the use for private gain of an institution which it supports. The University of Illinois further declines to permit the use of data so secured in advance of the complete publication of the results, unless it be with the understanding that any individual or corporation interested may have equal rights with the cooperating agency."

This circular also set forth a sample cooperative agreement, patent and publication policies, and a very clear statement of policy that research was a vehicle for graduate work. "One of the most important functions of the Engineering Experiment Station is the training of men in the methods of scientific research, and possibly the chief advantage in organizing the Station at the University, rather than as an independent research bureau, lies in the fact that the institution attracts many brilliant young men who desire to pursue graduate studies and become prepared to do scientific work. A commercial laboratory or public research bureau cannot offer equal opportunities for advanced study under direction. They must, therefore, look to the universities for their supply of research men. The stimulus to those professors who are directing the work of graduate students, working with them in the solution of scientific problems is doubtless responsible for the fact that very many of the most important discoveries in science have emanated from institutions of higher education.

"The Station employs a considerable number of full-time research assistants, of various academic rank, each of whom is attached to some one of the departments engaged in research. Many of these men are experts in research; others are beginners who need to be developed. In addition to this full-time research staff, the University of Illinois now maintains fourteen Research Graduate Assistantships in the Engineering Experiment Station, together with two Research Graduate Assistantships in Gas Engineering which have been established under the patronage of the Illinois Gas Association. These Assistantships are open to graduates of approved American and foreign universities and technical schools who are prepared to undertake graduate study in engineering, physics, or applied chemistry. Each assistantship carries a stipend of six hundred dollars, and freedom from tuition, incidental, and laboratory fees. Appointment to these positions must be accepted for two consecutive collegiate years, at the expiration of which period, if all requirements have been met, the degree of Master of Science will be conferred. Not more than half of the time of these assistants, during ten months of each year, is required in connection with the work of the department to which they are assigned; the remainder is available for graduate study.

"Information concerning the opportunities for graduate study in engineering will be found in the circular of the Graduate School of the University of Illinois, which may be obtained upon request to the Dean of the Graduate school. Further details regarding appointment to the Research Graduate Assistantships will be supplied upon request by the Director of the Engineering Experiment Station."

Direct industrial support of research and testing began to flow. Unlike earlier reports, acknowledgements would contain recognition of funds supplied by individuals and associations of companies. During the 20's and 30's, research support was received from such Illinois based companies as the Utilities Research Committee of Northern Illinois, The Consolidated Coal Company, The Western Brick Company, Chicago Bridge and Iron Works, Allis-Chalmers, Western Electric, Commonwealth Edison of Chicago, Zeigler Coal and Coke Company, The Illinois Master Plumbers Association, Illinois Central Railroad Company, and the Carnegie-Illinois Steel Corporation, to name a few of the Illinois industries which supported research at Illinois during this era. It appears quite clear that the Engineering Experiment Station was meeting a great industrial need during this early era. Industry had little competence or facility for research during this period. The University could supply both.

It is important to note that, from its very inception, the Station staff looked upon itself as providing a management function in the expansion of an overall research effort. All of the research was conducted in the academic departments. The Station staff and its executive committee aided in the coordination of multi-departmental projects, passed upon proposals and publications, but tended generally to adhere to a decentralized program.

This characteristic of decentralization is one which tends to portray the entire history of the Station down to today. There are several other characteristics of the early history which should be discussed and contrasted with present conditions. There was a distinct separation between the formal instruction of the College and its research. Titles such as Research Professor began to creep in and staff holding such titles were contrasted from teaching professors. Staff were hired for the specific purpose of devoting their energies and efforts to research alone. Supportive staff without professional titles were hired to work on specific projects. Research facilities, laboratories, and buildings were created and in many cases quite separated from the teaching activities of the department. It would be an error to leave the impression of a thick wall between teaching and research; however, the two efforts were regarded as substantially different. This is perhaps thoroughly understandable during the early period and up to 1945. Little importance was given to graduate education in engineering. Research as the prime vehicle for advanced work was given only a passing acknowledgement.

It is perhaps now well to contrast what went on at Illinois during the war years and most importantly the decade that followed the war. It is also important to try and put into some perspective what effect the earlier history of the station had upon the development of the post-war era.

The war itself, of course, had the singular and most profound result of creating a federal commitment to scientific and technological progress. The war was one not only of ideologies but a war of technologies as well. Many faculty members were drawn away from the Urbana campus to devote their talents to war research, particularly those who had built a research reputation. When the war ended, faculty returned, having been exposed to and participated in large federally sponsored research efforts. Many of them, such as Wheeler Loomis and William Everitt, were called upon to continue their relationships with the federal government in advisory capacities in aid of forming federal postwar policy in support of basic research. The stage was set for the College to conduct a wide spectrum of investigations of interest to both the institution and the federal government. This was because of its long history of research involvement, its faculty participation in war-related research, and its many faculty relationships with newly created federal research support agencies. There was a critical point in history in which it was clear that the College was able to respond actively and productively to the post-war federal commitment to research. It was a fortuitus blending of the history of conducting research fostered by the Station and the personalities who emerged in leadership roles during the post-war era. The sciences were represented by Loomis, Adams, and Ridenour soon to be joined by Sietz and Bardeen. Engineering was represented by Everitt, Jordan, Newmark, and the many faculty who saw very early that the dynamics of research support were changing dramatically.

Engineering education was also going through a dramatic change. Much greater emphasis was being given to science-based curricula, graduate education, and the intimate relationship which existed between research and post-graduate work. The federal government with its commitment to the advancement of science and technology viewed its role as a purchaser of scientific and engineering progress. This role, stemming primarily from the war years, sought out the most qualified sources of talent and expertise. The best sources tended to be ten or a dozen large universities of which Illinois was one. This federal commitment, again stemming from the war years, was most evident in the Department of Defense and its service agencies and the AEC. The federal role was not viewed as a support of higher education. However, with the intimacy that had emerged between graduate education and research, the effect upon the universities was the same as if the support had been specifically for the advancement of higher education. There was perhaps one significant difference. This form of federal commitment preferred those institutions who had the managements, facilities, and talent to conduct large scale, fundamental research. Only those institutions that had a history for substantial amounts of research were able to participate effectively. For those who were able to participate, however, it permitted a very large expansion of graduate education and research. At Illinois, the four-or five-decade experience with the Station had provided the means to participate and to go "big time" into a predominantly federally supported research program. Once the College began to move in this direction, the phenomena began to feed on itself and escalate. The Station still followed a pattern of decentralization. However, by developing an awareness where research support opportunities existed within the federal establishment, it provided a coupling between the research investigators and the federal support agencies. Faculty became aware of the business of doing research and began to understand very well the means by which federal support was generated. The professional enterpreneur emerged during this period as a distinct and recognizable faculty phenomena. It is extremely difficult to quantify or give reasons why the University of Illinois made such great progress during this period at a far different rate than many sister institutions who had also early created experiment stations. It became quite clear that Illinois was operating in a league occupied by MIT, Cal Tech, and Berkeley. In the entire mid-continent of the United States, the University of Illinois College of Engineering had outranked all of its sister institutions. Partly the differences relate to the leadership and faculty which it attracted to the University. Partly it relates to a very long history of encouraging research. Partly it relates to enlightened state support of research. One cannot discount the management of indirect cost funds during practically the entire history of research conduct at Urbana. The fact that engineering departments were substantial contributors to and recipients of indirect cost recovery enabled departmental administrators to initiate many research activities which would not have otherwise been started because the program lacked the maturity to warrant outside funding. In recent times, this source of funding for departmental use reached levels in excess of $1 million per year.

When one distinguishes the pre-war versus the post-war era down to the present date, a conclusion might appropriately be drawn that the College of Engineering went through a transition from being a state resource with support flowing primarily from the private sector to a national resource with federal support being dominant. Such is probably the case. The question that might be posed is whether this change has been at the expense of the College's ignoring its land-grant history and a consequent ignoring of its economic role in the state. A more appropriate way of putting it perhaps is that it is more a matter of scale and balance than an ignoring. Prior to the war, support from the private sources represented 100% of external support. Until a few years ago, the absolute amount of private sector support was at least as much (excess adjusted for cost increases) if not more than the pre-war period, but in the post-war period it represented about 5-10% of our total support.

Certainly, as one views the sources of support which the College has enjoyed over the past 20 years, federal sources have dominated. In the early 1960's there began to emerge at state and federal levels a substantial inquiry into the role of a college of engineering in regional economic development. This was the era in which many universities attempted to repeat the experience of MIT and Stanford. This was the era when many research parks close to universities were created. This was the period of time when university and government leaders were asking whether the new knowledge which had been developed in a nation-wide federally supported research effort could be translated for application in the private sector. The State Technical Service Act came into being in 1964 with the promise that federal support would be available to create new mechanism of technology transfer. Even the most generous assessment of all of these efforts would draw the conclusion of, at best, modest success. The effect upon the College of Engineering at the University of Illinois has been to sustain a concern about the ways in which we can relate effectively to and serve industry. The Station created in 1964 an Office of University-Industry Relations. It became clear to the College of Engineering that in many Midwest-based industries, research capabilities had been created and expanded within those companies during the preceding 20 years. Industries had become more self-reliant and did not need to look to universities for project- type research. However, there was still a need to provide industry with a translation of frontier research as it was being conducted by our College and underwritten by the federal government.

There has emerged a technique which provides interface. Within the past four years, three new programs have been created. The Industrial Affiliates Program in Electronics, the Industry Affiliates

Program in Electro-magnetics, and the Industry Affiliates Program in Civil Engineering. All of these programs require financial participation by individual companies for which they receive summaries of research results and annual visits by appropriate faculty. These programs, one of which is still in the early stages of formation, involved the participation by approximately 20 companies and the order of one-third of a million dollars of support. Two other programs which have been developed involve very specialized technical interests by a collection of companies having related interests. The Program on Fracture Control involves a consortium of ten companies who are interested in furthering their own research efforts in product reliability and fracture control. The second similarly structured program involves six companies interested in supporting research leading to the attenuation of noise in off-road equipments.

Yet another program involves the Civil Engineering Systems Laboratory program which is partially supported by 10 private contracting firms. These firms are provided with advanced computer methods of construction projects planning and management. The conclusions which one must draw from the success of these programs is that the College has now begun again to understand and develop mechanisms for relating to the private sector of our region. Several distinguishing characteristics of these newer programs should be highlighted as an indication of how future and expanded relationship with industry might be structured.

1. These newer relationships often involve a consortium of companies with a common set or sub-set of fairly specific interests. This observation is certainly true in the case of the Civil Engineering Systems Laboratory, the Fracture Control, and Noise Programs.

2. These new programs often draw upon existing technology with the main thrust directed toward translation for possible real world application.

3. Continuing communication and access to individual problems are provided by joint university and industry advisory committees. Communications are further enhanced by scheduled visits by selected faculty to the plants of sponsors.

4. The individual proprietary interests of participating companies must be acknowledged. New and more enlightened (flexible) patent policies still await development.

5. These new programs tend to be interdisciplinary in faculty composition thus providing a broader base of expertise consistent with the

interdisciplinary nature of many industrial problems.

6. Dominant industrial participation tends to be from Illinois or at least the region.

7. It must be recognized that enormously greater amounts of time and energy must go into the creation and sustaining of such programs than in the creation of other forms (public) of support. Professional staff must often be identified to work full-time in developing very complicated relationships—relationships not only with industry, but among and between faculty as well. This function at Urbana is performed most conveniently under the auspices of the Engineering Experiment Station.

Appendix

A Selected Sample of Engineering Experiment Station Publications As They Relate to Regional Economic Interests

COAL

Fuel Tests with Illinois Coal, L. P. Breckenridge and P. Diserens. 1908.

The Economical Purchase and Use of Coal for Heating Homes, with Special Reference to Conditions in Illinois. 1917.

The Utilization of Pyrite Occurring in Illinois Bituminous Coal, E. A. Holbrook. 1917.

The Storage of Bituminous Coal, H. H. Stoek. 1918.

The Economical Use of Coal in Railway Locomotives. 1918.

The Analysis of Fuel Gas, S.W. Parr and F. E. Vandaveer. 1924.

Papers Presented at the Short Course in Coal Utilization. 1938.

A Survey of Sulphur Dioxide Pollution in Chicago and Vicinity, A. D. Singh. 1938.

Papers Presented at the Fifth Short Course in Coal Utilization. 1939.

Papers Presented at the Sixth Short Course in Coal Utilization. 1942.

Hand-Firing of Bituminous Coal in the Home, A. P. Kratz, *et al.* 1942.

Papers Presented at the Seventh Short Course in Coal Utilization. 1946.

Papers Presented at the Eighth Conference on Coal Utilization. 1949.

Fuel Tests with Illinois Coals, L. P. Breckenridge, et al. 1906.

How to Burn Illinois Coal Without Smoke, L. P. Breckenridge. 1907.

The Weathering of Coal, S. W. Parr, *et al.* 1907.

The Modification of Illinois Coal by Low-Temperature Distillation, S. W. Parr and C. K. Francis. 1908.

The Occluded Gases in Coal, S. W. Parr and P. Barker. 1909.

Unit Coal and the Composition of Coal Ash, S. W. Parr and W. F. Wheeler. 1909.

The Weathering of Coal, S. W. Parr and W. F. Wheeler. 1909.

Tests of Washed Grades of Illinois Coals, C. S. McGovney. 1909.

The Spontaneous Combustion of Coal, S. W. Parr and F. W. Kressman. 1910.

The Coking of Coal at Low Temperature, with a Preliminary Study of the By-Products, S. W. Parr and H. L. Olin. 1912.

The Steam Consumption of Locomotive Engines from the Indicator Diagrams, J. P. Clayton. 1913.

Coal Washing in Illinois, F. C. Lincoln. 1913.

The Analysis of Coal with Phenol as a Solvent, S. W. Parr and H. F. Hadley. 1914.

The Coking of Coal at Low Temperatures, with Special Reference to the Properties and Composition of the Products, S. W. Parr and H. L. Olin. 1915.

Dry Preparation of Bituminous Coal at Illinois Mines, E. A. Holbrook. 1916.

Specific Gravity Studies of Illinois Coal, M. L. Nebel. 1916.

Subsidence Resulting from Mining, L. E. Young and H. H. Stoek. 1916.

Effects of Storage upon the Properties of Coal, S. W. Parr. 1917.

Percentage of Extraction of Bituminous Coal with Special Reference to Illinois Conditions, C. M. Young. 1917.

Comparative Tests of Six Sizes of Illinois Coal on a Mikado Locomotive, E. C. Schmidt, *et al.* 1917.

A Study of the Forms in Which Sulphur Occurs in Coal, A. R. Powell with S. W. Parr. 1919.

Panel System of Coal Mining; A Graphical Study of Percentage of Extraction, C. M. Young. 1919.

Bituminous Coal Storage Practice, H. H. Stoek, *et al.* 1920.

The Distribution of the Forms of Sulphur in the Coal Bed, H. F. Yancey and T. Fraser. 1921.

The Ignition Temperature of Coal, R. W. Arms. 1922.

A Study of Coal Mine Haulage in Illinois, H. H. Stoeck, et al. 1922.

Power Studies in Illinois Coal Mining, A. J. Hoskin and T. Fraser. 1924.

A Study of Skip Hoisting at Illinois Coal Mines, A. J. Hoskin. 1925.

The Measurement of Air Quantities and Energy Losses in Mine Entries, A. C. Callen and C. M. Smith. 1926.

Tests on the Bearing Value of Large Rollers, W. M. Wilson. 1927.

The Measurement of Air Quantities and Energy Losses in Mine Entries, Part II, A. C. Callen and C. M. Smith. 1927.

The Classification of Coal, S. W. Parr. 1928.

Friability of Different Coals, C. M. Smith. 1929.

Combustion Tests with Illinois Coals, A. P. Kratz and W. J. Woodruff. 1930.

Washability Tests of Illinois Coals, A. C. Callen and D. R. Mitchell. 1930.

Accidents from Hand and Mechanical Loading in Some Illinois Coal Mines, A. C. Callen and C. M. Smith. 1931.

RAILROADS

The Economical Use of Coal in Railway Locomotives. 1918.

The Railroad Dynamometer Car of the University of Illinois and the Illinois Central Railroad, J. K. Tuthill. 1947.

Holding Power of Railroad Spikes, R. I. Webber. 1906.

The Effect of Scale on the Transmission of Heat Through Locomotive Boiler Tubes, E. C. Schmidt and J. M. Snodgrass. 1907.

High Steam Pressure in Locomotive Service, W. F. M. Goss. 1908.

Freight Train Resistance, E. C. Schmidt. 1910.

Resistance to Flow Through Locomotive Water Columns, A. N. Talbot and M. L. Enger. 1911.

Superheated Steam in Locomotive Service, W. F. M. Goss. 1912.

The Effect of Cold Weather upon Train Resistance and Tonnage Rating, E. C. Schmidt and F. W. Marquis. 1912.

The Tractive Resistance of a 28-Ton Electric Car, H.H. Dunn. 1914.

Laboratory Tests of a Consolidation Locomotive, E. C. Schmidt, *et al.* 1915.

Some Graphical Solutions of Electric Railway Problems, A. M. Buck. 1916.

The Tractive Resistance on Curves of a 28-Ton Electric Car, E. C. Schmidt and H. H. Dunn. 1916.

Comparative Tests of Six Sizes of Illinois Coal on a Mikado Locomotive, E. C. Schmidt, *et al.* 1917.

Passenger Train Resistance, E. C. Schmidt and H. H. Dunn. 1918.

An Investigation of the Properties of Chilled Iron Car Wheels, Part I, Wheel Fit and Static Load Strains, J. M. Snodgrass and F. H. Guldner. 1922.

An Investigation of the Properties of Chilled Iron Car Wheels, Part II, Wheel Fit, Static Load, and Flange Pressure Strains; Ultimate Strength of Flange, J. M. Snodgrass and F. H. Guldner. 1922.

An Investigation of the Properties of Chilled Iron Car Wheels, Part III, Strains Due to Brake Application; Coefficient of Friction and Brake-Shoe Wear, J. M. Snodgrass and F. H. Guldner. 1923.

A Study of Fatigue Cracks in Car Axles, H. F. Moore. 1927.

Freight Train Curve Resistance on a One-Degree Curve and a Three-Degree Curve, C. E. Schmidt. 1927.

Heat Transmission Through Boiler Tubes, H. O. Croft. 1927.

Effect of Enclosures on Direct Steam Radiator Performance, M. K. Fahnestock. 1927.

A Study of Fatigue Cracks in Car Axles, Part II, H. F. Moore, *et al.* 1929.

Tests of a Mikado-Type Locomotive Equipped with Nicholson Thermic Syphons, E. C. Schmidt, *et al.* 1930

POWER

Fuel Economy in the Operation of Hand Fired Power Plants. 1918.

The Analysis of Fuel Gas, S. W. Parr and F. E. Vandaveer. 1924.

An Electrical Method for the Determination of the Dew-Point of Flue Gases, H. F. Johnstone. 1929.

Condensation of Moisture in Flues, W. R. Morgan. 1934.

A Survey of Sulphur Dioxide Pollution in Chicago and Vicinity, A. D. Singh. 1938.

An Economic Study of Fuels in Manufacturing, W. H. Voskuil. 1952.

Lighting Country Homes by Private Electric Plants, T. H. Amrine. 1908.

On the Rate of Formation of Carbon Monoxide in Gas Producers, J. K. Clement, et al. 1909.

Tests of Two Types of Tile-Roof Furnaces Under a Water-Tube Boiler, J. M. Snodgrass. 1909.

Tests of a Suction Gas Producer, C. M. Garland and A. P. Kratz. 1911.

Street Lighting, J. M. Bryant and H. G. Hake. 1911.

Inductance of Coils, M. Brooks and H. M. Turner. 1912.

Mechanical Stresses in Transmission Lines, A. Guell. 1912.

Starting Currents of Transformers, with Special Reference to Transformers with Silicon Steel Cores, T. D. Yensen. 1912.

Characteristics and Limitations of the Series Transformer, A. R. Anderson and H. R. Woodrow. 1912.

Thermal Properties of Steam, G. A. Goodenough. 1914.

A Study of Boiler Losses, A. P. Kratz. 1915.

Corona Discharge, E. H. Warner with J. Kunz. 1919.

Some Conditions Affecting the Usefulness of Iron Oxide for City Gas Purification, W. A. Dunkley. 1921.

A Study of Explosions of Gaseous Mixtures, A. P. Kratz and C. Z. Rosecrans. 1922.

An Investigation of the Maximum Temperatures and Pressures Attainable in the Combustion of Gaseous and Liquid Fuels, G. A. Goodenough and G. T. Felbeck. 1924.

A Thermodynamic Analysis of Gas Engine Tests, C. Z. Rosecrans and G. T. Felbeck. 1925.

The Effect of Temperature on the Registration of Single Phase Induction Watthour Meters, A. R. Knight and M. A. Faucett. 1926.

The Cause and Prevention of Embrittlement of Boiler Plate, S. W. Parr and F. G. Straub. 1926.

An Investigation of the Mechanism of Explosive Reactions, C. Z. Rosecrans. 1926.

A Thermodynamic Analysis of Internal-Combustion Engine Cycles, G. A. Goodenough and J. B. Baker. 1927.

Heat Transfer in Ammonia Condensers, A. P. Kratz, *et al.* 1927.

An Investigation of Checkerbrick for Carburetors of Watergas Machines, C. W. Parmelee, *et al.* 1928.

The Corrosion of Power Plant Equipment by Flue Gases, H. F. Johnstone. 1931.

CONSTRUCTION

A Simple Method of Determining Stress in Curved Flexural Members, B. J. Wilson and J. F. Quereau. 1927.

Tests of Welds, W. M. Wilson. 1931.

Simplified Computation of Vertical Pressures in Elastic Foundations, N. M. Newmark. 1935.

Contributions to Proceedings of the Second International Conference on Soil Mechanics and Foundation Engineering, 1949.

Lectures on Foundation Engineering, A. E. Cummings. 1949.

What Do We Know About Diagonal Tension and Web Reinforcement in Concrete? E. Hognestad. 1952.

Tests of Reinforced Concrete Beams, Series of 1905, A. N. Talbot. 1906.

Tests of Concrete: I, Shear; II, Bond, A. N. Talbot. 1906.

Tests of Concrete and Reinforced Concrete Columns, Series of 1906, A. N. Talbot. 1907.

Tests of Reinforced Concrete T-Beams, Series of 1906, A. N. Talbot, 1907.

A Study of Roof Trusses, N. C. Ricker. 1907.

Tests of Cast-Iron and Reinforced Concrete Culvert Pipe, A. N. Talbot, 1908.

Tests of Brick Columns and Terra Cotta Block Columns, A. N. Talbot and D. A. Abrams. 1908.

A Test of Three Large Reinforced Concrete Beams, A. N. Talbot. 1908.

A Study of Base and Bearing Plates for Columns and Beams, N. C. Ricker. 1909.

Tests of Timber Beams, A. N. Talbot. 1909.

An Investigation of Built-up Columns Under Load, A. N. Talbot and H. F. Moore. 1910.

Tests of Columns: An Investigation of the Value of Concrete as Reinforcement for Structural Steel Columns, A. N. Talbot and A. R. Lord. 1912.

Tests of Reinforced Concrete Buildings Under Load, A. N. Talbot and W. A. Slater. 1913.

Reinforced Concrete Wall Footings and Column Footings, A. N. Talbot. 1913.

The Strength of I-Beams in Flexure, H. F. Moore. 1913.

Tests of Bond Between Concrete and Steel, D. A. Abrams. 1913.

Wind Stresses in the Steel Frames of Office Buildings, W. M. Wilson and G. A. Maney. 1915.

Influence of Temperature on the Strength of Concrete, A. B. McDaniel. 1915.

Tests of Reinforced Concrete Flat Slab Structures, A. N. Talbot and W. A. Slater. 1916.

The Strength and Stiffness of Steel Under Biaxial Loading, A. J. Becker. 1916.

The Strength of Webs of I-Beams and Girders, H. F. Moore and W. M. Wilson. 1916.

Tests to Determine the Rigidty of Riveted Joints of Steel Structures, W. M. Wilson and H. F. Moore. 1917.

Test of a Flat Slab Floor of the Western Newspaper Union Building. A. N. Talbot and H. F. Gonnerman. 1918.

Analysis and Tests of Rigidly Connected Reinforced Concrete Frames, M. Abe. 1918.

The Relation Between the Elastic Strengths of Steel in Tension, Compression, and Shear, F. B. Seely and W. J. Putnam.

The Thermal Conductivity and Diffusivity of Concrete, A. P. Carman and R. A. Nelson. 1921.

Studies on Cooling of Fresh Concrete in Freezing Weather, T. Yoshida. 1921.

Effect of Moisture Content upon the Expansion and Contraction of Plain and Reinforced Concrete, T. Matsumoto. 1921.

The Strength of Concrete: Its Relation to the Cement, Aggregates, and Water, A. N. Talbot and F. E. Richart. 1923.

An Investigation of Web Stresses In Reinforced Concrete Beams, F. E. Richart. 1927.

The Effect of Climatic Changes on a Multiple-Span Reinforced Concrete Arch Bridge, W. M. Wilson. 1927.

An Investigation of Web Stresses in Reinforced Concrete Beams, Part II, Restrained Beams. F. E. Richart and L. J. Larson. 1928.

A Study of the Failure of Concrete Under Combined Compressive Stresses, F. E. Richart, *et al.* 1928.

The Failure of Plain and Spirally Reinforced Concrete in Compression, F. E. Richart, *et al.* 1929.

Laboratory Tests of Reinforced Concrete Arch Ribs, W. M. Wilson. 1929.

Dependability of the Theory of Concrete Arches, H. Cross. 1929.

The Torsional Effect of Transverse Bending Loads on Channel Beams, F. B. Seely, *et al. 1930.*

Column Analogy, H. Cross. 1930.

Movement of Piers During the Construction of Multiple-Span Reinforced Concrete Arch Bridges, W. M. Wilson. 1931.

MANUFACTURING

High-Speed Tool Steels, L. P. Breckenridge. 1905.

Repeated Stress (Fatigue) Testing Machines Used in the Testing Laboratory of the University of Illinois, H. F. Moore and G. N. Krouse. 1934.

Papers Presented at the First Short Course on Industrial Packaging and Materials Handling. 1949.

Proceedings of the Special Symposium on Behavior of Welded Structures, W. H. Munse. 1962.

Tests of High-Speed Tool Steels on Cast Iron, L. P. Breckenridge and H. B. Dirks. 1905.

The Strength of Chain Links, G. A. Goodenough and L. E. Moore. 1907.

Tests of a Liquid Air Plant, C. S. Hudson and C. M. Garland. 1908.

The Strength of Oxyacetylene Welds in Steel, H. L. Whittemore. 1910.

Magnetic Properties of Heusler Alloys, E. B. Stephenson. 1910.

Tests of Nickel-Steel Riveted Joints, A. N. Talbot and H. F. Moore. 1911.

An Investigation of the Strength of Rolled Zinc, H. F. Moore, 1911.

A New Analysis of the Cylinder Performance of Reciprocating Engines, J. P. Clayton. 1912.

Magnetic and Other Properties of Electrolytic Iron Melted in Vacuo, T. D. Yensen. 1914.

The Effect of Boron upon the Magnetic and Other Properties of Electrolytic Iron Melted in Vacuo, T. D. Yensen. 1915.

Magnetic and Other Properties of Iron-Silicon Alloys Melted in Vacuo, T. D. Yensen. 1915.

A Preliminary Study of the Alloys of Chromium, Copper, and Nickel, D. F. McFarland and O. E. Harder. 1916.

The Embrittling Action of Sodium Hydroxide on Soft Steel, S. W. Parr. 1917.

Magnetic and Other Properties of Iron-Aluminum Alloys Melted in Vacuo, T. D. Yensen and W. A. Gatward. 1917.

Twist Drills, B. W. Benedict and W. P. Lukens. 1917.

Tests to Determine the Rigidity of Riveted Joints of Steel Structures, W. M. Wilson and H. F. Moore. 1917.

Hydraulic Experiments with Valves, Orifices, Hose, Nozzles, and Orifice Buckets, A. N. Talbot, *et. al.* 1918.

The Orifice as a Means of Measuring Flow of Water Through a Pipe, R. E. Davis and H. H. Jordan. 1918.

Dissolved Gases in Glass, E. W. Washburn, F. F. Footitt, *et al.* 1920.

An Investigation of the Fatigue of Metals, H. F. Moore and J. B. Kommers. 1921.

An Investigation of the Fatigue of Metals, Series of 1922, H. F. Moore and T. M. Jasper. 1923.

Investigation of Antennae by Means of Models, J. T. Tykociner. 1925.

An Investigation of the Fatigue of Metals, Series of 1925, H. F. Moore and T. M. Jasper. 1925.

Tests of the Fatigue Strength of Cast Steel, H. F. Moore. 1926.

An Investigation of Twist Drills, Part II, B. W. Benedict and A. E. Hershey. 1926.

Tests of the Fatigue Strength of Cast Iron, H. F. Moore, *et al.* 1927.

Surface Tension of Molten Metals, E. E. Libman. 1927.

A Metallographic Study of the Path of Fatigue Failure in Copper, H. F. Moore and F. C. Howard. 1928.

Tests of the Fatigue Strength of Steam Turbine Blade Shapes, H. F. Moore, *et al.* 1928.

Rolling Tests of Plates, W. M. Wilson. 1929.

Acid Resisting Sheet Iron Cover Enamels, A. I. Andrews. 1929.

The Hydroxylation of Double Bonds, S. Swann, Jr. 1929.

A Study of the Ikeda (Electrical Resistance) Short-Time Test for Fatigue Strength of Metals, H. F. Moore and S. Konzo. 1930.

Studies in the Electrodeposition of Metals, D. B. Keyes and S. Swann, Jr. 1930.

Study of Slip Lines, Strain Lines, and Cracks in Metals Under Repeated Stress, H. F. Moore and T. Ver. 1930.

Tension Tests of Rivets, W. M. Wilson and W. A. Oliver. 1930.

The Effect of Furnace Gases on the Quality of Enamels for Sheet Steel, A. I. Andrews and E. A. Hertzell. 1930.

Treatment of Water for Ice Manufacture, D. Burks, Jr. 1930.

The Effect of Smelter Atmosphere on the Quality of Enamels for Sheet Steel, A. I. Andrews and E. A. Hertzell. 1931.

The Microstructure of Some Porcelain Glazes, C. L. Thompson. 1931.

The Effect of Thermal Shock on Clay Bodies, W. R. Morgan. 1931.

HEATING AND VENTILATION

The Warm-Air Heating Research Residence in Zero Weather, V. S. Day. 1927.

Papers Presented at the First Conference on Air Conditioning. 1936.

Papers Presented at the Second Conference on Air Conditioning. 1939.

Simplified Procedure for Selecting Capacities of Duct Systems for Gravity Warm-Air Heating Plants, A. P. Kratz and S. Konzo. 1942.

Papers Presented at the First Short Course on Steam and Hot Water Heating Systems. 1948.

A Comparison of the Performance of Various Room Heating Units Used In the I=B=R Research Homes, W. S. Harris and L. N. Montgomery. 1960.

A Study in Heat Transmission, J. K. Clement and C. M. Garland. 1909.

A Study of the Heat Transmission of Building Materials, A. C. Willard and L. C. Lichty. 1917.

Report of Progress in Warm-Air Furnace Research, A. C. Willard. 1919.

Investigation of Warm-Air Furnaces and Heating Systems, A. C. Willard, *et al.* 1921.

Investigation of Warm-Air Furnaces and Heating Systems, Part II, A. C. Willard, *et al.* 1924.

Investigation of Warm-Air Furnaces and Heating Systems, Part III, A. C. Willard, *et al.* 1929.

Factors Affecting the Heating of Rooms with Direct Steam Radiators, A. C. Willard, *et al.* 1931.

Humidification for Residences, A. P. Kratz. 1931.

Illinois Institute of Technology Research Institute

Thomas W. Knowles

The Illinois Institute of Technology Research Institute is a not-for-profit research foundation which was incorporated on April 6, 1936 "to promote, encourage, maintain and aid scientific investigation and research in affiliation with Armour Institute of Technology by the faculty, staff and alumni thereof, and others associated therewith." The informal organization originated on September 18, 1933 when Armour Institute of Technology President Willard E. Hotchkiss appointed Professor Joseph B. Finnegan director of "all technical work of a non-academic nature carried on by faculty members in the Institute laboratories." Prior to this time professors at AIT often performed industrial research privately.

The research contracts in the first year, 1936, totaled $20,000 and consisted of a coal washability survey for a coal company and a study of the electrical and magnetic characteristics of a kind of wire rope for a steel company. These reflect the emphasis on industrial research during the early years.

"In the early days, the director and staff all were part-time researchers and full-time teachers. The research institute was secondary to the academic work of the school. The teachers' first interest was in teaching and their first loyalty to the classroom."[1]

In 1938 the name was changed to the Armour Research Foundation. Simultaneously, Harold Vagtborg became the director and the

Thomas W. Knowles is an Assistant Professor in the Department of Industrial and Systems Engineering at Illinois Institute of Technology in Chicago, Illinois.

transition from a teaching-research staff began. The Foundation established the principle of entrusting fundamental research to the academic group and building up a large, full-time staff of industrial research men that handle primarily applied research.

In 1940, AIR and Lewis Institute merged to form Illinois Institute of Technology. The name of the foundation was changed to Armour Research Foundation of Illinois Institute of Technology. In 1963, the name of the foundation was changed to its present name, IIT Research Institute.

IITRI exhibited a phenomenal rate of growth for an institution without any endowment or source of funds other than research contracts and licensing fees generated from the research (patents). Table 1 shows this growth in terms of dollar volume and number of employees.

Table 1 Fiscal Year $ Volume and Average Employees

Year	*$ Volume*	*Employees*
1936	$20,000	NA
1937	$45,240	10
1938	$70,750	23
1939	$118,200	43
1940	$264,730	97
1941	$608,390	192
1942	$1,094,520	258
1943	$1,314,820	294
1944	$1,525,000	348
1945	$1,750,000	378
1946	$2,100,000	445
1947	$2,550,000	448
1948	$3,378,379	597
1949	$4,116,500	630
1950	$4,614,958	683
1951	$6,225,302	879
1952	$8,106,877	1,074
1953	$10,011,741	1,135
1954	$11,118,114	1,112
1955	$10,756,764	1,121
1956	$11,910,042	1,237

Table 1 (Continued)

Year	*$ Volume*	*Employees*
1957	$14,067,534	1,290
1958	$14,647,873	1,236
1959	$15,430,827	1,267
1960	$16,190,123	1,260
1961	$16,200,782	1,238
1962	$20,306,808	1,537
1963	$23,551,789	1,734
1964	$24,329,084	1,628
1965	$25,481,862	1,780
1966	$27,323,182	1,842
1967	$28,741,095	1,859
1968	$29,404,335	1,696
1969	$27,229.986	1,503
1970	$26,536,558	1,419
1971	$24,170,944	1,265
1972	$25,366,438	1,277

Source: IIT Research Institute Annual Reports.

During the first years of the ARF, almost all of the research was industrial and contracted by industrial firms and trade associations. Very little government sponsored research was done. In fact, government sponsored research was not considered in the province of the ARF during its early years. The 1939 annual report states, "Three years have passed since the organization of the Research Foundation of Armour Institute of Technology for the exclusive purpose of rendering a *research and experimental engineering service to industry* on a not-for-profit basis." (Italics added). However, this began to change during World War II, when governmental contracts with the military took precedence over industrial research.

That the view towards government sponsored research changed markedly during the war is reflected by the following statement from the 1947 annual report: "The character of the work we are continuing to take for the armed forces and various other government agencies is becoming steadily more desirable from a research point of view. . . ." Unfortunately, no information is available giving the relative industrial and governmental research dollar contract volume over the years. Table 2 gives the breakdown in number of projects, governmental and industrial, and dollar volume of internally funded research.

Table 2 Analysis of Research Projects by Type and Number

Year	*Total*	*Government Sponsored*	*Industry Sponsored*	*Internally Funded*
1948	191	62	129	NA
1949	243	74	169	NA
1950	276	94	182	NA
1951	304	146	158	NA
1952	381	216	165	$112,697
1953	438	158	281	$138,484
1954	538	203	335	$173,278
1955	512	170	342	$320,235
1956	691	298	393	$324,684
1957	734	300	404	$580,754
1958	780	375	405	$548,642
1959	695	381	314	$572,385
1960	700	357	343	$878,237
1961	661	363	298	$1,045,338
1962	727	402	325	$1,342,305
1963	718	361	357	$1,723,514
1964	1,058	531	527	$1,515,753
1965	841	411	430	$1,475,304
1966	1,060	436	624	$1,537.072
1967	991	401	590	$1,424,310
1968	897	369	528	$1,215,684
1969	936	430	506	$1,437,861
1970	768	377	391	$1,525,937
1971	610	293	317	$1,432,685
1972	594	262	332	$1,231,721

Source: IIT Research Institute Annual Reports

In terms of dollar volume, governmental research accounted for 78.5% of the research in 1959 and in recent years averaged approximately 80%. In terms of number of projects, government projects averaged 42.5% and 45.4% during the periods 1948-1952 and 1968-1972, respectively. The fundamental change in the industrial-governmental mix appears to have occurred during and after World War II.

The organizational relationship of IITRI to Illinois Institute of Technology has remained the same since 1936. The Board of Trustees and President of IITRI are the same as that of IIT. However, IITRI is separately incorporated in the State of Illinois.

There is still a sharing of skills between the staffs of IITRI and IIT. IITRI has a tuition program under which many of their employees enroll in courses at IIT. During the 1970-71 academic year there were approximately 200 course registrations by IITRI employees. In addition, some IITRI staff teach evening courses at IIT. IIT faculty participate in IITRI projects at a much lower rate than in the earlier years. Unfortunately, data concerning the amount of days IIT faculty devote to IITRI projects was not available. However, the utilization is still significant.

REFERENCE

1. *The Frontier*, Armour Research Foundation of Illinois Institute of Technology, Chicago, Illinois, Vol. 18, No. 2, Summer 1955, p. 21.

Statement of the Unique Historical and Geographic Factors Which Impinge on the Regional Role of the Engineering Programs at Southern Illinois University–Carbondale

Thomas B. Jefferson

HISTORICAL FACTORS

Engineering at Southern Illinois University—Carbondale is young and developing, the first engineering degrees having been awarded in 1966. "Steady-state" impact on the region has therefore not yet been experienced.

A change in organization and name of the former School of Technology was effected in April, 1971. The School of Engineering and Technology comprises five departments:

Thomas B. Jefferson is the Dean of the School of Engineering and Technology at Southern Illinois University at Carbondale, Illinois.

Department of Electrical Sciences & Systems Engineering
Department of Engineering Mechanics and Materials
Department of Thermal and Environmental Engineering
Department of Occupational Education
Department of Technology

The first-listed three departments, acting in concert with the dean, are responsible for B.S. and M.S. programs in engineering. The Department of Occupational Education has responsibility for B.S. and M.S. programs which produce teachers in a number of occupational specialties. The Department of Technology operates two B.S. programs, Engineering Technology and Industrial Technology. In addition to the programs listed here, an M.S. program in Applied Science is offered in the School of Engineering and Technology.

SPECIAL FEATURES OF ENGINEERING PROGRAM

The baccalaureate program in engineering is a single-degree program; the degree is Bachelor of Science with a major in Engineering. The curriculum is a unified one, with the objective of graduating a broadly-based engineer who is well prepared either to enter the field immediately or to pursue graduate studies. The program of study therefore features a large common core of subjects, while providing the student with the opportunity to elect one of three 26-quarter-hour optional course groupings at the senior level. The student is thus given sufficient depth that he has the confidence and ability to solve real problems in a selected area; and he is given sufficient breadth to have an appreciation of the nature of problems in other areas. The capstone of the curriculum is a 6-quarter-hour sequence in Engineering Design, which is taught by a team of professors. Teams of three or four students each design a device or system requiring the integrated use of principles and experience gained in earlier stages of the program.

The baccalaureate in engineering thus produces an engineer whose elective program is biased toward a specific area, but who is not a specialist. For example, the student who selects the optional group in Electrical Sciences and Systems Engineering will not be a traditionally trained Electrical Engineer, Systems Engineer, or combination of the two. He will however be able to deal with real engineering problems requiring a reasonable depth of knowledge of electrical matters and systems approach.

Since the baccalaureate engineering program does not follow the conventional specialty pattern, it is possible that significant differences in regional impact will emerge as the study progresses.

Graduate study programs leading to a Master of Science degree with a major in Engineering are available in the following functional areas of engineering:

Electrical sciences—
circuits, electronics, fields, controls

Fluid mechanics—
viscous and inviscid flow, compressible flow, turbulence, wave motion

Solid mechanics and structures—
continuum mechanics, dynamics and vibrations, experimental stress analysis, stability, structural analysis and design

Transfer processes and environmental engineering—
heat transfer, mass transfer, chemical processes, water resources development, air and water pollution control

Systems engineering—
bio-engineering, hybrid computation, control and instrumentation, computer applications, operations research and mathematical model formulation

Some 82% of the engineering faculty hold doctors' degrees; these were awarded by eighteen universities. Some 33 universities are represented in degrees held by engineering faculty members at all three degree levels.

GEOGRAPHIC FACTORS

S.I.U.—Carbondale is located in extreme Southern Illinois in a predominantly rural area. Mining, fruit farming, and small industry characterize the immediate surroundings.

Carbondale is 18 air miles from the Mississippi River and 38 air miles from the Ohio River. St. Louis, Missouri is 85 air miles to the Northwest; Paducah, Kentucky is 58 air miles to the Southeast; and Evansville, Indiana is 95 air miles away in a generally Easterly direction. These constitute nearby industrial centers.

A recent survey of engineering graduates over a five-year period, 1966-1970, indicates locations of graduates. We attempted contact with the list of 131 graduates for that period and got 78 responses. For these, the following breakdown applies:

Table 1 Location of 1966-1970 Graduates of the SIUC School of Engineering and Technology

	Number		Percent of Sample	
Location	Industry	Graduate School	Industry	Graduate School
Illinois	25	14	32	18
Ohio	4		5.1	
California	4		5.1	
Kentucky	3		3.8	
Florida	3		3.8	
Missouri	2	1	2.6	1.3
Pennsylvania	2		2.6	
Michigan	2		2.6	
Maryland	2		2.6	
Okla., Tex., Tenn., Wash., N. J., D. C., N. Y., Ind., Wisc.l Colo. — 1 each	10		12.8	
	57	15	73.0*	19.3*

*Armed Forces accounted for the remaining 6 graduates, or 7.7% of those who responded.

The above table indicates rather heavy settling in Illinois, with adjacent states accounting for fewer graduates than some more distant locations. Informal information from Placement Office personnel suggests that shifts in national emphasis and related shifts in recruitment may be reflected substantially in the data. At the beginning of the five-year period studied, aerospace activity was strong and centers in California, Washington, Florida, and elsewhere were recruiting vigorously. It is felt that perhaps 90% of S.I.U.—Carbondale's present technical recruiting area would be encompassed by a 200-mile circle around Carbondale, plus the remainder of Illinois (see following Figure).

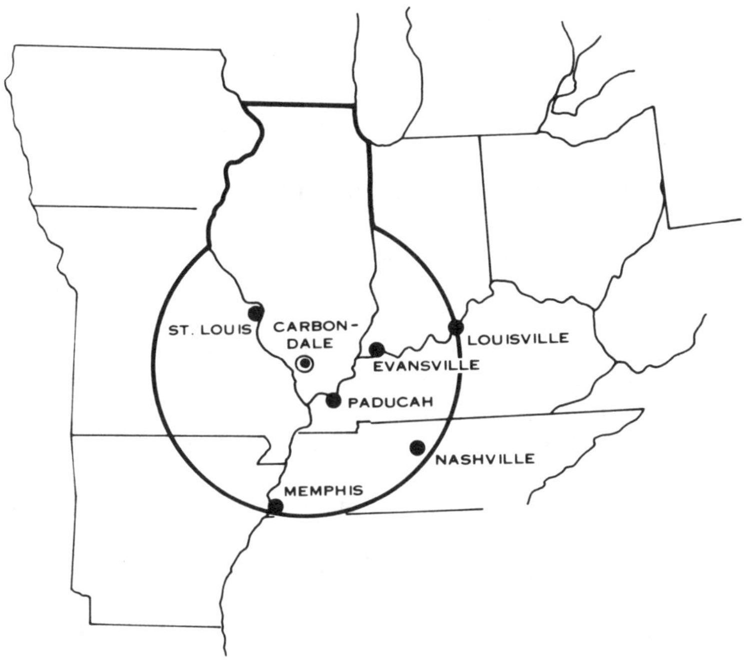

Fig. 1 Southern Illinois University Recruiting Area for 90% of the Students.

Geography affects professional ties of engineering faculty:

1. A number of faculty members are affiliated with the Illinois Society of Professional Engineers, through the Egyptian Chapter of I.S.P.E. This constitutes a combination of State-wide contacts and more localized contacts within a part of the State. The S.I.U. Engineering Club is a student chapter of I.S.P.E.

2. Engineering technical society dealings involve both Kentucky and Missouri, and faculty members active in these societies are in general members of sections headquartered either in St. Louis, Missouri or Paducah, Kentucky. Activities include section or chapter memberships held by faculty, attendance of some section officers at S.I.U.'s annual Engineer's Week Banquet, support of the S.I.U. Engineering Club, student scholarships, and attendance at meetings on the part of S.I.U. personnel. Technical society chapters with which we have had dealings of this sort include:

American Institute of Chemical Engineers	—Western Kentucky Section
American Society of Civil Engineers	—St. Louis Section
American Society of Heating, Refrigeration and Air Conditioning Engineers	—St. Louis Chapter
The American Society of Mechanical Engineers	—Paducah Section
Institute of Electrical and Electronics Engineers	—Both St. Louis & Paducah Sections

3. In the American Society for Engineering Education, our affiliation is with the Illinois-Indiana Section, as is that of other engineering schools in Illinois.
4. Contacts with industries develop in many ways. One important vehicle for meaningful interchange between industry and S.I.U. engineering faculty was the recently terminated State Technical Services program. In connection with this program, some 10 faculty members from S.I.U.—Carbondale were involved in visits to some 80 industries for the purpose of identifying problems in computer applications or industrial waste handling with which we could assist. The area of Illinois covered was roughly that portion of the State south of a line through Alton and Salem. Included were visits to industries in the Paducah, Kentucky area, Evansville, Indiana area, and Cape Girardeau-Sikeston, Missouri area. S.I.U. staff worked closely with about 12 of the industries in relation to problems in which a mutual interest developed.
5. Engineering courses have been offered by extension from time to time in Metropolis, Illinois, which is on the Ohio River and a part of the industrial area surrounding Paducah, Kentucky. We are en-

tering our third year of an Engineering M.S. program which is taught at a Resident Center in Sauget, Illinois, adjacent to East St. Louis.

APPENDIX II

CONFERENCE PROCEEDINGS*

*The complete Proceedings include the Project Reports marked by asterisk in Appendix I contents at p. 91.

Regional Role of Engineering Colleges Conference

INTRODUCTION

George Bugliarello

The proceedings of this Conference are an integral part of our study of technology, the University and the community. The Conference has provided an exchange of views among persons and institutions that have been grappling with the problem of analyzing and guiding the interaction between the university and the surrounding community. Sufficient information is emerging, through the proceedings of this Conference, as well as through the other phases of this study, and through other investigations, to indicate that this is an area in which society fails to take full advantage of the complex resource that a university represents for its community. At a moment in which our society seems increasingly reluctant to support a costly system of higher education, it is important that the role of higher education institutions be understood in all of its facets and that actions be taken to develop potentially beneficial functions of these institutions.

If we focus specifically on the interaction between an engineering college and the community, this interaction could be fruitfully viewed in terms of what are—or ought to be—the basic tasks of the college.

In the first place, the college ought to be involved in communicating the meaning of technology to the rest of society. At this moment if one were to use a scale between 0 and 1 to assess how an engineering college succeeds in this task of communicating to the rest of society the meaning of technology as a revolutionary step in the evolution of life, we would have to assign a mark very close to zero. The colleges have not succeeded in communicating on this subject because at present they are not very concerned with it. Yet it is a most vital and crucial issue. It is not a matter of public relations, of glamorizing what we are doing as engineers. It is rather a matter of developing an understanding of what machines are about—the machine as an extra corporeal organ, the machine as a substitute for some of our biological drives, the machine as a substitute for our social drives, and so on.

Secondly, colleges ought to be pushing the boundaries of technological knowledge. Everybody accepts this point. However, there has been relatively little appreciation in our colleges of the fact that technology is the combination of devices and organizations; technology is both hardware and social technology. We have been able to do the first but we have not been able to do the second.

In the third place—and this is the most obvious—the colleges are involved in the training of personnel. Again, on a scale from 0 to 1, one could give a mark of only about two thirds for what colleges of engineering have been able to do. We are not recycling or retraining our faculties, and we are not recycling the overwhelming majority of engineers in the country. The process is very much hit and miss. We have not taken the view that our province is all the engineers that we have produced, that our province is all the technological activities around us; we have only been concerned with those that come to us.

A fourth important, but debated, role is to respond to other needs of the community—direct operational needs, such as the need for assistance in the formulation of legislation, the need for various facilities, or the need for assistance in the operation of some critical programs. In the national context, on our 0 to 1 scale, we have done this perhaps for two thirds of a point; but in the regional context, we have done this to a very limited degree, as we have seen throughout this study.

Indeed, if we look at what engineering colleges are doing or could do in response to regional needs (the region being defined broadly to mean a state, a multi-state region, a county or a city), it is clear that also in the other three role areas the colleges have been considerably less effective than at the national level. For instance, when it comes to pushing the boundaries of technological knowledge, we have not been able to push effectively a regionally oriented technology simply because we do not know the meaning of technological knowledge in the regional context. When it comes to the training of personnel, we find that we cannot respond effectively in the regional context until we develop regional manpower systems and regional manpower models.

Some of the problems in response to regional needs are exemplified by the difficulties encountered when a college endeavors to respond to the needs—stated or unstated—of a state legislature for technical advice. Aside from the lack of formal channels of communication, there is the serious problem of lack of continuity in state policies. At this moment it is very difficult to anchor the regional role of an engineering college to state mechanisms that see very rapid changes in their advisory groups, particularly scientific advisory groups. We have experienced this in the present study. We make contacts with state officers, they give us support, then at a later date we find that our contacts have changed jobs and their successors are not interested any more in the same problem. There are very few science advisors that survive in office more than one term. Frequently they change even more often than governors, so there is no stability—not the same kind of stability that exists at the Federal level. Therefore, we must find some mechanism that offers a means of achieving the necessary stability.

However, the crucial problem remains inside the university. It stems from the disagreement as to what the very goals of the university should be. Part of the problem, as it concerns the ability of the university to respond effectively to some needs of its environment, derives from the inevitable conflict between the democratic, consensus operation that is so characteristic of the academic enterprise and the need for a more hierarchical approach, which is necessary if the university is to respond rapidly.

The conflict can be resolved if the view is accepted that a university must operate in a sense like an amoeba. An amoeba can respond to its environment. Certain portions of the amoeba are inner directed and other portions adapt with pseudopodes to the outer environment. Similarly a certain portion of the university remains inner directed and another portion is outer directed. In today's climate it is to a considerable extent the outer directed portion of the university that makes the life of the inner directed portion possible. But we have no good theory as to how we can develop an effective university organism with these two components.

Another unresolved question in this respect is the role of research institutes that were created originally with a fairly close coupling to the university but have increasingly acquired a life of their own—for example, Illinois Institute of Technology Research Institute or the Stanford Research Institute. We must really ask ourselves what the role of the university is if these institutes that were created to respond to needs of the environment simply respond less and less to the need of the parent institution.

These are some of the very basic philosophical issues that remain unresolved in this study. They imply value judgments and research much beyond that contained in these proceedings and in the other portions of our study.

Higher Education in the United States

(Luncheon Talk)

Warren B. Cheston

I want to speak to you briefly today on a number of interrelated topics concerning the immediate future of higher education in the United States, with particular reference to publically supported institutions. Whenever possible I shall illustrate my general remarks with allusions to engineering education.

Universities apparently have survived the sometimes bloody, quasi-political battles of the period 1968-1970, sparked by those who would radically change the form and content of higher education and its relation to a politicized society. Engineering colleges were little affected in this period, except insofar as the institutions of which they were a part underwent significant change. True, there emerged a more concentrated focus on the relationship between technology and its fruits, on the one hand, and the remainder of societal concerns on the other; much of this change of emphasis was long overdue and had virtue in itself, but by and large engineering faculties and the students which they serve are still deeply immersed in Reich's Consciousness I and II and there is little significant interaction with those issues paramount in the gestalt of Consciousness III society. The extra-educational events of 1970-71, which caused vast and significant changes in the employment pattern in the United States, have probably had more serious impact on engineering colleges than the campus upheavals of the previous two years.

Warren B. Cheston is Chancellor of the University of Illinois at Chicago Circle.

However, this is not my primary message. The focus of my remaining comments will be on the subject, "Who decides the programmatic development of higher education in the United States?" Although this subject has a certain pressing immediacy to those of us in publically supported higher education, the subject has great relevance to the private sector as well (conversations with the chief executive officers of the major private institutions in metropolitan Chicago indicate that they are truly aware of this relevance). All of us who have had some administrative responsibilities within colleges of engineering know that resource allocation is strongly coupled to the analysis of quality of programs. We all think we know how to judge the quality of a given department within a college. Let me list some of the quality criteria which I consider relevant—each of you has a similar list:

(a) How easy is it for your baccalaureate candidates to obtain fellowships and assistantships in prestigious institutions?
(b) How easy is it to recruit the best baccalaureate candidates from prestigious institutions into your graduate programs?
(c) How easy is it to place your Ph.D.'s in positions in prestigious institutions?
(d) How easy is it to recruit onto your faculty the Ph.D.'s from prestigious institutions?
(e) How successful is your faculty in competing for Federal and foundation support for their programs of research or instructional innovation?
(f) How successful is your faculty in obtaining Guggenheim, senior NSF fellowships and the like?
(g) Where does your department rank in the Cartter survey or similar studies?

These criteria of quality are, by their very nature, qualitative —which simply means that those who apply these criteria, most usually deans and academic vice presidents and vice chancellors, need to meet certain minimal standards of competence and accrued experience. These senior academic officers have realized for a number of years that this qualitative judgment scheme needs to be supplemented by some sort of quantitative analytical scheme to increase the efficiency of internal management. Many such management tools have been in use for some years. Most of you are familiar with those developed by former Provost Terman of Stanford. However, most university administrators have realized the limitation of these quantitative management tools and indices and have applied them with great caution. These indices are many in number and I shall mention only a few.

(a) FTE students/FTE faculty.
(b) Dollars expended/SCH produced, etc.

To these have been added primitive measures of "productivity;" that is, it has been realized that a department which devotes X-SCH on a student who never gets a degree should somehow be differentiated from another department which devotes X-SCH on a student who does get a degree. In addition, it has been realized that it is necessary to distinguish between departments who require Y credits for a degree from those who require (Y) credits (≠1) for a degree—i.e., educational philosophy is a parameter which needs to be quantified. No matter what specific quantitative management tool is developed, the output is usually stated in terms of "resource entitlement" for a given program. In the end, however, the dean or the academic vice-chancellor must make a decision and he makes it, in a certain sense, by the seat-of-his-pants; namely, his application of qualitative criteria tempered by his quantitative analysis.

However, as the demand for higher education has increased and as new institutions have proliferated, those who provide the funds for public higher education have felt the need to establish mechanisms whereby decisions on resource allocation among the many institutions which constitute the higher education system of the state are made. These mechanisms take the form of coordinating bodies which attempt to make the same kind of decision among institutions as deans, etc., make among departments, within an institution. To carry out their charge, these coordinating bodies develop an insatiable appetite for data and statistics. Formulas are developed into which these data are fed and resource allocation based, at least in part, on the outcome. Realizing that such a scheme for resource allocation leaves much room for criticism, the coordinating bodies employ in addition qualitative measures. Unlike the "quality" measures used for internal university management, we have instead "goals criteria" which have a high political content. For one fact is certain, these coordinating bodies are political bodies appointed and (or) elected by governors, legislatures, etc.; these bodies are intended to produce results which are at least consistent with the political strategies of those political agencies which appoint them. If they fail to fit into the overall strategy for the attainment of political goals, their advice and council will be ignored and they will be impotent. There is nothing basically wrong with this in a democratic society.

That which we in universities must contend with in the 1970's appears to be an increase in management at levels far removed from those who produce: management decisions and resource allocations based more and more on the attainment of political goals of those outside the university framework—a reliance on measures for resource allocation which are more and more quantitative but which relate less and less to the traditional goals of the universities themselves, which include a strong component of the search for quality.

Although the Federal government has greatly influenced the development of institutions of higher education (when Federal funds are available for aeronautical engineering and not for philosophy, it is quite obvious that the goals and mission of the university are thereby altered), the agencies distributing largess to the universities have by and large used criteria of quality, although no one can deny the influence of certain strategically placed political figures on the determination of the direction in which the largess shall flow. However, those at the working level in the Federal agencies have been sensitive to quality criteria and have been some of the most outspoken and eloquent defenders of quality in higher education of any segment of society in the United States. At present, with my limited experience with coordinating agencies, I am pessimistic that we shall find within these agencies the same dedication to quality that we have experienced in Federal agencies. I hope that I am wrong in my present view.

With the increasing reliance on such management and resource allocation data as unit costs, engineering colleges shall be obliged to be increasingly adroit at explaining their high unit cost figures. There is some hope within the university management scheme since quality plays such an important role there, as I have stressed previously. I know that deans of colleges of engineering are very clever in explaining the high costs of engineering programs, and when searching for evidence of efficiency within their operations compare themselves with other colleges of engineering.

However, with limited resources available to higher education and the increase in emphasis on egalitarianism in higher education, high cost professional and semi-professional programs are going to come under increased scrutiny. The 1970's, as well as being characterized by the aforementioned, will also see available funds channeled preferentially into medicine and the health-related disciplines.

My own view is that the large comprehensive engineering colleges should look more and more towards cooperative efforts to reduce costs. Through agencies such as the CIC (Committee for Institutional Cooperation), such cooperative efforts should be developed. There is an obvious example where cooperation is certainly called for—natural resource engineering (mining engineering, etc.) Most programs in this discipline have sub-marginal enrollment. In some the number of faculty is not much different from the number of students. Here is an area ripe for regional cooperation—every state in the upper midwest has an important stake in this field, but the region itself probably cannot support more than one or two quality programs at a reasonable cost. It is probably cheaper for the state of "X" to send its students to the state university of "Y," paying all costs including housing, than to attempt to mount a program of its own. It is undoubtedly better for the colleges of engineering involved to develop and prepare such a plan rather than leave the initiative to a cabal of state coordinating agencies. Coordination and cooperation fostered by the institutions with a steady eye on institutional autonomy is far preferable to forced cooperation, with coordination provided by an agency external to the universities involved. I contend that such cooperative programs are both desirable and possible and my one example does not exhaust the field for cooperation.

Needed: An Interface Between Science and Politics

(Dinner Talk)

John H. Conolly

Ever since science and technology put men on the moon, people have been asking, if we can put men on the moon, why can't we solve the traffic problem?

Or: Why can't we provide decent housing for everyone?

Or: Why can't we clean up the environment?

There are probably several answers to these questions, but in my line of work—as an elected public official—I am very keenly aware of a major difference between landing on the moon and solving these other problems.

The difference, to use a nice sounding expression for it, is "intergovernmental relations." Or, there is a shorter word—a word that doesn't always sound so nice to everyone, even though there is no reason why it should be a dirty word. The word is "politics." There is no government on the moon, no taxpayers there, no constituents—so when the United States sent a space ship to the moon, we didn't have to coordinate the decision with any other government.

But the situation is vastly different when we start to deal with domestic problems. In my home county, Lake County, Illinois, we have more than 150 different local governments. About a third of these are special districts—the kind of local governments that typically deal with sanitation, water supply, drainage, conservation, all of the

John H. Conolly is State Representative from Waukegan, Illinois and Chairman of the Illinois Legislative Council. Mr. Conolly has also served at President of the National Legislative Council.

problems that today involve the quality of the environment. So when we start to deal with a problem that goes beyond an immediate neighborhood, there's much more involved than the knowledge of how to solve the problem. There is the politics of the problem—deciding what government or governments can best deal with the problem. This involves many people. The local governments of Illinois have more than 35,000 *elected* officials—and many times that number of appointed officials. We might propose to abolish townships. Do you know how many elected officials townships in Illinois have? Almost 10,000. Abolish 10,000 elected offices—propose that and you know what political problems are.

Unfortunately, to many people that word "politics" has an unpleasant odor. They would like to get along without politics. But politics, according to my dictionary is, "The art or science of political government . . . The conducting of or engaging in political affairs, often professionally . . . The business, activities, or profession of one so involved . . . The methods or tactics involved in managing a state or government." The word "politics" is ultimately derived from the Greek "polis," meaning city, and from the associated Greek word, "polites," meaning a citizen. And the word scientist comes from a Latin root that means "to know."

Today, if we want to use science and technology to solve our domestic problems, we must bring our scientists and our politicians together.

It may be helpful to keep in mind Dr. Daniel P. Moynihan's[1] description of the two kinds of problems we face in our society.

First, there are political problems.

Second, there are knowledge problems.

A political problem is, how do you get something done through government? For example, how do you pass a law? Parenthetically, I might say that "passing a law" is only the first step to getting something done through government. But the fact that the law is on the book signals to everyone one important fact—that there is an important consensus of opinion in support of this objective.

A knowledge problem is quite different. Most of you are familiar with knowledge problems and how you go about solving them. The process is considerably different from the political process.

In the case of the successful program to put men on the moon, we first had to have a consensus that this was desirable. Congress had to agree to appropriate the funds to finance the program. This was the political problem. The knowledge problem came next. Or did it? I understand, on fairly good authority, that we knew how to get men to the moon before we started out on the Apollo mission. After the money was appropriated, it was a matter of working out the details.

Going back to Moynihan again, he has something very interesting to say about the relationship between knowledge problems and political problems. He says of these two kinds of problems:

> There is an increasingly dense mutuality between the two, it being a defining characteristic of technological society that *what can be done must be done, so that where knowledge comes into being on its own*, as it were, *it commonly produces pressures within the political system to adopt the social goals made possible by such knowledge*.[2]

If Moynihan's view is correct—then we whose primary profession is politics are eventually compelled to adopt the advances made possible by science and technology.

It is easy to say this—but not always easy for us who are engaged in politics to do so. For example, technology has developed an interesting little device called the "breathalyzer." Blow into it after you have been drinking, and it measures the amount of alcohol in your blood and therefore your intoxication level.

The value of this device in connection with the arrest and punishment of drunken drivers is obvious. It was so obvious that every state in the Union has required such tests except Illinois. Here the Administration caused the introduction of such legislation, to require tests, in January 1971.

In view of the previous action of the other 49 States, you might have expected this bill to pass easily in Illinois. If you followed this legislation, you know it did not. The Federal Secretary of Transportation had to threaten to reduce Federal grants for highway purposes to get some action. Finally, at the fall session of the Legislature, on November 13, ten months after the bill was introduced, it was passed and sent to the Governor for his approval.

No, it is not always easy to solve the political problems, even when we know what to do at the technological and scientific end of the problem.

However, activity is going forward in several areas which we hope will overcome some of the political problems in putting technology to work for us.

The National Legislative Conference, for example, has created a

Committee on Science and Technology. This group is studying the ways in which science and technology can be made directly available to state legislatures. We are seeking to learn how to create an interface between the scientific community and the legislature. This particular committee is an outgrowth of a more extensive survey of the uses of science and technology in state and local government conducted by the Council of State Governments.

In this connection, Dr. M. Frank Hersman is chairing a Committee on Intergovernmental Science Relations which is about to report their recommendations to the Federal Council for Science and Technology—recommendations that will have a good deal to say, I am sure, about the interface between men of science and men of government.

Perhaps of even more basic importance is strengthening the communications between state legislatures and the Federal government, particularly Congress. The National Legislative Conference Committee on Intergovernmental Relations, which for some years I chaired, has been particularly active in presenting the needs of the state and local governments to the White House and the Congress.

The most visible part of our activity has been our concern for revenue sharing and convincing the Federal government that it should assume the basic financial responsibility for public aid programs. The states must have more funds—or must be relieved of a substantial portion of the public aid burden—if we are going to meet our needs in the field of education, including higher education.

In particular, if the state and local governments are going to utilize science and technology for new approaches to problems of social policy, they are going to have to receive substantial infusions of new revenues. The most likely source for such revenues is shared funds from the Federal government. Without money of this kind, our states are not going to have the "venture capital" that is needed for research and development.

More than money is involved in intergovernmental relations. Just as in the Intergovernmental Science Program of the National Science Foundation, an important element is to bring state and Federal officials and the scientists together. The significance of this is that we are learning to talk together—scientists and politicians.

Politicians, I hope, are going to understand better the needs of the scientists and the universities (because that is where most of the scientists are—in the universities). And the scientists are going to understand the dimensions of the political problems that must be dealt with if we are going to put scientific knowledge to work for the common good.

This conference is an important part of that process. Part of the knowledge problem in higher education is, how do we organize higher education to get the most out of it? What are we putting into higher education? What goes in terms of faculty? Students? Input from other schools? From industry? From government? What is coming out? Where do the students go? What—besides teaching—do the faculty members contribute to the community? Is the community failing to utilize what is available to it from higher education?

This conference is taking questions like these and examining them in the context of the regional role of the engineering colleges.

This conference is dealing with what Moynihan terms a "knowledge problem," and because it concerns engineering colleges, the source of so much of our technological knowledge today, your findings are going to be highly important for the political realm as well. I, for one, await your findings with a great deal of interest because—if Moynihan is right—the answers you and your fellow scientists come up with in conferences such as this, may determine the shape of our political life for the next decade.

REFERENCES

1 Moynihan, D. P., "To Solve Problem, First Define It," *New York Times*, Annual Education Review, Jan. 12, 1970; see also Moynihan, "The Knowledge Gap in Education," *Chicago Tribune*, Jan. 10, 1972.

2 Moynihan, D. P., "To Solve Problem, First Define It," *New York Times*, Annual Education Review, Jan. 12, 1970.

The Evolution of Engineering Education

(Luncheon Talk)

Steven B. Sample

The State of Illinois Board of Higher Education was formed about ten years ago for the purpose of coordinating all of the publicly-supported colleges and universities in the State, including junior colleges. It is a lay board comprising ten members appointed by the Governor, plus the Superintendent of Public Instruction and the chairmen of the five college and university governing boards. The Board of Higher Education hires and fires its own administrative staff, which presently consists of roughly thirty-two professional and clerical people, headed by the Executive Director, Dr. James B. Holderman.

The Board is charged with preparing and implementing the State's master plan for higher education. It must approve all new programs, new degree offerings, buildings, and campuses at the various colleges and universities, and it must advise the Governor and General Assembly on all matters relating to the higher education budget. It must also make recommendations directly to the governing boards concerning the elimination of existing programs that are no longer economically or educationally justifiable. It reinforces these recommendations through its power of budget review.

The BHE, as it is called, does *not* set campus policy, nor does it control university appointments, nor does it control the disbursement of funds once they have been allocated by the General Assembly; these

Steven B. Sample is Deputy Director for Programs, State of Illinois Board of Higher Education.

powers are strictly the prerogative of the various governing boards. Thus, the ultimate control of higher education is *shared* between the governing boards on the one hand, and the BHE on the other. This situation often leads to rather raucous differences of opinion between the two sides, as some of you may be aware from a number of recent news articles.

Before discussing any long-term evolutionary trends in engineering education, let me first recount some of the dramatic changes I have witnessed in the decade or so that I have been associated with the engineering profession. When I enrolled at the University of Illinois as a freshman in 1958, the college of engineering was entering a truly golden era. Engineering was highly respected on the campus as a tough, glamorous, no-nonsense discipline in which only the very best students could survive. Expansion and growth were everywhere evident in the technical disciplines—in fact, as I remember, the staff of the Electrical Engineering department alone numbered over 100 professors by the time I graduated.

Research was king. Federal dollars flowed ever so freely, especially to support research in applied science and mathematics within the engineering departments. As a consequence, such mundane concerns as teaching and the practical problems of industry were often ignored, and at times even treated with contempt. In spite of the huge professional staff, it was not unheard of for a student to earn his E.E. degree at Urbana without ever seeing a professor in the classroom, his education having been acquired completely at the hands of graduate students.

Now, roughly a decade later, the picture is considerably changed. Freshmen engineering enrollments are down sharply—as much as 20% at some schools—and the decline is even more marked at the graduate level. Engineering is often denigrated on the campus as the tool of the military-industrial complex—the very handmaiden of those who would destroy the world through war, pollution, and dehumanizing materialism. Federal support for research, graduate education, and equipment has been substantially decreased. Moreover, engineering now has a much lower priority in the competition for State funds, as evidenced, for example, by the recent recommendation of the Board of Higher Education to discontinue the separate engineering school at Southern Illinois University at Edwardsville. And finally, the engineering job market has gone from one of insatiable demand just a few years ago, to gross oversupply in some areas today, especially at the doctoral level.

In order to really understand these changes, and to infer any long term trends from them, they must first be placed in a larger historical framework.[1]

Formal engineering education in this country is a relatively recent phenomenon. Prior to the nineteenth century engineering was taught almost completely on an apprentice basis. Then, with the establishment of the military academies in the early 1800's, and with the subsequent formation of a number of technical institutes, engineering instruction began to move from the field into the classroom. However, it is important to remember that these early centers of technical education were in no way connected with the academic colleges of that period.

By the middle of the nineteenth century a few colleges began to permit the establishment of scientific schools on their campuses. However, these schools enjoyed at best a loose affiliation with the colleges, in that the scientific students were segregated from the regular students and pursued a separate curriculum. Moreover, even this minimal acceptance within the confines of academe depended upon the scientific schools being concerned with the teaching of natural philosophy, as opposed to the practical exploitation of physical laws such as would characterize an engineering curriculum.

Finally, with the establishment of the land-grant colleges in 1864, engineering found a comfortable legitimacy in schools dedicated to teaching agriculture and the mechanic arts. However, these institutions were not really colleges in the traditional sense, but rather trade schools established for the benefit of the industrial and agrarian classes. The curriculum was heavily laced with laboratory investigations, field work, and shop practice, as opposed to the normal collegiate curriculum of Greek, Latin, mathematics, and moral philosophy. In addition, the faculty of the land-grant colleges were often rough-hewn men who had little or no teaching experience and who possessed precious little classical knowledge themselves.

Over the next half century the classical colleges and the land-grant colleges began to more closely resemble each other. The traditional academic establishment was undergoing what has come to be called the academic revolution,[2,3] in which research and graduate education were to transform the colleges into universities, in accordance with the model developed in Germany. This revolution opened up the established private colleges to the point that science and the scientific method became an integral part, if not the corner stone, of these institutions. Some of the private universities even began to permit the development of engineering schools on their campuses. The land-grant colleges, for their part, introduced more and more of the humanities into their curriculum, and also evolved into universities through the establishment of graduate schools. By the 1930's, except for the presence of agriculture at the state universities, there was little difference in kind between these and the private universities, although the latter perceived themselves to be enormously superior in quality.

During this period engineering education remained essentially unscathed by the academic revolution, in that it retained its traditional emphasis on teaching its students a relatively fixed body of technology. This is not to say that the engineering curriculum failed to reflect advances in technology or that engineers were not introduced to the fundamentals of mathematics and science. But the faculty and profession still fervently believed that an engineer could be given a set of basic *technological skills* that would stand him in good stead for a lifetime of professional practice.

Then came World War II, and the engineering profession was found sorely wanting. Radar, rockets, the atomic bomb, sophisticated ship and aircraft design, the computer, synthetic materials—all these *practical* developments were not the product of engineers, but rather of scientists and mathematicians pressed into service by the exigencies of the war.

In response to this intellectual impotence on the part of the engineering profession, a number of young engineering educators began to feel that the time had come for their profession to undergo its own academic revolution. Men like Bill Everett, formerly Dean of Engineering at the University of Illinois, began to acquire control of the larger engineering schools, and under their patronage emerged the research laboratories, the emphasis on graduate education, and the requirement of the doctor's degree for faculty appointment that characterize the modern college of engineering. The curriculum became heavily oriented toward science and mathematics and much less so toward the teaching of technological skills. Engineers would not only be able to follow the rapid advances in basic technology, but in increasing instances they would lead these advances. This philosophy undergirded the golden era from the late 50's through the 60's, which I described earlier in my talk, and culminated with the most monumental engineering achievement in history—the landing of men on the moon in 1969.

What can we conclude about engineering education from this brief historical perspective? First, in slightly over a century it has evolved from an apprenticeship in which various arts and skills were learned by emulating a practicing professional, to a body of technology learned at the hands of professors who were themselves closely identified with the engineering profession, to a scientifically based curriculum taught by men who are in many cases indistinguishable from applied scientists and mathematicians. Moreover, a discipline that once trained men to take independent responsibility for engineering projects early in their careers, now educates men who for the most part will work in giant corporate labyrinths and who may never have independent responsibility for anything.

Then, too, because of its having participated (albeit somewhat tardily) in the academic revolution, engineering education occupies a special position in the academic hierarchy, distinct from the other professions. Unlike law and medicine, for example, the highest degree in engineering is the traditional academic doctorate, the Ph.D. And unlike agriculture, teacher education and business, engineering is accepted as an *academic* discipline, as evidenced by its inclusion in surveys of the comparative academic strength of universities.

As I pointed out earlier, with decreasing enrollments and lessening Federal and state support, engineering education today is in serious trouble. However, its history of adaptability to change and its position as an integral part of academe will, in my opinion, help it to successfully weather the storm. This is not to say that reason and order will prevail at every turn as we enter the new reality of the penurious seventies.

While I do not pretend to have any special insights, I do believe that definite trends are discernible amongst the myriad and diverse proposals being put forth for solving the current dilemma of engineering education. First, I think it's safe to say that engineering education is becoming much more flexible in its curriculum. Three years ago a number of my Purdue colleagues and I began an honors program in electrical engineering in which each student followed a plan of study that was tailored to his individual needs.[4] The only requirements in this program were that each student had to complete 132 total hours, of which 24 hours were to be in the humanities and social sciences, and 32 hours were to be in electrical engineering; otherwise, the student and his advisor were completely free to work out the plan of study however they might choose. Moreover, if the student wished he could combine his undergraduate and master's work into an integrated five-year program, thereby allowing him to work on his master's thesis for two to three years while delaying his advanced course work in the humanities until he had achieved greater technical and intellectual maturity. Initially the faculty were somewhat reluctant to approve this program on even a limited basis; however, I am happy to say that many of the basic features of this plan have now been extended to *all* electrical engineering students at Purdue.

This program was built on a number of philosophical premises that are relevant to the current crisis in engineering education. First, in an era in which technology pervades every facet of society, a wide variety of students might be attracted to study engineering as a background for some other profession, such as law, medicine, or business, provided the engineering curriculum were sufficiently flexible. This hypothesis has indeed proven correct, to the point where I am convinced that providing pre-professional training for other disciplines, particularly medicine and business, could become a major market for engineering education in this decade.

Second was the realization that there are many different jobs and responsibilities within the electrical engineering profession itself, and, as a corollary, that there is an equal diversity of people who wish to become engineers, and that therefore no single, rigid curriculum will best prepare all people for all electrical engineering positions.

Finally, in an era of rapid, even violent, change it was felt that the most important things a student could learn are the techniques and the *need* for continued learning once he leaves school. To illustrate this point, let me define a ratio, which for lack of a better name we might call the knowledge perception ratio, or KPR. The KPR is the ratio of what you think you know, to what you think there is to know. It has been well documented that this ratio is identically equal to one upon graduation from high school, and decreases exponentially thereafter. By allowing the honors student to take a significant part in, and to share responsibility for, the design of his own professional education, we found we could increase the attenuation rate of the KPR substantially.

In addition to becoming more flexible, engineering education will in my opinion become much more closely allied with the social and human sciences. I foresee strong interaction over the next two decades with such disciplines as law, medicine, criminology, sociology, urban planning, political science, and public administration. To support my contention, let me read you an excerpt from a paper by Roger Levien of the Rand Corporation:[5]

> "The fault, you see, lies not with our technology, but with our systems. When I say "system" in this context, I mean the combination of technology, institutions, and persons intended to serve some purpose. The problem is that as engineers we have specialized in only one aspect of the design of these systems—the technology. Yet it is now evident that the proper functioning of our public systems is not principally a question of competent technological design—classical engineering. Better weapons, patrol cars, and computer systems will not, by themselves, significantly improve public safety and the quality of justice. To achieve such improvements will require changes in institutions as well. We as engineers have little experience with and little inclination toward such activities. For many of us, institutions are for the social scientists and professors of business administration. However, for those of us who were drawn to engineering neither by a love of resistors and condensers nor by an infatuation with LaPlace transforms, but by the desire to create things that serve a useful purpose, there is no escaping the need to expand our concerns and analyses to include institutions. And there is no reason the engineer cannot be trained to bring to the design process a comprehension of social and institutional phenomena as well as a command of physical ones."

In my opinion this increasing flexibility and expanded concern with the social sciences will be accompanied by a greater involvement of engineering schools with the communities in which they are located. Not only will the technical expertise of colleges be increasingly directed toward the practical problems of their surrounding region—as evidenced by the growth of industrial research parks on university campuses—but local society will play an expanded role in the planning and development of engineering education.

For example, Garner and Shaw of the California State Technical College recently described an experiment in industrial participation in engineering education.[6] Far from the cooperative work-study programs of the last two decades, the Garner-Shaw experiment involves bringing technical problems and engineers from industry directly into the college classroom. In this way students are provided with concrete, challenging design problems, and at the same time become acquainted with the engineering profession in their own community.

Increasing flexibility, increasing interaction with the social sciences, increasing involvement with the community—these are a few of the evolutionary trends I foresee for engineering education over the next decade or two. One might well ask whether these trends do not really presage the demise of engineering education in the classical sense, and with it the demise of the noble, disciplined profession with which each of us is so intimately familiar. I cannot really answer that question. But I can say that, as a profession, our first duty is not to preserve a particular technological or educational heritage, but rather to meet the practical needs of society in the most efficient manner possible. I believe John Truxal put it well when he said:[7]

"We must respond to the nationwide anti-technology sentiment and to the young people's accusation that engineering education is irrelevant with two broad statements:

(1) Engineering education is an essential element of a truly liberal education, especially in an era when technology interacts with so much of our daily lives, and

(2) While human desires, social pressures, and technology have combined to aggravate many of our national ills, the only real hope for the improvement of the quality of life lies in the *humane* use of new technology."

REFERENCES

1. Rudolph, F., *The American College and University* (New York: Alfred A. Knopf, 1962).
2. Jencks, C. and D. Riesman, *The Academic Revolution* (Garden City, New York: Doubleday, 1968).
3. Veysey, L. R., *The Emergence of the American University* (Chicago: University of Chicago Press, 1965).
4. Sample, S. B. and H. W. Thompson, "An Honors Program Leading Directly to the Master's Degree," *Engineering Education*, Vol. 60, 1970, p. 472.
5. Levien, R. E., "Institutions, Innovation, and Incentives," IEEE *Transactions on Education*, Vol. E-14, 1971, p. 148.

6. Garner, E. R. and A. C. Shaw, "Industrial Participation in Engineering Education," *Engineering Education*, Vol. 62, 1971, p. 114.
7. Truxal, J. G., "Toward a More Humane Use of Technology," *Engineering Education*, Vol. 62, 1971, p. 93.

*Regional Economic Effects of Higher Education Institutions**

Barclay Hudson

Higher education plays many roles, and its benefits can be assessed in a variety of ways. In this paper I wish to look at the effects of institutions of higher learning (IHL) from a rather special standpoint: the focus will be on *regional* effects, which are quite different from the often-studied effects that would accrue either to individual students or to the nation as a whole; the emphasis will also be on *economic* benefits, or monetary revenue that IHL might bring to a region, as distinct from cultural advantages or particular types of manpower development.

The analysis is based on a conceptual model which describes five separate economic mechanisms which transmit the effects of an IHL to its local region. Most of what follows deals with the nature of these five links. In the concluding section, I will try to focus on policy im-

Barclay Hudson is an Assistant Professor in the Urban Planning Program of the School of Architecture and Urban Planning at UCLA.

*This analysis was made possible by a grant provided by the Ford Foundation and the Rockefeller Foundation under their joint Program of Social Science and Legal Research on Population Policy. The purpose of this grant is to examine urban growth policies for the United States, focusing on the role of higher education and other professional services as possible instruments for deconcentration of urban populations into new growth centers. Besides the author, the other investigators in this program are John Friedmann and Leland S. Burns, all from the Urban Planning Program of the School of Architecture and Urban Planning at UCLA. The author accepts sole responsibility, however, for the contents of this paper.

plications, though any recommendations would have to be considered simply working hypotheses at this state in the research. Nevertheless, a few tentative points brought out by the exercise can be stated at the beginning.

1. A region is very much an "open" system, whose boundaries are easily crossed by capital and human resources: thus there is no guarantee that the resources produced by a local IHL will stay in the locality; they can easily "migrate" out with no compensation to the region. On the other hand, a regional IHL can serve as a means to *attract* resources to the locality, and possibly in greater amount than it might produce on its own. In this way the analysis of IHL economic effects at the regional level differs significantly from the calculation of IHL effects for individuals or for the nation as a whole, which represent relatively "closed" systems from the standpoint of recapturing the benefits of educational investment.

2. It follows that IHL policy—to the extent that it is geared to the goal of local *regional* development—should begin with a consideration of the specific mechanisms which transmit IHL economic effects. These need to be examined not only from an *empirical* standpoint to estimate the usual direction and magnitude of effects, but also from a *political and administrative* standpoint, to suggest possible means of intervention in the economic mechanisms. In general, our empirical work shows rather weak magnitudes of IHL effects on regional development. Our analysis also suggests, however, that somewhat unorthodox policies in higher education might create a significantly stronger role for IHLs within a local economy, based on evidence from "deviant" cases of higher education practices in the past.

3. Intervention strategies are difficult to prescribe in the abstract, because appropriate policy requires special attention to the nature of each particular region. This is inherent in the nature of exploiting comparative regional advantages. Nevertheless, certain generalizations can be offered:

—guidelines to success and failure are likely to be more easily found in *case studies* of IHL success and failure, than in aggregate statistical analysis (we have attempted both types of investigation in the research project described here);

—*entrepreneurship* appears to be a major factor, whether undertaken by persons within the IHL itself or by regional officials interested in enhancing the local role of educational facilities;

—by itself, an IHL would seem to have fairly small leverage on regional development but the effects are likely to be *least* significant in less-developed regions where they are easiest to measure. For example, in a "college town" the effects of an IHL are isolated and relatively easy to document, but the benefits tend to diffuse out of the local economy rather quickly. In contrast, the marginal effects of an IHL are probably greater in larger, more economically diverse regions which already have something going for them, though the problems of ascribing effects to the IHL factor become increasingly difficult in these cases.

The conceptual model shown in Fig. 1 illustrates five major links or processes through which an IHL might affect the economy of the surrounding region.

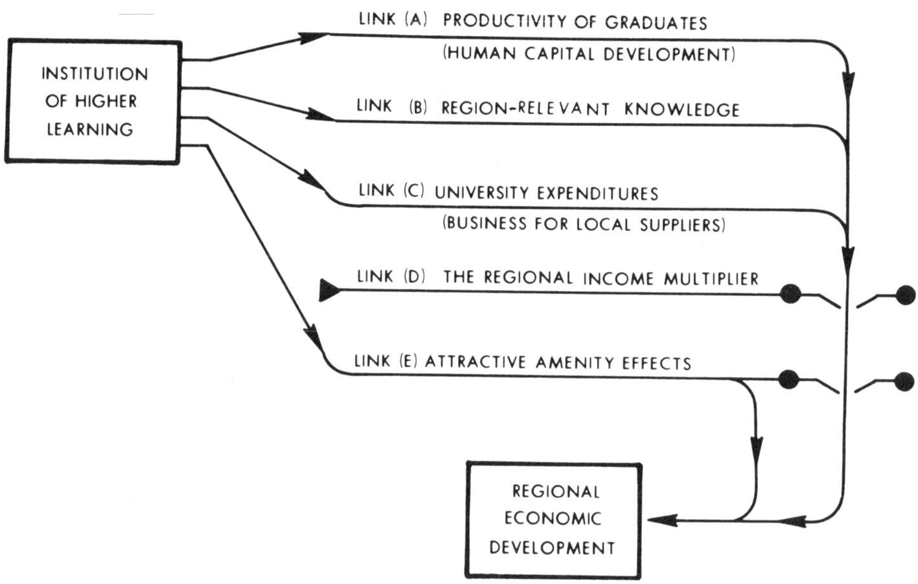

Fig. 1 The Five Link Model of the Economic Role of a Regional IHL
Note: the "gates"(⊸⟜) connected with Links D and E indicate that these mediate (either enhance or dissipate) the effects of Links A, B, C.

Link (A) refers to the production of trained graduates. There is a fairly extensive literature on the economic benefits of training from the standpoint of either manpower development policies (the calculation of skills needed to attain sectoral growth targets), or the analysis of human capital investments (which measures the economic value of education based on the added income it confers on individuals who have more advanced levels of schooling). Such studies almost exclusively refer to Link (A) in our model (See Hüfner, 1968).

So called "employment multiplier" effects may enhance the regional economic value of training programs beyond the direct contribution of the graduates themselves. For example, a University of California study group has asserted, "Many industries know that one good Ph.D. can provide jobs for anywhere from 10 to 100 less highly educated employees. For this reason they are . . . locating new plants near university campuses to take advantage of the talent available . . ." (University of California, 1967 p. 21).

This kind of claim for higher education effects has to be qualified, however. If the "10 to 100 less highly educated employees" are attracted from other jobs within the local region, net gains will be marginal. It is fairly easy to measure new employment created, but less easy to measure the cost of diverting labor from earlier jobs. For this reason, the "employment multiplier" can easily be exaggerated (See Hardin and Borus, 1971; Somers and Woods, 1969). There are two ways in which new employment can be created without the offsetting costs of diverting manpower from other regional uses: one is to bring local unemployed people into the labor market; the other is to attract labor from outside the region—an effective and legal form of piracy, since human capital can be imported across most regional boundaries with no compensation to the exporting region.

A strategy that focused on hiring of the unemployed might have some extra side benefits, such as reduced welfare payments, a fairer redistribution of income, and a sharper sense of regional identity by the local inhabitants. To be sure, past experience has shown that regional unemployment is often a function of many factors besides the unavailability of jobs and lack of training (Cumberland, 1971; Somers and Woods, 1969, pp. 119-51). In addition, job-creation and training programs have to be carefully integrated in order to prevent trainees from taking their skills elsewhere (U.S. Department of Commerce, 1964, p. 31; Luytjes, 1971). This is not to suggest that full employment strategies are impossibly difficult; on the contrary it is to suggest that a regional university, more than almost any other regional industry, is in a position to consciously orient its employment-creation effects in directions consistent with regional goals. This is one of the major points I wish to make in this paper.

The other strategy—pirating manpower away from other regions—is consistent with traditional economic philosophy which says that full mobility of resources will help to insure their most efficient use. Here, an important point needs to be made about the quality of labor imported. The motivation to import labor often stems from the fact that it is cheap. The question is, in whose interests? Employers of imported labor will benefit, but average productivity levels will decline to the extent that the marginal productivity of newcomers is below the average level of the region's initial labor force. This would be contrary to any regional goal seeking to raise average levels of income. It follows that when a regional university trains a Ph.D. who will attract "10 to 100 less highly educated employees" into the region, a truly regional role for the university would demand that it undertake an entreprenurial role in developing the kind of economic activities that are both consistent with regional goals and attractive means of employment for local graduates.

The idea of "pirating" human resources works both ways. It may be true that an IHL can create significant employment multiplier effects this way, but it can also lose many of its own graduates through the same process in reverse. This will be a problem particularly in lesser-developed regions, where the idea of a "regional" university is most attractive from the standpoint of stimulating growth (Messner, 1969; Klaassen, 1968).

Few studies adjust the calculated value of higher education for the out-migration factor, because migration represents no loss from the individual's standpoint (he takes his human capital with him), and migration is relatively unimportant from a national standpoint, because there is relatively little movement across national boundaries. In smaller, developing countries, however, outmigration of the most highly trained cadres can be a major problem. The same is true of *regional* level development within the United States, where movements of manpower across regional boundaries is even less restricted.

A study of migration among graduates of a large university was conducted by myself and Marshall Feldman, using current locations of present and past alumni from UCLA.* Migration patterns were

*The UCLA Alumni Association has unusually complete data on its graduates based on active follow-up of alumni in selected field visits. Data is computerized, facilitating

analyzed in terms of six concentric definitions of the UCLA region, the smallest being Westwood (the immediate campus surroundings) and the largest being the U.S.A. outside of California. The "target" region chosen for analysis was "West Los Angeles", (including Westwood), comprising concentric regions members one and two, as shown in Table 1. The table shows an expected diffusion of alumni out of the region over time, but as people get older they tend to settle down (see Lowry, 1966, p. 27), and the rate of diffusion seems to diminish until it settles on a steady state after about 15 years. Table 2 shows that the proportion of alumni settling within West Los Angeles is the same for those who graduated in 1955-59 (15-17 percent) as it was for those who graduated nearly fifty years ago.**

Over time, the rate of outward diffusion may change, but the dominant changes tend to cancel each other out: each generation is more mobile than the last, and the general sprawl of greater Los Angeles means a growing proportion of students (and jobs after graduation) within accessible commuting distance *outside* of the immediate West Los Angeles "region" as we have defined it. On the other hand, there is increasing development of business in the Westwood area and a trend toward higher residential densities in West Los Angeles, reflecting both national shifts of population into the Southern California area, and the relative attractiveness of West Los Angeles within the larger area, with local access to beaches, the Santa Monica foothills, and entertainment, not to mention the University itself. (The amenity effect of IHLs will be discussed later).

If we take ten years as a reasonable span for looking at where a "regional" university's alumni are, the UCLA data tells us that less than three percent are in the immediate locality (see Table 1, Col. 1, for the 1960-64 cohort). About a fifth are in the easily commutable region (in this case, West Los Angeles with a population of about one-half million). Two thirds are in the larger metropolitan region (Los Angeles County—population about 7 million); eighty percent in the State (California, with about 20 million); and the other twenty percent elsewhere.

analysis. Our study traced alumni movement using zip codes, with names and streets removed to protect individual identies. Our data excluded a small fraction of alumni living outside the continental U.S.A.—about one percent.

**This long-run "stable" pattern probably conceals offsetting secular trends: earlier alumni were likely drawn from a mere local population, before UCLA's national reputation developed; this would make them settle locally in greater proportion, hiding the fact that they may have continued their secular draft away from West Los Angeles.

Table 1 Percentage Distribution of UCLA Graduates by Concentric Regions as of 1972.*

	REGION					
	(1) Westwood	(2) West L.A.	(3) L.A. City	(4) L.A. County	(5) California	(6) Total USA
Population of Region (1970) (Cumulative)**	3.6×10^4	5.1×10^5	2.8×10^6	7×10^6	2×10^7	2×10^8
1970-71 *Alumni Distribution* –cumulative –marginal*** (n=6848)†	6.4 6.4	35.2 28.8	47.4 12.2	78.1 30.7	94.4 16.3	100.0 5.6
1960-64 *Alumni Distribution* –cumulative –marginal (n=11,980)	2.5 2.5	19.5 17.0	31.1 11.6	64.2 33.1	81.4 17.2	100.0 18.6
1950-54 *Alumni Distribution* –cumulative –marginal (n=9,423)	1.8 1.8	15.6 13.8	24.4 8.8	59.1 34.7	81.8 22.7	100 18.2

*Hudson, Appendix II-5.
**"Cumulative" denotes the entire alumni population in the region.
***"Marginal" refers only to the increment in the number of alumni added by moving from one region to the next larger region.

†The sample(n) does not include persons with unreliable addresses. This may bias the sample slightly in the direction of exaggerating the proportion of graduates living in the local area, but the size of error is unknown.

Two things must be kept in mind when interpreting UCLA as a "regional" university. First, for *any* university, the definition of regional role clearly depends a great deal on how the boundaries of the region are to be defined. This is especially clear in the case of alumni distribution but the problem applies to other university linkages with

Table 2 Location of UCLA Graduates by Year of Graduation as of 1972 (Percentages)

	Graduating Class (Grouped Cohorts)							
	1920-29	30-39	40-49	50-54	55-59	60-64	65-69	70-71
No. in sample	2,110	5,768	9,696	9,423	11,666	11,980	19,427	6,848
Present location:								
West Los Angeles	15.5	17.3	17.9	15.6	16.7	19.5	25.6	35.2
Outside West L.A.	84.5	82.7	82.1	84.4	83.3	80.5	74.4	64.8
Total	100.0	100.0	100.0	100.0	100.0	100.0	100.0	100.0

the economy as well. Secondly, UCLA is not typical of many regional universities in that it is located within a fast-growing metropolitan area.

In carrying out our investigation into the spatial distribution of UCLA graduates, we were interested not only where alumni ended up, but in the *net* shift of residence that took place, taking into account the fact that some students did not come originally from the local area. We had two competing hypotheses about the effect of the university as a "portal" through which young people would pass and consequently change their mind about place of residence: on the one hand, the IHL could introduce students from other regions to the locality where they might settle, bringing with them the human capital they had accumulated in their earlier schooling. On the other hand, a university might have the effect of providing a portal *out* of the region, by broadening the horizons of local residents and enhancing accessibility to the outside world.

Our data showed that neither effect seemed significant, or to the extent they were present, they cancelled each other out. This was true for undergraduates as well as graduates. In the late 1960's, for example, about 77 percent of the students accepted to UCLA graduate studies were from Los Angeles County, while a roughly equal number—75 percent—of the alumni from the same period were living in the County. Among graduate students, the figures were 64 percent in each case.

There is reason to believe that this finding would not be typical for many regions. Particularly among advanced students in slower-growing regions, there may be a strong tendency for the "portal" effect to work in the direction of out-migration. We know that at the Ph.D. level, there are significant net shifts at the national level between the location in which a person receives his doctorate and the place where he gets his first job. Data from the early 1960's shows that the West had about 20 percent more people in first post-doctoral jobs than numbers of Ph.Ds it produced; the South had at least 80 percent more; New England, on the other hand, *exported* about a third of its Ph.Ds and the Midwest about a quarter (See Shimshoni, 1968 p. 24).

Perhaps, then, the most important lesson from the UCLA data is not to be drawn from the numbers themselves, but from the factors that need to be taken into account for their interpretation.

Two types of considerations emerge. *First*, the effect of an IHL probably cannot be generalized unless explicit allowance is made for the size and economic diversity of the region it is intended to serve and the size and diversity of the IHL itself.

Secondly, if a university begins to think about a specifically regional role and does not remain indifferent to the problems and opportunities of resources flowing in and out across the region's boundaries, then it clearly implies the need to think in terms of competition with the rest of the world for the things it produces and for the things it can attract, apart from the resources that keep it going.

In these ways, regional higher education policy has dimensions quite different from policy conceived at higher levels and different from the sum of individual perceptions about what is to be gained from higher education. The regional role is distinguished by the fact that the region is an open system. Link A in our model suggests the need for policy intervention to keep graduates in the region. It also requires strategies to maximize the employment multiplier effects in ways that reduce unemployment and promote the in-migration of higher—rather than lower—productivity skills. In short, in a regional role, an IHL takes on the job of planning and entrepreneurship for economic activities, not simply the production of skills.

Link (B) refers to the knowledge generated by a local IHL and its effects on raising the productivity of the regional economy. The economic value of knowledge has been estimated in several studies, but findings are somewhat unreliable because they require some tenuous assumptions about the way "knowledge" is defined, the processes behind its diffusion, and the way its effects are valued in monetary terms.*

A crude measure of the effect of knowledge would be the increase in economic productivity of a region, after holding certain other things constant. Michael Rodell and I have attempted to measure the overall effect of an IHL in regional productivity at the county level, using 24 cases of large universities and a control sample of similar counties in the same States.** To the extent that IHL presence is associated with higher regional productivity, we can say only very tentatively that this was due to the diffusion of locally generated knowledge. The effect may have come through the IHL, but through other links than knowledge—for example human capital (Link A), the attraction of progressive-minded businessmen seeking advancement of their children's education, or other amenity effects (Link E). Nevertheless, to the extent that any relationship can be found, it makes the existence of Link B—the contribution of knowledge—somewhat more plausible. An even more basic question this addresses is whether an IHL seems to have any effect at all.

We included in our sample only IHLs with enrollments greater than 14,500 in 1970, and excluded cases where more than one major university was present in the county, in order to isolate the effects of single IHL campuses. We took as a proxy for the IHL's contribution of knowledge the *change* in student enrollments during the decade of the 1940's, and measured the impact in terms of changes in productivity during the subsequent decade of the 1950's.

Productivity effects were assessed both in the agricultural sector (changes in sales per acre of farmland) and in the manufacturing sector (changes in value added per worker). Separate linear regression equations were measured for agricultural and manufacturing, with results shown in Table 3.

*For example, Denison (1968) estimated that about 20 percent of the increase in U.S. national product (and 40 percent of the per capita income gain) between 1929 and 1957 was due to the "advancement of knowledge". But he notes that a broader definition of knowledge would suggest that it plays an even greater role in economic development (1968, p. 319).

**Two control counties were used for each test case. Counties were matched on the basis of population size and income level in the initial period. The test and control groups were scanned for other statistically significant differences besides the presence of an IHL: counties with universities tended to have more rapid growths of urban population in the 1940's (r=.34). The university counties also had a greater proportion of adults with a college education, and fewer acres in farming in 1949; neither of these two variables entered the regression equations with any degree of significance, however. See Table 3.

Table 3 County Productivity and Regional IHL*

Agricultural Equation

$$(1)\ \Delta A = 3.7 + \underset{(.04)}{.31}\,A + \underset{(3.9)}{3.6}\,IHL + \underset{(.0006)}{.0004}\,\Delta IHL - \underset{(.07)}{.04}\,\Delta U$$

Degrees of freedom = (4,67), F = 10.29 (significant at P = .05)

Manufacturing Equation

$$(2)\ \Delta M = 2746 + \underset{(.10)}{.19}\,\Delta IHL - \underset{23.4}{22.6}\,\Delta U - \underset{(17.4)}{15.1}M + \underset{(252)}{238}\,IHL$$

Degrees of freedom = (4,63), F = 0.699 (not significant)

Dependent variables

Δ A = Change in agricultural sales per acre in farmland between 1949 and 1959, in 1949 dollars.

Δ M = Change in manufacturing value added per production worker between 1947 and 1958, in 1949 dollars

Independent variables

Δ IHL = Change in student enrollments from 1939 to 1949 (This is the policy variable of major interest, a proxy for the degree of knowledge transmitted to the local region)

IHL = a dummy variable, given a value of *zero* for non-university counties and *one* for university counties

A = initial (1949) level of agricultural sales per acre

M = initial (1947) level of value added per worker in manufacturing

Δ U = changes in urban population between 1940 and 1950. (This was the only "contextual" variable that entered significantly in any of the regression equations estimated.

*The numbers in parentheses are the standard errors. When these exceed the value of the coefficient, results are unreliable.

Our analysis suggested only a weak link between the presence of an IHL and the productivity of the county region. The regression equations shown in Table 3 represent a pooled sample of data for both the counties with and without a local IHL. Figures in parenthesis are standard errors. In the *agricultural equation,* the signs were in the expected direction, the equation as a whole was significant at the .05 level, and the IHL variables entered early in the stepwise regression. Nevertheless, the standard errors of the IHL variables were slightly greater than the value of the coefficients themselves, making inference risky. A more refined model might show the IHL variables as significant in their relationship to growth of agricultural productivity but the data used here does not establish that fact.

In the *manufacturing equation*, the growth of enrollments was more significant in terms of having a relatively low standard error for the coefficient but the overall equation was not significant, again failing to substantiate our hypothesis about IHL effects.

The weakness of these relationships could conceivably be explained by our study design. We chose large universities and looked at their impact in fairly local terms, at the county level. Far from providing a "strong" input, the size of the IHLs sampled may have meant that the county comprised only a small faction of their service area, providing a weak measure of their true effect. If this is true, it presents a dilemma, because it becomes increasingly difficult to measure the effects of higher education in broader-defined regions. We infer this from two supplementary studies carried out in connection with this research. In the first study, we surveyed both State and local Chambers of Commerce throughout the country in regard to their *perceptions* about the economic role of IHLs: it clearly emerged that IHL effects—to the extent they were discerned as effective promoters of economic activity—were seen to operate almost entirely at the local level and very little at the State level.*

In the second study, we used multiple regression analysis to measure the relationships between higher education inputs and economic activity at the State level. Although we found some significant relations, we found it impossible to draw any meaningful inferences from them. In the first place, higher education expenditures represent only about one percent of economic activity in the nation at large, and it would require an extremely delicate instrument to distinguish these effects from the other possible determinants of State-level economic activity.

Secondly, even if we could find a statistically significant relationship *with everything else held constant*, we would likely eliminate from consideration the very processes by which IHL makes its effect. For example, we would want to control for levels of investment by non-IHL industries, but a local IHL may operate on the local economy precisely

*While they are not usually staffed by social scientists, Chambers of Commerce are in the business of promoting economic activity, and assessing what contributes to it. We therefore regard their opinions as useful. Our Chamber of Commerce survey is described more fully in connection with Link E.

in terms of stimulating such investment (as the University of California Report suggested in connection with Link A). In short, *regional infrastructure* is both an effect of an IHL (a dependent variable) and at the same time something we want to hold "constant" in order to measure the "pure" effect of higher education. There seems little way of handling this difficulty, which is a conceptual problem apart from being a statistical one.

A third problem is the likelihood of reverse causation: even if a reliable measure of association could be established between higher education and State-level economic activity, cause-and-effect probably works in both directions: a rich State generates better education as well as vice versa. It does not help to employ lead-lag analysis to sort out these effects, when both variables are growing together in fairly linear fashion over time. Here is a typical case of so-called "second cybernetics," involving deviation-amplifying mutual causal processes. Where such causal loops occur, it is no longer possible to statistically isolate the magnitude of one variable's effect on another (Maruyama, 1963; Wright, 1971, p. 105).

These three difficulties involved in measuring the one-way effect of an IHL upon a regional economy can be made more graphic by presenting a revised version of our five-link model to include three additional mechanisms which are implicit in the entire analysis. Figure 2 includes an additional Link (F) representing the effects of *regional infrastructure* as a mediator in the transmission of IHL effects to the local economy: initially, an IHL may have a real effect in promoting the development of an infrastructure, but once in place, the infrastructure has an effect of its own that cannot be attributed to higher education; for in subsequent periods the IHL could conceivably be removed without dimishing the infrastructure's effects.

An additional mechanism is shown in Link (G), representing the *reciprocal of effects* of regional development upon IHL growth. As already mentioned, this makes it difficult to sort out statistically the opposite effect of higher education investments upon regional development, which is the critical question for regional development strategy.

A third link (H) suggests the effect of spurious factors. There are really two problems here. One is the difficulty of real IHL effects being statistically swamped by other factors affecting regional development, given that IHL activities are such a small portion of the whole and are easily hidden by the random "noise" and by problems

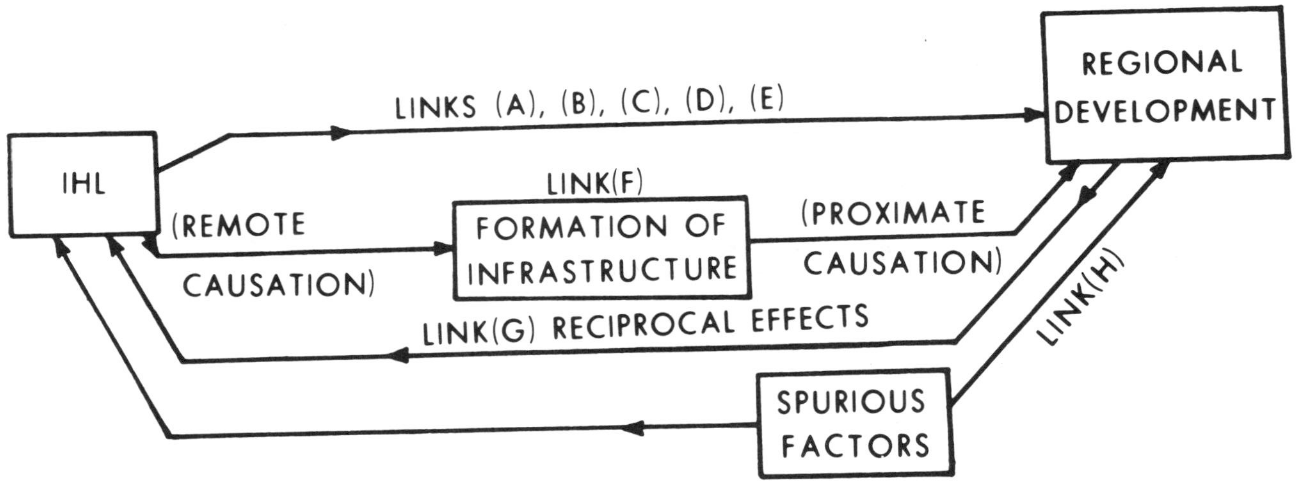

Fig. 2 Processes Implicit in the Five-Link Model

of measurement elsewhere in the system.* The other problem suggested by Link (H) is the possibility of spurious correlation between IHL growth and regional development, involving no causal link between the two, but a mutual relation to a third set of variables. For example, the statistical association between higher education and regional development may stem from the fact that they are both influenced by such things as a high level of "need for achievement" in the local population, or by initial endowments of wealth in the region, or by local amenities that attract higher income people into the area. The first problem means that real effects cannot be measured in the presence of excessive noise; the second problem raises the opposite possibility of wrongly inferring causality to relationships that *can* be measured. Link (H) depicts both of these problems created by spurious variables that make it difficult to measure IHL effects on regional development at the State level.

The less we can take data results at face value, the more we have to rely on a strong theoretical framework for interpreting empirical findings. This brings us back to the two equations from Table 3. One thing the equations suggest is that IHL effects, though weak, seem to be somewhat different in the agriculture and manufacturing sectors. This gives rise to the possibility that perhaps knowledge serves these sectors in different ways. Or rather, the type of knowledge involved may be different.

In the case of manufacturing, knowledge requirements are basically related to the science and technology of the production process, with little consideration needed of environmental variables. In this case, a local IHL would need to have a particular strength in some technical area in order to give the region a comparative advantage over other regions. It is unlikely that many universities could provide a significant competitive edge on this basis.

In agriculture, however, enviromental variables play a much greater role, and a different sort of knowledge is required. Differences in local climate, geological conditions, soil types, farming traditions, personal wealth, and other factors could have a major effect in determining "optimal" farming practices. Raising agricultural productivity would take not only general knowledge of a production technology,

*Morganstern's famous study *On the Accuracy of Economic Observations* (1963) points out that data on levels of economic activity are generally in error on the order of 10-15 percent (1963, p. 9).

but *specific knowledge about the locality,* as might be provided by a regional IHL or by a system of extension agents.

This suggests a possible distinction between two types of knowledge that a regional IHL might provide. Each would act in a different way on regional development and imply quite different academic functions and structures for a regional IHL. (1) *Sector-specialized knowledge* would consist of R&D activity within a particular field, providing a competitive advantage in local industry and support for an export base in the regional economy. (2) *Regionally-specialized* knowledge, on the other hand, relates to development programs and strategies outside the export-base sector. This would imply a different type of role for the IHL, emphasizing the generalization of knowledge about the region itself in terms of its particular needs, development potential, and eligibility for public and private investments from outside the region, as well as collation, interpretation and synthesis of knowledge at large for purposes of application to specific development programs.

Sector-specialized knowledge may have limited applications for regional development, despite some dramatic examples of knowledge-based industries growing up around major universities, for example in the cases of Route 128 and Palo Alto. In the first place, academic knowledge diffuses quite readily across regions, so that the local presence of an IHL would provide only a marginal comparative advantage for local industry in this respect. Secondly, closer examination of development experiences like those at Route 128 and Palo Alto show that other relatively unique factors were at work. In Boston, for example, the concentration of massive federal R&D programs during World War II and a banking community sympathetic to research ventures were significant determinants of the Route 128 development (Shimshoni, 1968). In Palo Alto, environmental amenities and some extraordinary leadership from within Stanford University played major roles. Third, where knowledge is very important to an industry's comparative advantage (as in defense research), economies of scale tend to foster agglomeration of efforts, putting a limit on the number of regional centers that one nation could support. Fourth, sectoral knowledge would have its greatest potential as an independent influence over location decisions in the case of those industries with rapidly developing technologies and a "footloose" independence from markets, suppliers, and labor markets. An example would be the pharmaceutical industry in Kalamazoo, which is closely associated with the local campus of the University of Western Michigan. It is important to note that an increasing portion of American economic activity seems to be in this "footloose" category (Stanbeck and Knight, 1970, pp. 229-40). On the other hand, a local IHL would have to provide a superior quality of specialized knowledge if it were to offer a substantial competitive advantage for the region vis-a-vis other IHL in other regions. This would be difficult for a younger university to provide.

More likely, this kind of symbiosis would tend to derive from the parallel evolution of certain lines of specialization between universities and nuclear industries already in place. It is less easy to conceive of an IHL *initiating* sector-specialized programs on the gamble that appropriate industries would show up to take advantage of this source of knowledge. Rather, the pattern would probably follow the example of Beersheba University in Israel. In this case, a major industrial complex was developed rapidly with the intention of deconcentrating economic activities in the country. Close ties presently exist between the University and local industry. The important thing to note, however, is that the University and its programs were the *consequence* of industrial development rather than its cause. It was founded after the initial starting-up phase of industrial development in order to hold high-level manpower in the region that might have otherwise drifted back to more established centers.

Regionally-specialized knowledge stands in contrast to sector-specialized knowledge in that it focuses on needs of the region, as opposed to the needs of a particular economic sector. This would imply a regional role for an IHL quite distinct from traditional university functions. Much of this knowledge would relate to activities in the not-for-profit sector: knowledge about the availability of government programs in health, education, and area redevelopment; knowledge about development experiences in other regions (or other countries) that have special applicability to local conditions; knowledge about emerging national priorities and their implications for spatial distribution of government spending, together with conditions for determining local eligibility for funded programs of this nature.

What is implied here is the enhancement of a region's capacity to attract outside resources, many of them in non-traditional programs associated with the public rather than the private sector. Past examples might be the federal sponsorship of agricultural sciences development through the land-grant colleges, or the public works programs of the 1930's, or the area redevelopment programs of the 1960's. Past efforts in publically-funded regional development in the

U.S. have been characterized more by physical than social planning (Cumberland, 1971). In future, one can anticipate a more sophisticated science of regional development based on systematic social experimentation as was done earlier in the agricultural sector.

The resulting policy implications for higher education were alluded to in the first section of this paper: (a) On the *research* side, it suggests the need for case studies of IHL regional roles, as opposed to aggregate models which tend to blur the importance of variables describing the regional environment. (b) On the side of *social action*, a greater emphasis seems warranted on the university's direct involvement in planning and executing regional development strategies, as opposed to passively responding to market demands. (c) Moreover, the University's *entrepreneurial role* in attracting economic activity by the development of sector-specialized knowledge would seem to be limited to cases where IHL resources were concentrated heavily in a particular speciality and where other regional amenities were present to attract footloose industries (something to be discussed later under Link E). (d) *Regionally specialized knowledge*, however, might provide the capacity to mobilize public funds for the region's use even in the absence of a clear economic advantage for location of private industry. (e) In view of the apparently weak effect of IHL's on local economies, fairly sharp *departures from past educational practice* would likely be needed in order for higher education to serve as a major influence on the productivity of a local region—at least through the mechanism described by Link B in our model. This is not to say that the link is inherently weak; it may simply mean that IHL policy makers have been slow to exploit its full potential.

Link (C) refers to effects of IHL-related expenditures on the local economy. These expenditures might consist of construction outlays, staff salaries, spending by staff and students, consumption of equipment, and other purchases of locally supplied goods and services. In economic terms, these are the institution's "backward linkages" to its sources of supply, in contrast to its forward linkages to the industries which consume the IHL's output. Insofar as the inputs to the educational production function will be very roughly equal to the outputs in terms of economic value, the backward and forward linkages will be in principle roughly of the same order of magnitude. In other words, the economic stimulus that an IHL produces in terms of business and jobs for suppliers (Link C) will be roughly the same as the value of the IHL's output represented by Links (A) and (B). Much of the recent literature on regional economic impacts of higher education institutions refer exclusively to Link (C), or the effects of an IHL as a source of business for local suppliers. Two studies which review the literature in this vein are Caffrey and Isaacs (1971) and Strang (1971). Most such studies give little recognition to the regional economic effects of Links (A) and (B), but give greater emphasis to (D)—multiplier effects, and passing reference to (E)—amenity effects.

From the regional standpoint, of course, it matters a good deal whether suppliers are local or outsiders: most salaries will be paid locally, but other operating costs may not. Just how much leeway university administrators have in choosing suppliers is unclear, but it probably depends on the institution's sense of identity with the region and its willingness to take an entrepreneurial role in the development of viable local suppliers. If this appears inconsistent with traditional university practices, it reflects the traditional lack of concern for specifically *regional* consequences of IHL operations. Once the region becomes a salient unit, the operational significance of backward linkages becomes clear, and they become more obviously a matter for policy intervention just as the forward linkages have always been.

In practice, backward linkages of IHLs seem to come to public attention mainly when they are operating contrary to local interest groups. For example, IHLs may in some cases consume local public services and utilities without adequate compensation to the community, particularly since educational property is often exempt from taxes, and because students pay little personal income tax. Another salient aspect of university backward linkage effects concerns the consumption of land by expanding urban campuses. The encroachment of the university upon low income neighborhoods was a catalyzing issue for student unrest in the late 1960's, notably at Berkeley, Columbia and Harvard. Moreover, the housing demand of a university population affects not only occupancy rates but prices for housing elsewhere, and this affects the social economic composition of the population far beyond the numbers directly displaced by IHL activities themselves. Whether these people-displacing effects of backward linkage are counted as benefits or costs depends on one's political priorities in regard to the raising of average incomes at the expense of the relatively poor.*

*Hansen and Weisbrod's study of the University of California (1969, Chap. 4) showed that contrary to the original intentions of providing greater equality of opportunity, the

On the more positive side, university-related expenditures have certain advantages over other industries, as an economic base for a locality. In the first place, for every dollar that goes into the university's operating budget, almost twice that amount flows into the local economy from student and faculty spending.** In addition, the demand for IHL services is generally more stable than for most other industries.

The net effect of IHL-related expenditures depends on the source of financing—a factor that few studies deal with very explicitly. If higher education is financed from non-local sources, and if students come from outside the locality, there is a clear net gain for the region. On the other hand, if spending comes out of local pockets, it is necessary to consider whether that money would have been spent locally just as well in the absence of an IHL. Would college-age students have taken their money and spent it elsewhere? Could the money invested in development of an IHL facility have been invested in some other local industry? How much income, earning power, and spending power were sacrificed by students who withdrew from the local labor market to further their education?

These questions suggest a rather narrow-minded view of education. But right now, we are only dealing with the economics of the case, from a regional perspective. And from an economic standpoint, the question of income foregone while studying is extremely significant. Blitz (1962) has estimated that income foregone by students at the higher education level are roughly 240 percent larger than the cost of education as measured in IHL ledgers themselves. In principle, the loss is recaptured in the long run by the greater productivity of graduates, but there are two major qualifications to this that each region must take into account in calculating the economic value of investing in higher education. First, future benefits have to be discounted, and the benefit-cost ratio is extremely sensitive to alternative assumptions about the proper social discount rate: above a 6 percent discount rate, the rate of returns to educational investment often diminish rapidly (Houthakker, 1959). Secondly, the assumption that present in-

system has a net regressive effect on income distribution, taking into account that subsidized students come from higher income families on average than those who supported the system through taxes.

**Based on a study by Ira Fink of the Berkeley and Santa Cruz campuses of the University of California, and their local economic impacts (1967, pp. 6, 8).

comes foregone in the locality will be locally repaid by higher future incomes depends on the willingness of graduates to stay in the region—which may not always be the case.

A final point about backward linkages is that their benefits for a local economy depend on the region's absorptive capacity, which depends on the area's size and diversity. *In the short run*, the effects of IHL expansion might well be stronger in less developed regions: whereas a mature region might absorb new business without much expansion of overhead, a newly developing region would show more pronounced economic "accelerator" effects, consisting of the build-up of new infrastructure for public utilities and commercial suppliers. A study of campuses in the University of California system, for example, showed that at least in the short run, the new campus in the sparsely settled area of Irvine brought about a larger local investment of commercial infrastructure than did the older urbanized campus at Berkeley.

In the long run, however, a large, diversified local economy—say, that of a middle size city—could provide for a large portion of an IHL's normal operating needs, and absorb more of its business. This may run counter to intuition, because the economic effects of an IHL are more visible and more easily documented in a small "university town." Nevertheless, college-dominated communities probably benefit less per dollar of educational expenditures than larger cities. This point should become clearer in connection with the discussion of income multiplier effects which comprise Link D.

Link D represents the economic multiplier effects associated with IHL-initiated investments. When an exogenous mechanism, including any of the Links (A), (B), (C), or (E), results in a net increase in local income, that increment will be respent, raising other people's incomes. The total gain in local income depends primarily on how much of the respending is absorbed by local suppliers versus the amount that leaks out of the region to business on the outside. In general, the smaller and less self-sufficient the region, the smaller the multiplier. The larger the region, the greater the total cycle of respending is recaptured by the region, and the larger the multiplier. In the largest of regions—say, the United States as a whole—the multiplier reaches a magnitude of about three fold the size of the initial investment; at the state level, the multiplier is lightly less, perhaps 2.5-3.01; in a small town the multiplier would approach 1.0, or simply the value of the initial investment itself, with virtually all of the responding taking place on outside commercial and industrial suppliers.*

As shown in Fig. 1, the size of the multiplier mediates the ultimate contribution of higher education to the regional economy and has an important bearing on measuring the total economic impact of IHL investments. Unlike the previous three links, the size of the multiplier is generally not under the control of IHL policy, being more a reflection of the economic context. On the other hand, it is conceiveable that in a relatively small region, a conscious policy of an IHL might be to build up the economic self-sufficiency of the locality, enhancing the multiplier in the same way as it might enhance the strength of Link (C) through the creation of local suppliers.**

Finally, Link (E) consists of the amenity effects of a local college or university. This refers to the attraction of human and capital resources into the region due to the presence of IHL facilities. This is different from the other four links described by the model, in that the resources generated for the region are entirely external to the IHL activities themselves. Links (A) and (B) transmit direct products of the IHL (human capital and knowledge), while (C) and (D) represent secondary effects of resources which flow through the IHL system (direct expenditures and subsequent recirculation of these resources released into the local economy). In contrast, the amenity effect (Link E) refers to resources supplies *outside* the IHL system, due to other activities locating nearby because of the presence of higher education.

Loosely speaking, the amenity effect is a "multiplier" of its own. The educational opportunities an IHL confers upon a college freshmen may be the factor that tips the father-businessman's decision to set up shop in the immediate area. A professional consulting firm

*The size of the multiplier depends on both the proportion of income spent outside the area, and also the marginal propensity to consume (as opposed to save) increments to income. See Caffrey and Isaacs, 1971, Appendix B, and Isard and Czamanski, 1965.

**The policy of "import substitution" is common among developing countries attempting to create their own industries to compete with foreign imports and thereby create jobs and retain wealth within the country's own boundaries. The basic principal also applies to regional development, but with certain qualifications: the strategy would have to concentrate on supply sectors which could operate fairly efficiently on a small scale (since local suppliers could not begin to match the efficiency of national industries where major economies of scale were operating); it would also make most sense to develop local suppliers in cases where otherwise unexploited regional resources could be mobilized—particularly unemployed or underemployed manpower.

may have a manager who likes the mountains, but could only justify a move to Boulder because of the intellectual community that the University of Colorado's campus confers on the area. A Federal Model Cities grant might go to a particular corner of urban America because a local graduate school has an interest and capacity for evaluative research on community development programs.

The very fact that an IHL has academic and cultural ties with a network of similar institutions around the country (including a good football team) may put on the map a local community that might have otherwise remained anonymous to potential investors. Professional service industries may develop faster in IHL communities because of the available labor pool of student and faculty wives; among other things, this skill pool supply could be tapped to upgrade the quality of education at primary and secondary levels, further enhancing the attractiveness of the region to outsiders.*

Amenity effects have been little studied in the literature (with the notable exception of Leo Klaassen's *Social Amenities in Economic Growth*, 1968). There are perhaps three major reasons for this. First, there are few amenity effects that occur in "pure" form, except perhaps the "cultural magnetism" of an IHL. This might be measured by numbers of visitors to the area (mentioned, for example, by Messner, 1969, p. 17; and by Caffrey and Isaacs, 1971, Appendix 3, pp. 67-70). But what about the effect of the cultural "magnet" on more permanent decisions for relocation? Here, it becomes difficult to separate amenity effects from the other linkages. For example:

—the "portal effect" we talked about in connection with Link (A)—the in-migration of non-local students who subsequently take up residence—represents an attractive amenity effect of the IHL as well as its capacity to produce human capital.

—the use of university faculty members in local business and public services—ordinarily considered under Link (B) as transmission of knowledge and skills—becomes an IHL-provided amenity if the services are uncompensated, or if they are instrumental in attracting other resources to the area.**

*These last points—which would also apply to husbands of students and faculty women—was suggested by Professor Frank Thomas, Chairman of the Department of Geography at Southern Illinois University at Carbondale. Thomas has been studying the regional influence of SIU, and feels that campus-associated wives in particular have had the effect of "very considerably upgrading the educational system" in Carbondale. (Personal communication, 23 August 1971).

**Outstanding examples of this are (a) Beersheba University, designed for the purpose

The location of business in a university town may partly reflect Link (C)—the backward linkages of IHL expenditures—but if the business is, say, the head office of a firm serving a much larger area, the effect of the IHL will have been to attract economic activity on a scale considerably larger than is needed to provide service for the IHL population itself; the effect of the IHL, then, will be partly that of an amenity, according to our definition (economic activities generated by the IHL presence, but not directly comprising the human and capital resource flows provided by the IHL itself).*

The *second* reason why amenity effects are generally not considered as a separate item for investigation is that the most powerful amenities are often unique to an area and shaky as a basis for drawing general conclusions. The quality of life in an intellectual community like Harvard Square, the Latin Quarter in Paris, Oxford, or Berkeley is unreplicable, or at least city planners are only beginning to think seriously about intellectual and cultural life styles as major variables in city design.** Similarly, the establishment of a college or university gives professional people *access* to amenities that cannot in any case be recreated, but are simply there, only waiting for the establishment of an IHL. Colorado, rural New England, Southern California, and Florida are conspicuous examples.

The *third* reason why amenities are often passed up for study is that their worth is highly subjective and variable—a matter of taste which cannot be given a dollar value in the same way as consumables like

of preventing the outflow of high-level manpower after the starting-up phase of Beersheba's industrial development, and largely relying on faculty who work part time in local industry; and (b) the case of SIU at Carbondale, which contributed the Chief Planner for the Carbondale Model Cities Agency, along with the services of graduate student interns (Monty and Thomas, 1970, p. 2-3).

*This principle operates on a grand scale in the case of Federal R&D allocations. In the mid-1960's, California was getting approximately 40 percent of the Federal Government's entire outlays for major R&D programs. A University of California study committee would like to interpret this as follows: "Undoubtedly, the research money goes where the brains are, and brains are where the educational system nurtures them." (University of California, 1967, p. 8). During this same period, over a span of two years, California received $1.5 billion more from the Federal Government than she paid out in taxes, while New York State (with roughly an equal income) lost $4 billion net to the government. (*Ibid.*) At that time, California was planning on spending nearly three times as much as New York State on current costs of higher education ($1.8 vs. $0.6 billion, compared to about $8.6 billion in current outlays for the entire country. These were 1965 estimates projected for 1970.) See *Ibid.*, p. 6.

**Thrusts in this direction can be seen in the work of such people as Elie Bensimon (1971); Ian McHarg (1971); and Christopher Alexander (1969).

entertainment, art, and high fashion. Amenities are too bound up in the "livelihood" issues of location decisions to be easily separated out. This is not just a problem of statistical discrimination but the basic intermingling of motives in persons actually making location decisions. Evidence on whether an IHL's presence is a "decisive" factor therefore tends to be most specific when it is contextual and anecdotal and relates to particular people, places, and moments in history.*

Referring back to Figure 1, the amenity Link (E) is less determinant than the others—not necessarily weaker, but less clear-cut as a cause-and-effect process. It does not represent a flow of resources through the IHL itself, and its effect refers not only to the offerings of higher education but to the presence of *prior* amenities in the local environment whose access is enhanced by the marginal addition of an IHL. In addition, the amenity effect of an IHL is not simply transmitted in terms of physical resources, but in terms of the shaping of people's perceptions about the quality of a local environment, and its *potential* for providing future resources to complement other economic activities. Here, we are dealing in the realm of people's beliefs and judgments, things which cannot be entirely reduced to objective reality.

We decided to explore this area of beliefs and informed judgment about IHL amenity effects by going directly to Chambers of Commerce. The idea was to systematically survey a set of institutions whose function is, precisely, to deal with the operational factors affecting location decisions.**

In January 1973, a survey letter was distributed to the offices of 67 metropolitan and 46 state Chambers of Commerce across the country. To date** we have had responses from 47 metropolitan Chambers of Commerce (70%), and 25 state Chambers of Commerce (56%). At the same time we continue to receive responses from various additional

*An outstanding and thorough study along these lines was carried out by George Feliz (1972) in connection with the role of Regional College campuses in outlying regions of Chile. Among other evidence he gathered, Feliz interviewed 65 "prominent citizens" in the various areas served by the campuses. One citizen summed up the local situation by saying: "The creation of the Center (regional college) has meant a change toward permanence among the population. The people want to stay here because there are educational opportunities" (1972, p. 314).

**Simon Fass, a graduate student in the Urban Planning Program at UCLA, has had a major hand in this part of the study.

***As of 1 March 1973.

local organizations as a result of some state Chambers passing our survey letter along to potential interested bodies.

In general, the contents of the response letters have been enthusiastic—and promotional material appended to the letters quite voluminous. Some tentative conclusions can be offered at this time, but we emphasize that these findings are very preliminary.

—Local (i.e. metropolitan, city and town) Chambers indicate a greater interest in the economic amenity effects of institutions of higher learning than state Chambers.

—Among the local Chambers, the largest (e.g., Jersey City) and smallest cities (e.g., Pocatello, Idaho) indicate lesser effects than intermediate size cities.

—Quantitative data has not been provided in support of any response opinions.

—In order of the frequency of appearance in the response letters, the amenity effects Chambers of Commerce perceive of colleges and universities are as follows:

1. The presence of a university permits the pooling of resources and knowledge between the academic community and the local private sector (e.g. economic development committees composed of university department heads, Chambers of Commerce executives, and leading business and community leaders) in order to arrive at *decision goals and methods of promoting local economic development*. This characteristic is implied or stated explicitly in almost every response—and often appears to be the most important identifiable characteristic.
2. Firms and branches specializing in *research and development* type functions are the ones for which the influence of a university is most important. Here, several roles are attributed to the university including the function of tapping university human and physical resources, or providing for periodic managerial retraining, and to a lesser extent, providing recreational, entertainment and other activities for the management employees of the firm. This is particularly true for the intermediate and larger cities.
3. For the *smaller cities*, the R&D orientation is important as well. However, in these cases the importance of advanced *vocational* and *junior college institutions* is equally and perhaps more important in providing skills and retraining programs for the management and employees of manufacturing firms and branches.

4. In all cases, the availability of *higher educational facilities* for management and employee families is of great importance in the attraction of economic activity. This is particularly true of the "professional" types of occupation.
5. The impact of universities is seen in most cases to be highly localized. Regional amenity impacts tend to be small (but this varies of course with particular regions). This we deduce from the general lack of interest on the part of state Chambers of Commerce. It is also consistent with statistical data provided in Stanbeck and Knight's work on urban and regional growth potentials (1970).
6. At the present, we are attempting to generate frequency counts on the contents of response letters in a detailed fashion. Simultaneously we intend to disaggregate the responses with respect to city size, rate of growth, and nature of the economic region.

Thus far, our investigation into Link (E) offers some tentative conclusions that are consistent with our remarks about the other links: *first*, the effects are difficult to measure; *second*, to the extent we do have indirect evidence about the magnitude of effects, their strength seems to vary considerably with the nature of the region; and *third*, it seems likely that the effects could be enhanced by higher education policy that departed somewhat from tradition. In other words, a stronger role for IHLs in regional development requires not only a better understanding of how these linkages operate, but a willingness to use higher education resources for economic purposes other than traditional education and research.

This much the 'amenity" link has in common with the other linkages. The amenity factor deserves special attention for future investigation, however, not only because so little is known about it, but because it is likely to grow in importance over time. In the first place, the American economy has an increasing proportion of its industries falling into the "footloose" category, meaning they are not tied geographically to particular raw materials, labor pools , or markets (Stanbeck and Knight, 1970, p. 230). The amenity factor will therefore become more decisive in location of economic activity. The same is likely to be true of individuals: as income rises, a lesser value is placed on further economic rewards, and greater importance is given to the quality of life style, the environment, and non-material considerations that are "beyond the reach of economics."* Amenities become corres-

*This is John Kenneth Galbraith's central thesis in *The New Industrial State* (Boston: Houghton Mifflin, 1967).

pondingly more decisive in individual location decisions.

Another way to look at this is to note that in the aggregate production function, *human* capital is becoming an increasingly large ingredient over time, for the U.S. as a whole (Denison, 1962). This has obvious implications for strategies of regional growth—specifically the need for less earth moving equipment and more skills and organizational capacity to mobilize human resources previously underutilized. It also gives greater plausibility to the statement that industry will go where the brains are—where higher education nourishes them. The point is, higher education not only *produces* skills; it also enhances the *demand* for cultural amenities and for better education for the next generation.

CONCLUSIONS

Our major conclusions were cited at the beginning of the paper, but they may be restated now in somewhat different form.

(1) The regional role of higher education involves a quite different set of economic mechanisms than the ones involved in the economics of education from an individual or national standpoint. This is because the region is an open system, with human and capital resources free to flow in and out of the local area. As a result, analysis and policy formation must focus on the specific nature of these flows and ponder ways to intervene in keeping benefits within the region. This is the rationale behind the development of the five-link model, and it explains the importance of the amenity link in particular, since this relates most directly to the spatial question of the location of economic activities.

(2) It is difficult to establish empirically that an IHL has a major economic impact on its surrounding region, although *a priori* arguments for it abound. This is partly a reflection of statistical difficulties in measuring the marginal contribution of education to the larger processes involved in regional development. In part, though, it reflects the point made earlier—that the benefits of higher education easily diffuse out of a specific locality. The specific ways this happens has been described in connection with each of the five links.

(3) It is not the traditional role of higher education to meddle directly in economic processes, but there are a few exceptions which may be the most useful prototypes for defining a specifically *regional* role for an IHL. We have the examples of Route 128 and Palo Alto, where sector-specific knowledge, combined with a strong entrepreneurial push from within major local universities, created dramatic

effects. Under most circumstances, sector-specific knowledge cannot provide a strong comparative advantage for a region, but *regionally*-specific knowledge could become an entrepreneurial focus for mobilizing local resources and attracting economic activity in the not-for-profit sectors based on Federal or State funding. This may include a stronger involvement in policy sciences, perhaps along the lines suggested by recent speculation on a "Fourth Branch" of government (see, for example, Golovin, 1969).

Another, even more attractive prototype for IHL regional roles is the land-grant college, dating from the last century. Starting with a virtually blank slate (agriculture was even less a science in those times than regional development is today), college administrators were forced to send their teachers out into the field to invent a curriculum: it was based on learning about problems first hand from people who lived them; it was based on research designed to yield enough information to have something useful to teach about; and it called on help from non-academic sources, including government at all levels and private business, which could offer investment capital and experience in selling new ideas. Persons trained were not set loose into society at large, but prepared for specific roles as researchers, specialists and extension agents in an environment familiar to them through direct contact in their fieldwork. (See Bowman, 1968).

This is not a model to be accepted at face value. There are major differences between a farm and a region in terms of the problems of development, the ways to examine these problems, and the appropriate scale of attack. The lesson is rather that higher education policy has feasible options of considering the full range of its economic impacts on a region and working to strengthen these various linkages by direct political and economic intervention.

Until now, the literature has given scattered attention to one or another of the links cited in this paper; but the notion of intervening in the mechanisms on behalf of the specific interests of a delimited region tends to be in conflict with the traditional IHL concerns about quality and efficiency throughout. The land-grant college was neither particularly efficient, nor likely to produce the "quality" of output associated with major institutions producing sector-specific knowledge and higher level manpower. At the same time, the land-grant colleges made a difference in the American rural landscape. The fact that our findings show such a weak effect of present-day IHLs on regional development may suggest above all that American higher education has failed to break the old molds and adapt as well to the needs of the Twentieth Century as it succeeded in doing in the past.

BIBLIOGRAPHY

Alexander, Christopher, "Major Changes in Environmental Form Required by Social and Psychological Demands," Second International Symposium on Regional Development (Proceedings) (Tokyo: Japan Center for Area Development Research, 1969), 66-79.

Bensimon, Elie, "Project d'une université urbaine dans le 'Pentagone' bruxellois," *Environment*, 12-1 (Décembre 1970-Janvier 1971), 55-59.

Blitz, Rudolph C., "The Nation's Educational Outlay" in Selma J. Mushkin, ed., *Economics of Higher Education* (Washington, D.C.: Office of Education, U.S. Government Printing Office, 1962), 147-69, 390-403.

Bowman, Mary Jean, "The Land-Grant Colleges and Universities in Human Resource Development," in Bowman *et.al.*, eds., *Readings in the Economics of Education* (Paris: Unesco, 1968), 634-50.

Caffrey, John, and Isaacs, Herbert H., *Estimating the Impact of a College or University on the Local Economy* (Washington, D.C.: American Council on Education, 1971).

Cumberland, John H., *Regional Development Experiences and Prospects in the United States of America* (Paris: Monton & Co., 1971).

Denison, E. F., "Measuring the Contribution of Education and the "Residual" to Economic Growth," in Mary Jean Bowman *et.al.*, (eds.) *Readings in the Economics of Education* (Paris: Unesco, 1968), 315-37.

Denison, E. F., *The Sources of Economic Growth in the United States and the Alternative Before Us*, A Supplementary Paper for the Committee for Economic Development (New York: Committee for Economic Development, 1962).

Feliz, George C., *The Regional University Centers: Innovation in Chile* (Berkeley: University of California Center for Research and Development in Higher Education, 1972).

Fink, Ira Stephen, *The Community Impact of the University of California's Berkeley and Santa Cruz Campuses* (Berkeley: Office of the Vice President, University of California, April 1967).

Galbraith, John K., *The New Industrial State* (Boston: Houghton Mifflin, 1967).

Hansen, W. Lee, and Weisbrod, Burton A., *Benefits, Costs, and Finance of Public Higher Education* (Chicago: Markham Publishing Company, 1969).

Hardin, Finar and Borus, Michael E., *The Economic Benefits and Costs of Retraining* (Lexington, Mass.: Heath Lexington Books, 1971).

Houthakker, H. S., "Education and Income", *Review of Economics and Statistics*, 41 (February 1959), 24-28.

Hüfer, Klaus, "Economics of Higher Education and Educational Planning—A Bibliography," *Socio-Economic Planning Sciences* 2:1 (October 1968), 25-101.

Isard, Walter and Stanislaw Dzamanski, "Techniques for Estimating Local and Regional Multiplier Effects of Changes in the Level of Major Governmental Programs" (Ithaca, New York: Cornell University Center for Housing and Environmental Studies, Article Reprints #23), (Reprinted from Peace Research Society Papers, III, Chicago Conference 1965).

Klaassen, Leo H., *Social Amenities in Economic Growth* (Paris: OECD, 1968).

Lowry, Ira S., *Migration and Metropolitan Growth* (San Francisco: Chandler Publishing Co., 1966).

Luytjes, Jan B., "An Note on the Impact of Increased Educational Funds in Lagging Areas", *Growth and Change* 3:1 (January 1971), 38-41.

Maruyama, Magoroh, "The Second Cybernetics: Deviation-Amplifying Mutual Causal Processes" *American Scientist*, 51 (1963): 164-179.

McHarg, Ian L., *Design with Nature* (Garden City, New Jersey: Doubleday and Co., Inc., 1971).

Messner, Stephen D., "Market Analysis for the Single Industry Community: An Illustrative Study of a University Town" (Storrs, Conn.: Real Estate Reports No. 6, Center for Real Estate and Urban Economic Studies, University of Connecticut, April 1969).

Monty, Donald D. and Thomas, Frank H., "The University as an Instrument of Regional Competition." Paper read at Conference on Competition Among Regions: Around the Metropolis (April 7, 1970).

Morganstern, Oskar, *On the Accuracy of Economic Observations*, 2nd. Ed., (Princeton, N.J.: Princeton University Press, 1963).

Shimshoni, Daniel, "Regional Development and Science-Based Industry" (Cambridge, Mass.: Program on Regional and Urban Economics, Harvard University, Discussion Paper No. 31, May 1968).

Somers, Gerald G. and W. Donald Woods, eds., *Cost-Benefit Analysis of Manpower Policies* (Kingston, Ontario: Industrial Relations Centre, Queen's University, 1969).

Stanbeck, Thomas M. and Richard V. Knight, *The Metropolitan Economy: The Process of Employment Expansion* (New York: Columbia University Press, 1970).

Strang, William A., "The University and the Local Economy: A Study of the Economic Interaction Between the University of Wisconsin and the Dane County Economy" (Madison: Bureau of Business Research and Services, Graduate School of Business, University of Wisconsin, Wisconsin Economic Studies No. 4, September 1971).

University of California, "The University as a Major Influence in the State" (Proceedings of the University of California Twenty-second All-University Faculty Conference, May 20-23, 1967, University of California, Santa Barbara), n.p., n.d. (1967).

U. S. Department of Commerce, Area Redevelopment Administration, "The Geographic Mobility of Labor: A Summary Report" (Washington, D.C.: Government Printing Office, September 1964).

Wright, Sewall, "Path Coefficients and Path Regressions: Alternative or Complementary Concepts?" Chap. 6 in H. M. Blalock, Jr., ed., *Causal Models in the Social Sciences* (Chicago: Aldine/Atherton, Inc., 1971), 101-14.

Environmental Factors Affecting Technical Employment

Eric S. Whitman

The State Science Advisory Council of Illinois was empowered by the State Legislature to discuss significant scientific and technological developments and their possible effects, and encourage an increased scientific climate within the State; suggest ways to improve the climate for the production and development of new ideas in science and technology; and seek ways to connect the present and potential capabilities of Illinois with emerging sciences and technologies having future important economic potential.

Since it was an accepted wisdom that Illinois was not the home of a large number of high technology industries and therefore the State was serving as a prime exporter of technical talent (about 50%), the Council undertook a project* which was intended to more closely define the factors which were influencing this migration. A representative sample of the members of the industrial and academic communities was asked to participate in the study. The participants were as follows:

1. Those currently employed in Illinois, but who, during the last

Eric S. Whitman is a Research Project Planner in Corporate Research and Development, Allied Chemical Corporation, Morristown, New Jersey.

*The project was sponsored jointly by the Office of State Technical Services and the State Science Advisory Council, April 1968-September 1968. Mr. Whitman served as Executive Director of the project.

three years**, made a decision to remain at their position as opposed to leaving Illinois.
2. Those who left Illinois within the last three years.
3. Those who were offered a position in Illinois, but turned it down.

It was hoped that this work would provide a factual data base for the development of a science policy for the State.

Preliminary results of the survey which are based upon discussions with university administration faculty members, industrial scientists, members of the financial community and members of the public sector indicate that the most important reason for the migration of scientists from Illinois is the lack of employment opportunities in the high technology areas in which they were trained. The quantitative or questionnaire portion of the study was not designed to amplify on the early conclusion, but attempted to ascertain what environmental factors peculiar to Illinois, if any, were important variables in an employment decision. The factors considered are shown in Table 1.

**1965 to 1968.

Table 1 Environmental Factors Studied

1. Climate and Related Factors
2. Geography
3. Educational Opportunities for Children
4. Recreational Opportunities
5. Entertainment Opportunities
6. Living Conditions
7. Professional Opportunities and Employment Conditions
8. Long Distance Transportation
9. Local Transportation
10. Personal Factors

METHODOLOGY

Participating universities and industrial organizations were asked to supply lists of names and addresses of people who had refused offers in Illinois or had left a position in the State. Respondents in the first category, those employed in the State but who had considered leaving, were obtained by a mass mailing of questionnaires (Fig. 1) to current employees. Each participant was asked to rank the factors as to their importance in the employment decision. Then each factor and subfactor was rated in terms of degree of preference for the Illinois location versus the alternate location. At the completion of the questionnaire the respondents were asked to rerank the factors. A breakdown by subfactor (Table 2) of the category dealing with professional opportunities reveals that many of these could be related to the availability of a technically oriented university.

Mark an X in the box showing the extent of your preference of this factor *at the time of your decision.*

Strongly favored the specific Illinois location	Moderately favored the specific Illinois location	Neutral	Moderately favored the specific alternative location	Strongly favored the specific alternative location

Was this factor an important consideration *at the time of your decision?*

☐ YES ☐ NO

Fig. 1 Survey Questionnaire—Professional Opportunities and Employment Conditions

Table 2 Subfactors Comprising the Professional Opportunity Factor

1. Reputation of Employer
2. Responsibility in Position
3. Salary
4. Fringe Benefits
5. Large Company or Institution
6. Small Company or Institution
7. Professional Freedom in Position
8. Local Sections of Professional Societies
9. Professional Colleagues in Area
10. Availability of Technical Library Facilities
11. Educational Opportunities
12. Convenience of Travel Facilities
13. Supporting Facilities

DATA ANALYSIS

The response rate to the questionnaire mailing was approximately 30% for each category surveyed. The researchers felt that the lack of characterization of the nonresponse category put any statistically inferred conclusions on a highly tenuous basis. Instead the numerical results deemed to be illustrative of trends were identified and organized in such a manner as to allow for nonstatistical comparisons of respondent categories.

RESULTS

A comparison of the importance that each respondent attached to each factor is contained in Tables 3, 4 and 5. In each case the professional opportunity was considered the most important consideration in the employment decision. Looking at the three categories deemed most important, one sees that those who left Illinois and those who refused a position selected the same categories in the same order. However, those who remained chose personal considerations as their third most important factor. (Table 6)

The data contained in Tables 7 thru 12 are a breakdown of how the respondents ranked each factor in comparison with the other factors. These ratings agreed with the importance attributed to each factor. (Table 13)

Table 3 Importance of each factor at the time a decision was made to remain at a position in Illinois.*

	Percentage		
	Important	Not Important	No Opinion
Factor			
1. Climate and Related Factors	29	70	1
2. Geography of Area	37	60	3
3. Educational Opportunity for Child	43	45	12
4. Recreational Opportunities	28	69	3
5. Entertainment Opportunities	17	80	3
6. Living Conditions	55	43	2
7. Professional Opportunities and Employment Conditions	94	5	1
8. Long Distance Transportation Facilities	16	82	2
9. Local Transportation Facilities	16	83	1
10. Personal Factors	52	45	3

*Number of respondents = 380.

Table 4 Importance of each factor at the time a decision was made to leave a position in Illinois.*

	Percentage		
	Important	Not Important	No Opinion
Factor			
1. Climate and Related Factors	48	51	1
2. Geography of Area	56	42	2
3. Educational Opportunity for Child	39	49	12
4. Recreational Opportunities	46	52	2
5. Entertainment Opportunities	19	78	3
6. Living Conditions	55	42	3
7. Professional Opportunities and Employment Conditions	90	8	2
8. Long Distance Transportation Facilities	9	89	2
9. Local Transportation Facilities	14	83	3
10. Personal Factors	38	58	4

*Number of respondents = 105.

Table 5 Importance of each factor at the time a decision was made to refuse a position in Illinois.*

	Percentage		
	Important	Not Important	No Opinion
Factor			
1. Climate and Related Factors	35	65	0
2. Geography of Area	44	52	4
3. Educational Opportunity for Child	39	59	2
4. Recreational Opportunities	32	65	3
5. Entertainment Opportunities	20	76	4
6. Living Conditions	68	30	2
7. Professional Opportunities and Employment Conditions	89	7	4
8. Long Distance Transportation Facilities	17	81	2
9. Local Transportation Facilities	4	93	3
10. Personal Factors	40	53	7

*Number of respondents = 93.

Table 6 Comparison by respondent group of importance attached to each factor

	STAYED IN ILLINOIS	LEFT A POSITION IN ILLINOIS	REFUSED A POSITION IN ILLINOIS
	PERCENTAGE RANKING AS IMPORTANT		
PROFESSIONAL OPPORTUNITIES	94	90	89
LIVING CONDITIONS	55	55	68
PERSONAL FACTORS	52		
GEOGRAPHY		56	44

Table 7 Ranking of factors influencing decision concerning employment in Illinois versus some other location by those who have decided to remain in Illinois.

% OF PARTICIPANTS RATING EACH FACTOR*

	1st	2nd	3rd	4th	5th	6th	7th	8th	9th	10th
1. Climate	3	10	13	11	16	11	9	11	8	9
2. Geography	2	8	12	14	17	15	9	6	10	6
3. Ed. Op. for Child	4	26	19	11	10	7	3	6	1	13
4. Rec. Op.	1	4	9	16	17	15	15	9	9	5
5. Ent. Op.	1	5	6	5	8	8	15	17	13	22
6. Liv. Cond.	5	16	17	17	12	11	11	3	2	5
7. Prof. Op.	76	10	4	2	4	1	2	1	1	1
8. L.D. Travel	1	2	6	6	8	8	11	22	17	16
9. Local Trans.	1	3	5	5	7	6	12	15	24	21
10. Personal	12	15	15	11	7	7	7	5	7	15

*Number of respondents = 380.

Table 8 Ranking of factors influencing decision concerning employment in Illinois versus some other location by those who have left positions in Illinois

% OF PARTICIPANTS RATING EACH FACTOR*

	1st	2nd	3rd	4th	5th	6th	7th	8th	9th	10th
1. Climate	6	17	20	8	12	12	8	6	6	5
2. Geography	8	18	14	23	15	12	4	2	1	3
3. Ed. Op. for Child	2	16	17	11	15	8	6	7	6	11
4. Rec. Op.	2	9	10	20	20	14	7	4	7	7
5. Ent. Op.	0	3	9	6	8	6	16	12	10	30
6. Liv. Cond.	3	22	15	12	7	15	10	11	1	4
7. Prof. Op.	73	6	5	5	4	2	4	3	0	0
8. L.D. Travel	1	2	5	4	8	3	11	20	23	23
9. Local Trans.	1	1	1	5	5	7	8	20	25	26
10. Personal	8	11	10	11	10	6	13	10	8	13

*Number of respondents = 105.

Table 9 Ranking of factors influencing decision concerning employment in Illinois versus some other location by those who refused a position in Illinois

% OF PARTICIPANTS RATING EACH FACTOR*

	1st	2nd	3rd	4th	5th	6th	7th	8th	9th	10th
1. Climate	5	11	14	15	15	15	10	7	6	1
2. Geography	5	12	19	10	23	15	12	3	0	3
3. Ed. Op. for Child	2	11	10	18	13	12	3	10	8	11
4. Rec. Op.	5	9	8	15	11	16	10	11	8	5
5. Ent. Op.	1	3	9	10	11	11	20	12	8	15
6. Liv. Cond.	16	22	19	11	8	9	10	3	0	3
7. Prof. Op.	60	13	10	7	4	1	3	1	0	1
8. L.D. Travel	0	3	1	4	2	0	13	19	45	13
9. Local Trans.	0	5	2	7	5	7	12	19	18	23
10. Personal	6	11	8	3	5	14	7	15	7	25

*Number of respondents = 93.

Table 10 Post-questionnaire ranking of factors influencing decision concerning employment in Illinois versus some other location by those who have decided to remain in Illinois

% OF PARTICIPANTS RATING EACH FACTOR*

	1st	2nd	3rd	4th	5th	6th	7th	8th	9th	10th
1. Climate	3	10	11	14	14	12	11	10	5	11
2. Geography	1	9	14	14	16	15	11	8	7	6
3. Ed. Op. for Child	4	19	21	11	10	7	4	6	3	13
4. Rec. Op.	1	6	8	14	19	15	16	9	7	5
5. Ent. Op.	1	4	6	5	8	6	14	18	14	24
6. Liv. Cond.	4	19	16	16	15	12	9	3	2	5
7. Prof. Op.	77	10	4	3	2	2	1	1	1	1
8. L.D. Travel	1	2	7	7	5	8	11	19	24	17
9. Local Trans.	1	2	4	7	9	8	11	16	21	23
10. Personal	12	21	11	9	7	4	8	6	10	11

*Number of respondents = 380.

Table 11 Post-questionnaire ranking of factors influencing decision concerning employment in Illinois versus some other location by those who have left a position in Illinois

% OF PARTICIPANTS RATING EACH FACTOR*

	1st	2nd	3rd	4th	5th	6th	7th	8th	9th	10th
1. Climate	6	22	16	9	14	11	6	6	5	4
2. Geography	7	21	15	22	16	9	5	2	0	3
3. Ed. Op. for Child	1	17	11	10	4	22	10	4	7	12
4. Rec. Op.	1	9	18	20	18	9	7	5	7	5
5. Ent. Op.	0	3	7	7	12	8	15	8	10	29
6. Liv. Cond.	6	19	18	10	16	9	8	9	2	3
7. Prof. Op.	76	4	5	3	4	2	3	1	1	2
8. L.D. Travel	0	1	2	12	3	4	18	23	25	11
9. Local Trans.	0	1	2	6	3	6	11	25	22	24
10. Personal	7	11	9	7	11	9	7	9	9	21

*Number of respondents = 105.

Table 12 Post-questionnaire ranking of factors influencing decision concerning employment in Illinois versus some alternate location by those who refused a position in Illinois

% OF PARTICIPANTS RATING EACH FACTOR*

	1st	2nd	3rd	4th	5th	6th	7th	8th	9th	10th
1. Climate	4	16	18	18	15	11	7	2	6	2
2. Geography	6	14	18	10	19	13	13	4	0	4
3. Ed. Op. for Child	4	15	4	15	15	8	8	5	4	25
4. Rec. Op.	6	9	10	11	18	18	9	8	6	6
5. Ent. Op.	0	1	10	14	14	14	10	15	4	20
6. Liv. Cond.	12	23	15	12	11	9	9	2	0	6
7. Prof. Op.	64	11	9	6	7	0	1	1	0	1
8. L.D. Travel	0	2	2	5	9	9	7	23	26	17
9. Local Trans.	0	1	1	5	3	4	10	20	34	23
10. Personal	12	8	7	7	8	7	13	9	7	21

*Number of respondents = 93.

Table 13 Comparison of ranked factors by respondent group

% OF PARTICIPANTS RATING EACH FACTOR*

	REMAINED			LEFT			REFUSED		
	1ST	2ND	3RD	1ST	2ND	3RD	1ST	2ND	3RD
1. CLIMATE	3	10	13	6	17	20	5	11	14
2. GEOGRAPHY	2	8	12	8	18	14	5	12	19
3. ED. OP. FOR CHILD	4	26	19	2	16	17	2	11	10
4. REC. OP.	1	4	9	2	9	10	5	9	8
5. ENT. OP.	1	5	6	0	3	9	1	3	9
6. LIV. COND.	5	16	17	3	22	15	16	22	19
7. PROF. OP.	76	10	4	73	6	5	60	13	10
8. L.D. TRAVEL	1	2	6	1	2	5	0	3	1
9. LOCAL TRANS.	1	3	5	1	1	1	0	5	2
10. PERSONAL	12	15	15	8	11	10	6	11	8

CONCLUSION

After analyzing the raw data and piecing together data presented here, it has become obvious that the single most important factor contributing to an employment decision is the professional opportunity itself. For those who are settled in an area personal considerations play an important part in any decision to leave the area. It is interesting to note that geography placed ahead of personal considerations for those who left or refused a position in Illinois. This is probably due to the high visibility of the differences between locations and the hesitancy of the analytically trained professional to allow or admit that a subjective consideration has greatly influenced a career decision.

In conclusion it can be said that it is the job opportunity that attracts the highly trained scientist or engineer. Other factors may play a part in his decision, but he or she will go where the work exists.

The Changing Demand for Scientists and Engineers

Betty M. Vetter

We are being asked, at this conference, to examine a small professional segment of manpower as it relates to a small geographic segment of the nation, in the hope that the smallness of the segments will allow us to build and test a model that will be applicable to the larger problem. While we work with the small pieces of our puzzle, it is occasionally necessary to step back and view it in a larger context so that we can make sure we are working with typical pieces.

I would like to expand your miniature geographically to include a whole nation; and to broaden your engineering population just slightly by including physical scientists and even an occasional technician as I look broadly at the employment picture today and set the focus of my crystal ball on the match or mismatch of supply and demand over the next decade.

This time period is a good one to choose for a number of reasons—pragmatically, because most of the applicable forecasts that have been made are for that nice neat packet of time that ends in 1980; and equally importantly because the time length of the academic pipeline, plus a little planning time which we have to offer to universities, also comes out pretty neatly in a package ending about 1980.

Because the future is partly based on the present, and the present is largely the result of the past, let's look backward long enough to see

Betty M. Vetter is Executive Director of the Scientific Manpower Commission, Washington D.C.

how we got here, and hopefully let some of our past mistakes give us some guidance for the future.

It was really only 3 or 4 years ago that scientists and engineers were riding on a wave of popularity, rapidly rising salaries, and constantly unfilled demand. The time must seem longer to some of the presently unemployed. Today these same heroes are castigated as the producers of pollution and the purveyors of destruction who fueled the fires of war with their hideous weapons. While not forgetting the marvels of the engineering feats that put men on their satellite moon, there is grumbling that they did not choose instead to fix our earthly transportation mess. Scientists and engineers are in the process of losing a public relations war they didn't even know they were fighting! More about that later when we look at the enrollment picture.

What happened to the jobs—those beautiful salary offers and the recruiters who wined and dined and vied with each other to capture even the graduates near the bottom of the class?

A number of things happened to make the bottom fall out. The most important was a change in the economy as a whole. The job market was affected strongly in all sectors by our retreat from a period of war in which we refused to give up butter while still producing guns. As in all previous wartime periods, unacceptably high inflation occurred, but unlike the immediate post war periods preceding, there was no great unfilled consumer demand waiting to take up the slack as plants converted from making tanks and trucks to automobiles, or from war planes to civilian aircraft. They made both this time, as they went along.

The winding down of the war and of the space program happened to occur about the same time. Thus, it is understandable that our unemployment percentages are worse for scientists and engineers under 26 than for those who are a little older. Among the 90,000 physical scientists and engineers who graduated in 1969, about 40% were employed by aerospace or defense contractors.

Even fairly small changes in government support for particular kinds of enterprises can cause severe fluctuations in the employment market for technological professionals. Between 1960 and 1972, the chief areas of Federal R & D funding have been national defense, space research and technology, and health. These functions together make up about 86% of the total Federal R & D expenditures estimated for 1972, but they represented 95% of the total in 1960.

By 1968, following some crash programs in educational recruitment, 235,000 scientists and engineers were trained and employed in aerospace. By June, 1972, that number will have dropped 38.5%, or 90,000 men. Meantime the tremendous increases in the Federal budget have been principally in those areas which do not produce many jobs—welfare and educational improvements.

How bad is the current employment picture among scientists and engineers?

The best data we have, which may not be very good by now, indicates that about 3% of the engineers were unemployed in June 1971, with an additional 1.7% of them unsuccessfully seeking engineering employment although in some cases they were employed part time or even employed full time in some non-engineering job. It is not clear whether this percentage has risen or fallen since the survey completed last June, although the National Science Foundation plans to survey the same group again this summer so that we may at least have some trend data.

Among scientists surveyed in April 1971, an unemployment rate of 2.6% was found. With the input of June graduates, the percentage certainly has risen and probably is around 3%. The additional rate for those who are under-employed, part time employed, and employed outside the field against their wishes probably is about the same as for engineering—another 1.5%. The best estimates I can make of the number of unemployed scientists and engineers, including '71 graduates, is about 60,000. In these estimates are included only those scientists and engineers that hold at least one degree in the field. If we use the generally accepted ratio of 7 technicians for each 10 scientists and engineers, an additional 40,000 technicians may be jobless.

Among our scientists and engineers, we may also add a third group of unknown magnitude who are underemployed in the sense that they are performing work in some way related to their training, but which could be done by persons with lesser training.

Our group of unemployed scientists and engineers was almost twice as large in the summer of 1971 as it was a year earlier. The bits and pieces of information we have obtained since last summer appear to show that the situation is somewhat more serious now than it was then although it certainly has not doubled again.

It is important that we understand as much as we can about this unemployed population beyond the obvious facts that aerospace engineers are in more trouble than civil engineers. Some of the things we know contradict the generally held beliefs. For example, we do know that recent and new graduates have considerable higher unemployment rates than older professionals. Scientists under 30 reported to NSF an unemployment rate of 5.3% compared to 2 to 3 percent for those who are over 30. Among engineers, where the bachelor's degree is more likely to be considered as professional qualification than in the sciences, men under 24 had a 5.5% unemployment rate compared to less than 3% for those who were between 30 and 50. While it is obvious that dislocation among older workers with more responsibility usually is more personally devastating than among those more recently out of school, a higher percentage of the younger engineers and scientists seem to be in difficulty. On the other hand, since the science and engineering population of the country is considerably larger in the 30 and up age group, even the smaller rate of unemployment produces larger numbers of unemployed men and women in their thirties and forties.

Although there are few women in engineering, we are concerned that the unemployment rate among women in the sciences is considered higher than among men although the group is smaller. NSF found a 5.2% unemployment rate among women scientists compared to 2.3% for the men. Among women chemists, a survey by the American Chemical Society reports a 6.3% unemployment rate compared to 2.4% for men.

Another factor that surely must concern us is the proportion of the unemployed whose last professional work was directly involved with federal activities. Among engineers previously employed in space and defense activities, the unemployment rates are 6.3% and 4.8% respectively, compared to 1.3% for those whose jobs were in public works. Among scientists, 45% of the unemployed reported that their last science-related employment was supported to some degree by government funds.

These same features are emphasized by the geographic distribution of unemployment—the two coasts, followed by the Gulf areas show far higher technical unemployment than the rest of the nation. As you are aware, our defense and space efforts have been largely concentrated in the areas now having the highest unemployment rates. The 15 local areas showing highest unemployment for both scientists and engineers do not include a single city in the Midwest. In this, at least, we may be sure that our model is atypical.

On the other hand, what Federal funds there are that are being devoted to employment opportunities for this unemployed group are available to any state or any locality which chooses to add a little of its own money for an opportunity to employ top quality people in state and local jobs.

Representing educational institutions as many of you do, you also are interested in participating in retraining efforts by which the skills of aerospace and defense engineers can be converted to meet needs in the civilian sector. Here, I would offer a great big warning: do not get involved in setting up programs of retraining in areas where needs exist until you have valid reason to believe that the needs will be translated into demand. I cannot emphasize too strongly that where educational institutions have done their job well in the first place in educating young scientists and engineers, there is very little need to bring these people back to school to teach them how to apply their training in a new area. Of course, transferring an aerospace engineer into an environmental job means he has to learn some new things. However, it is not formal education he needs—it is some on-the-job training while he learns how to apply his skills in a new dimension. Fortunately, the number of "retreading" programs has been fairly small, since most of them turned out to be failures. They failed because the jobs for which they were retraining the engineers did not exist. They would have been a failure anyway in the sense of being wasteful even if jobs had existed, since most of the retraining needed can be done better on the job with perhaps a few specialized short courses offered on or off campus.

Formal retraining programs for converting aerospace engineers should start off by asking themselves "convert to what?" Not only do we lack substantial programs in environmental control, urban renewal, and all the other "hard" science areas, but we also face a saturated job market. While it is certainly possible that the job market for converted engineers will improve in the future as today's small programs in housing and urban development and environmental cleanup gain momentum, it is perfectly obvious that such improvement will not result from the reconversion of engineers but from a growth in demand for engineering services. Good engineers can and will reconvert themselves from aerospace or defense if a market for their talent develops. Many of them already have done so as openings occurred in miniaturized electronics, nuclear engineering and systems analysis.

It is vital that we differentiate between "need" and "demand". We need all the trained specialists we can get just to clean up the environment, produce abundant power, reconstitute our molding cities, and a host of other desirable things. But until the decision is made at all levels to budget one or more of these activities at the level needed to solve the problem, we have not translated our "need" for specialists into "demand", which means jobs.

There is considerable discussion in Washington today on how to maintain America's technological lead, and how to foster research and development programs that would serve the triple purpose of maintaining that superiority, improving our technological export business, and providing employment for our trained technologists; but there is little concrete action. The President has noted our nation's basic dependence upon a healthy technological R & D program and promises a more detailed message to Congress in the near future. His fiscal 1973 budget, in one of its few new programs, includes an increase of $700 million in civilian research and development spending, up 15% over last year and 65% over 1969. The President's expanded State of the Union Message promised new emphasis on cooperation with research and development in private industry, including new experimental programs for cost sharing and for technology transfers from the public to the private sector. The President specifically urged support for the space shuttle program which would offer employment to about 50,000 of our unemployed aerospace professionals. However, the final shape of the package he is asking Congress to approve is not yet outlined, nor can Congressional reaction be forecast yet.

Congress too has expressed concern with these things. The Economic Conversion Bills offered by Senator Kennedy and by Representatives Giaimo and Davis in the House specifically ask Congress to commit a fixed percentage of the gross national product (about 2%) to Federal support of science and technology. Although hearings have been held in both the House and Senate on these bills, they have never come up on the floor.

There is no universal agreement in Washington as to how to encourage an industrial research and development base. The one thing on which there does seem to be general agreement is that the emphasis should be on applied research directly leading to solution of stated problems. Funds for basic research conducted either in government laboratories or in universities are down very substantially. Since you are concerned with engineering which is in itself the application of research in a very practical way, the trend away from basic research may not yet concern you. I believe that it should. To solve our technological problems, we need new science, which can come only from basic research. For example, we must find new ways of harnessing energy sources. We are still generating power by very primitive means. The technological problems of achieving fusion power are so formidable that experts do not see their solution in less than twenty years or so. But new scientific knowledge might show us a way to convert mass into energy without having to create and then contain a major catastrophe.

Meantime, there is no common agreement in Washington as to the best method of Federal intervention to stimulate the technological community. Tax credits to private industry for R & D efforts are not universally approved.

Given the pessimistic view of current unemployment that I have given you this morning, you might well expect me to applaud the 17% drop in undergraduate engineering enrollment this year and to express my horror that enrollments in freshmen and sophomore chemistry appear to be rising all out of proportion to total enrollment at universities across the country. However, my view of these phenomena is just the opposite—I am disturbed by the drop in engineering enrollments and encouraged by what appears to be an increased interest in chemistry even if I think it is going to lead to some problems. In the case of chemistry, it appears that the enrollment results from a combination of interest in getting into medical school and perhaps interest in working in environmental areas. As far as I can see the students interested in becoming M.D.'s will face some severe disappointment—not because we do not need the doctors and could not find work for them if we had them but because the medical schools cannot absorb them. If their interest is in the environmental area, perhaps there will be jobs for them if at the same time they exercise their newly won franchise to encourage governmental efforts in this area.

For the engineers, I find the drop in enrollment figures deeply disturbing. We are losing the public relations battle in which we should be demonstrating that technology is the essential part of repairing the wounds of the past on our planet and providing a livable environment for its burgeoning population. In spite of all I have said today about unemployment among young engineers, recent graduates in engineering have found jobs—and found them a great deal easier than their cohorts with majors in teaching, the humanities or the social sciences. By comparison with all other groups of graduates, engineering graduates did extremely well in 1971 both in terms of finding employment and in the salaries they were offered. The annual survey of job prospects for 1971 engineering graduates conducted by the Engineering Manpower Commission found that at the time of graduation 9% of the engineers were without job offers or other firm plans. This figure is double that for 1970 and is in sharp contrast to the boom years of 1965 to 1969 when any man who could earn his sheepskin had a choice of more than one job. However, follow-up studies by various placement offices at several universities indicate that these graduates did find jobs. For example, Ohio State's College of Engineering granted 519 bachelor's degrees in engineering and in computer and information science in 1970-71. By mid November, the engineering placement office reported only 8 or 10 of these graduates were without jobs. Lehigh University reported that all but 5% of its graduating class had jobs while the unemployment figure for its engineer-graduates was less than 3%. Following a 12% unemployment rate among the 1,000 seniors at Indiana University at graduation time, September checkup found all but three or four persons had obtained jobs. Among Stanford's 909 engineering graduates at all degree levels, only 32, most of them at the master's level, were still looking at the time of graduation.

Frank Endicott's annual survey of prospective job opportunities for the class of 1972 shows better promise than for the class of 1971 with a 5% increase anticipated in engineering hires at the bachelor's level. As should be obvious, the engineering market is dependent most highly on the general economic situation. If there are jobs for anybody, there will be jobs for engineers. Most economists predict an improved job market by June 1972.

Meantime, although total undergraduate enrollment this fall was up 2.4%, total undergraduate enrollment in engineering dropped 17.1%. While graduate enrollment dropped 7.8% in engineering and 2.7% in the physical sciences, total graduate school enrollment was up 5.0%. The Carnegie Commission report on fall 1971 enrollment shows increases in almost every field except science and engineering.

An added factor of concern comes from an NSF survey which shows a drop in enrollment in graduate science departments at the top rated private graduate schools, but a rise in enrollment at the less prestigious ones.

Both engineering and physical science have been dropping steadily since 1960 in the percentage of all students that they attract. The Office of Education predicts a continued decline in the percentage for both baccalaureate and master's engineering graduates through 1980. The percentage of students majoring in the physical sciences is projected to continue to decline at all three degree levels. Enrollments certainly tend to confirm these projections. In actual numbers, the Office of Education shows a slight rise in engineering degrees at all three levels to 1980 and an actual drop at all three levels in the physical sciences.

If we are to solve the problems that await our solution, we must have new engineers and scientists. The Department of Labor forecasts annual openings between 1968 and 1980 for 73,400 engineers. The anticipated number of graduates at the bachelor's level is only 42,000 per year during the 1968-80 period but the number needed to meet requirements would have to rise 3,000 a year above that level. If these projections prove accurate, we will continue to be dependent on non-graduate engineers as we have in the past. We might note here that the unemployment rate last summer among engineers with less than a bachelor's degree was 4.4%, compared to a 2.8% rate for those with at least a bachelor's.

Without in any way downgrading the importance of the present unemployment problems among scientists and engineers, I believe that we should not be so obsessed with that problem that we fail to look ahead a few years and note that given present trends, we will again find ourselves formulating crash programs to train enough engineers and scientists to meet our needs.

The factors that determine supply and demand cannot be predicted with absolute accuracy. How will potential students react to today's employment market? What direction will the government go in working to solve its problems? What scientific breakthrough comparable to the transistor or the laser will affect our technology in the next ten years? I don't know. However, after working laboriously through the statistics that measure both supply and demand, I am left with a feeling of optimism by indications that there will be no substantial surplus of scientists and engineers in another year or two, and that the shortages that appear to dominate the second half of the decade will not be serious enough to jeopardize national progress.

Public Service at the University of Wisconsin-Madison

C. A. Coberly, L. Jedynak, J. A. Duffie
(Presented at the Conference by J. A. Duffie)

We can define public service (in an admittedly unsatisfactory way) as the application of the professional competence of the university staff off campus and through other means than teaching of resident students. That is to say, we can ask, how does the College of Engineering professionally impact the community outside of the campus? We will address ourselves to activities of the faculty in consulting, research and other kinds of cooperation with outside institutions (which might be industry, government agencies, private organizations, or even individuals).

At the University of Wisconsin-Madison, there are three organizations that have to do with engineering public service. One of us (J. A. Duffie) is an associate dean of the Graduate School and is responsible for the University-Industry Research Program, which is part of the Graduate School. The College of Engineering, through its Engineering Experiment Station, is a major mechanism by which public service

Camden A. Coberly is Associate Director of the Engineering Experiment Station.

Leo Jedynak is Associate Professor of Electrical Engineering.

John A. Duffie is Associate Dean of the Graduate School and Director, University-Industry Research Program. All authors are at the University of Wisconsin—Madison.

activities as defined above are carried out. We also have Engineering-Extension, which is responsible for continuing education. A clarification of the activities of these three organizations may be useful.

The UIR Program (University-Industry Research Program) in the Graduate School is a small program which functions with the cooperation of the colleges and provides time for a few selected faculty members to get out and get acquainted around the state. Their job is to find out what is going on and see how the activities of their colleges can relate to the activities of industry and government. This is in support of the colleges, including engineering, which are the organizations in which cooperative programs are carried out.

University Extension at Wisconsin is a separate unit of the University which includes an Engineering Department. Its major concern is with continuing education of a formal sort, with credit or non-credit offerings. (The Engineering Extension Department, in cooperation with the College of Engineering, has started a Professional Development Degree Program which recognizes the completion of a planned program of continuing education).

The Engineering Experiment Station is an integral part of the College of Engineering. It is not a separate research institute, as are some experiment stations. The research that goes on in our experiment station is an integral part of graduate education. The staff of the Experiment Station is the staff of the College. The laboratories of the departments of the College are the Station. In addition, the Experiment Station provides a useful mechanism for organizing, housing, and conducting interdepartmental and interdisciplinary programs which do not logically fit in a single department.

The College of Engineering (i.e., the Engineering Experiment Station) together with the Graduate School and the University administration sets policies and constraints under which the faculty carries out its public service activities. Our policies in one sense are conservative and in another sense they are quite flexible. A main element of the policy is to keep our research that we do in the University very closely related to our education programs. We do not have research that is classified (in any sense) and we do not undertake research where the sponsors of the research are not willing to accept the pace at which research involving graduate students can progress.

We differentiate very carefully between an institutional activity —and what the faculty do as individuals in their consulting. Some of the problems which come to us in the public service category are institutional problems and some are more properly handled by individual faculty members as consultants (working within the constraints that the University sets).

Another fact that we are careful to specify in our relations with outside agencies is that there are very distinct limitations on what the University can do. Limitations arise out of policy, but they also arise out of the fact that even though we are a big university, there are many areas in which we do not have competence. Also, some of our people get fully committed and don't have any further time for additional activities. Thus, there are limits and we try to make them very clear.

Within this framework, we can talk about several activities. First of all, we do research. We have institutional funds, and we direct some of those funds toward research that is specifically aimed at the short and long range well-being of the region, particularly the State. For example, we are supporting a set of programs on the geology of Wisconsin which may have long range economic impact. We have research that is supported by outside agencies and it is often done in cooperation with those agencies. For example, the city of Madison, the Federal government, the Heil Corporation of Milwaukee, and the University have a four-way program on evaluation of a solid waste disposal system. The research provides projects for our graduate students, services for the city, evaluations for the Federal agency, and maybe some sales of process equipment for the industry.

There are additional comments that need to be made about consulting. We usually think of consulting as paid consulting for industries, when the faculty goes out and works for X dollars a day. Equally important and perhaps more widespread than most of us realize is what we can call non-paid consulting. We find that a very large number of our faculty work for State or local agencies. They work as members of boards, such as the Board of the Department of Natural Resources, and the Pesticide Review Board; they serve on civic committees and perform other formalized services that are rendered as individuals. These are the formalized part of the non-paid consulting activities.

On a more informal basis, there exists a host of personal contributions of people on the faculty to outside organizations. For example, members of our faculty meet frequently with people from the State Departments of Natural Resources and Agriculture to discuss problems of those Departments. We know that at least a quarter (and possibly more than a half) of our faculty at Madison have inputs to State government one way or another. This is a very significant part of the public service activity of the University which is not formally recognized and is not widely appreciated.

These informal activities are characterized for the most part by one-to-one relationships. People who have the problems and the people who know something about their solutions get together to discuss a situation. These relations are diverse in nature and they are hard to identify. We seldom find out what happens and the contributions are hard to evaluate. On the other hand, we think they are important and an integral part of our public service activities. They cannot be separated from teaching and research and are an integral part of the "three-legged" stool on which our State University stands, i.e., teaching, research and public service.

We can describe our relations with outside agencies as being either active or passive. They are *passive* in the sense that we respond to questions and problems that come to us from outside and which we select as being appropriate for institutional or individual attention. To do this, we need convenient ways for outside people to find their way into the university. Our Engineering Experiment Station and the UIR Program provide this access. *Active* relationships with outside agencies are those we originate, in anticipation of problems or opportunities. We then need to appropriately design our activities in the university to improve our services to the region on the basis of these anticipations.

We do have some problems and conflicts in the University which arrive out of the nature of the institution and its many responsibilities. There are conflicts in a university like ours between the national and international interests of the faculty on the one hand and regional and state interests on the other. The maintenance of a reasonable balance among these various activities, to adequately serve a variety of needs and interests, is not easy.

We have another dilemma in deciding how responsive we ought to be to problems of the region. The university in many respects is a unique institution in society. It is the main place where people can presumably sit back without too much vested interest in any particular proposition and hopefully evolve concepts that will impact on society years hence. A classic example from our campus is the concept of Social Security which was devised by a professor of economics years ago and started during the Roosevelt administration; this was a concept that came out of the campus and is today a part of the fabric of our society. So, we must weigh this facet of the University's life on the one hand against responsiveness to short range problems on the other. We think we have to have some of each of these kinds of activities in the University. The conflict is not new. One group says we are "purists," that we are not going to bother with short-range problems and another groups says the University is built on public service and we had better get with it. There seems to be room for both of these attitudes.

The question must be asked, "Are we adequately organized to encourage an appropriate level of public service activities such as has been discussed?" We think that our job is to encourage these kinds of off-campus and public service activities and make them possible. We need to recognize that these services are a part of the University faculty's responsibility and perhaps adjust our thinking as far as promotions and so on are concerned to take these contributions into account.

We feel strongly that it is not possible to "organize" all these public service activities. Over-organization will render less effective many of the most valuable contributions of the faculty (i.e., the one-to-one informal, and formal, ties that need encouragement). These activities defy organization, and our problem is to stimulate and encourage them without making them complicated and difficult.

Regional Role of Engineering Colleges: Delaware

Edward W. Comings

Delaware, the next to the smallest State in the United States, has a population of about half a million. In addition to the University, there are two other State supported educational institutions—Delaware State College and Delaware Technical and Community College. The latter has existing or contemplated campuses in each of the three counties. The College of Engineering at the University of Delaware conducts the only professional engineering education program in the State. The University of Delaware in 1971-72 has an undergraduate enrollment of 10,400, a graduate enrollment of 2,400, extension enrollments of over 3,900 and in its Summer Session 8,000 students. The University receives approximately 1/3 of its operating budget from State appropriations. The total budget is $46,700,000. The College of Engineering is one of eight colleges and conducts undergraduate and graduate programs. A number of aspects which help reveal the role of the College of Engineering in the State and region will be discussed.

The University has recently completed three planning studies made by commissions consisting of faculty, administrators and students. The first of these, the Community Design Planning Commission, was appointed by the President and has been active for about two years. It received and reviewed plans prepared by each of the academic divisions (departments and colleges) for the next 2, 5 and 10 year

Edward W. Comings is Dean of the College of Engineering at the University of Delaware, Newark, Delaware.

periods. The Commission held public hearings in which each unit presented and defended its plans. The hearings were open to faculty and students and, to some extent, the public at large. The College of Engineering plan will be considered in part.

Engineering Education is truly a community activity involving students, faculty, university administration, the State, federal agencies, professional societies and industry. In an effort to serve each other, it is not "Who is serving whom?"—their interests are enmeshed.

Engineering and Society. The most striking feature of this century is accelerating change which affects all aspects of our lives. Change has occurred: in large increases in scientific knowledge and its application, in population density, in standards of living and affluence, in communication media, and in the average level of education of our citizens. These changes have never been adequately predicted nor their effects understood. One little understood influence is the mismatch between education and the degree to which it prepares people for the realities of modern life. The disparity occurs at all levels. It is most evident in the ghettos where squalor and rioting make it apparent to all. However, a significant degree of mismatch also occurs among the affluent, young and old. The young are aware of this and in frustration too often seek scapegoats rather than real improvements. Mismatch in education is leaving pools of citizens undereducated relative to their potential to play a role in the community and for their own self-fulfillment.

Engineering education has supplied the trained manpower to solve the problems of technological change. To this extent, it is partially responsible for their effect on society. It must now, and in the future, help educate citizens to assess the choice between the benefits from technology and its adverse effects on society. It must educate the manpower to bring the maximum benefit to mankind from technological changes.

Engineering and the State of Delaware. Delaware employs more scientists and engineers than it educates. Its economic health depends upon the discovery and production of new materials and methods to make them competitive. This enterprise is challenging and offers satisfaction to those who make constructive contributions. Delaware's young people should be encouraged to pursue careers in these fields if they have the capabilities and interest to do so. They should seek the preparation. The deficit in their numbers should be partially made up by inviting qualified students from other states to enroll in engineering education in Delaware. Whether graduates from the University of Delaware remain in the State or go elsewhere, the State should aim to educate a number equal to those employed here. More and more Delaware, together with some other progressive states, is viewing the graduate programs in engineering along with the undergraduate as a foundation for desirable industrial growth and security for jobs for a skilled work force. Quality graduate and extension programs generate new contributions, maintain old competence and develop new breadth among employed professionals.

History and Current Status of Engineering Education at Delaware. Baccalaureate production in engineering at the University increased at 5% per year. This parallels the growth in numbers of Bachelor's degrees in engineering for the United States and for all fields since 1900; national growth in Master's and Doctorates at 11% and 12% per year also lie within the range of fluctuating experience here.

Functions of the College of Engineering. The College of Engineering offers Bachelor's degrees which provide flexibility for careers as professional engineers or for relating technology to business. Its Master's and Doctoral programs are also varied. The College thus offers its students preparation for specific careers, educating new engineers and providing advanced training for experienced engineers employed in this area. It depends upon the College of Arts and Science for many of the courses taken by its students and has an obligation to show the relevance of such courses to the objectives of its own students. It should also teach the relevance of technology to society to non-engineering students.

The College has recently accepted commitments to serve the State in areas recognized by the University. These commitments are in Environmental Engineering, Ocean Engineering and more tenuously in Bio-Medical Engineering. It is difficult to apply engineering in these areas and the practitioners will require the best education in the sciences and humanities as well as the best possible in engineering to be successful. The wider range of interests and competence in ample depth to meet this challenge can be developed in the existing departments but will need careful nurturing.

It improves the attractiveness of the State to professional and pre-professional recruits through the quality and reputation of its programs and services: its research, its graduate and undergraduate degree programs, its impartial technical advice to government and industry. In meeting its more recent service commitments to the State, it is cooperating with the other Colleges in education and research.

The Programs Offered. There are four separate Bachelor of Engineering degrees with elective options in each, nominally requiring four years. The Bachelor of Arts degree may be obtained simultaneously with each of these in five years.

General Goals of the College. The College aims: (a) To meet the reasonable demands of individuals and institutions in the State for professional programs in engineering and/or research, including quantitative type service courses to give the non-engineering student an insight into the role of technology in society; (b) At the *undergraduate* level, to accept as students all qualified Delaware residents, to plan special means for the potentially qualified students from Delaware who need assistance in making the transition into engineering programs and to invite a number of the more qualified students from outside the State in order to add to the quality of the student body and to increase the talent pool available within the state for its development. At the *graduate* level the number of students is likely to be determined by the funds available to support the program and by quality limits set for their admission. Increase in quantity without comparable quality is undesirable. On the other hand, to be precluded from accepting well qualified students by a shortage in funding is not likely to be in the best interest of the State; (c) To maintain the quality of research efforts and the flexibility to adapt to new fields by:

Regional Connections. The following information concerns relations, within the nearby geographical region, with other educational and research institutions directed toward strengthening its own research and educational programs.

A. With educational and research institutions

(1) U. S. Army Aberdeen Proving Ground—briefly, this had included graduate education in engineering for APG's Exterior Ballistic Laboratory's professional staff, and graduate student theses have been conducted by students registered at the University of Delaware using the extensive research facilities at APG. Their installation includes

wind tunnels and shock tubes that represent a major national investment and resource. For example, W. B. Sturek has completed a Ph.D. degree in the Department of Mechanical and Aerospace Engineering using APG facilities and received his degree in June 1971. Since 1967, there have been at least eight such graduate students involved in engineering programs at the University of Delaware. A number of APG staff members teach extension courses at APG. APG was the technical monitor of the University Themis I Grant from the Department of Defense. They currently support a consulting and summer employment arrangement with one member of the College of Engineering faculty and there is a potential for more activity of this type.

(2) Civil Engineering students have used laboratories of the State Department of Highways and Transportation.

(3) The potential for cooperative educational programs and transfer arrangements with the Delaware Technical and Community College are under consideration.

(4) Contacts have been made with some nearby institutions —Delaware State College, Cheney State College and Lincoln University—for example, with an eye towards increasing the number of successful minority group students in engineering.

(5) A few years ago, a faculty member in Mechanical Engineering and Mechanics at the University of Pennsylvania was employed on a part-time basis for a period of time. Civil Engineering arranged for Dr. R. Richards to serve as an Extension lecturer for one semester while he was a regular faculty member at Princeton. Also, Dr. Tschebatarioff came as a consultant and special (or extension) lecturer from the same University.

(6) The College has been active in preparing to cooperate with Jefferson Medical University and the Wilmington Medical Center in pre-med and biomedical engineering programs.

(7) Civil Engineering is now working on a consortium arrangement in Air Pollution Control education involving universities from Delaware to New York.

(8) Other nearby engineering colleges include those at the University of Pennsylvania, Drexel, PMC Colleges, Villanova, Princeton and Johns Hopkins. Reflecting on the great diversity of types of university funding and political connections reveals the hurdles to be overcome in developing closer cooperation. Exchange type programs are not so easily arranged, as say, between the various campuses of Pennsylvania State University which are knit together by much in common.

B. With industry

(1) Preliminary contacts and arrangements had been made to utilize the subsonic velocity wind tunnel and associated facilities at the Vertol Division of the Boeing Company in Chester just prior to the time that part of this Division was shut down in the general economy program.

(2) There have been various arrangements with the Du Pont Company. They have supplied personnel at little cost to the College to assist in the instructional programs in design in the Department of Chemical Engineering and also in the Department of Mechanical and Aerospace Engineering. The use of their equipment has been possible on occasion for Civil Engineering, for example, but is frequently limited by their proprietory requirements.

(3) Regular graduate extension courses as well as continuing education programs have been offered and this activity is described in the College plans as an area where the largest growth is expected.

Interdepartmental and Interdisciplinary Programs. Increased use is being made of all degree programs, undergraduate and graduate, to provide more flexible career objectives. These programs permit different combinations of mathematics, science and engineering science to be brought to bear on the solution of problems in technology, including those with implications for the well-being of society.

Engineering in the United States has bred scores of subject areas. Many have administrative counterparts in universities. Some areas show great permanence and stamina, symbolized by stable professional societies. Others are more transitory, serving limited or local needs.

To keep its primary administrative units to the current four departments while providing a selection of the other programs, the College of Engineering conducts interdepartmental and interdisciplinary area studies through committees, with the faculty members assigned to one of the existing departmental units. The choice of programs and numbers of units is governed to some extent by the size, breadth of interest and communications problems of faculty and student body as well as by the service of the area to engineering education as a whole. It is sometimes advantageous to give some of the areas considerable visibility to attract support from currently fashionable (and funded) federal programs.

In addition to educating its share of engineers, it must be recognized that there is an increasing demand for individuals with a variety of types of education, each resting largely on some of the principles of engineering but which do not include the full coverage of a professional engineering education. The recent program in Engineering Administration is an example which has attained success. A comparable program in engineering and the arts is likely to hold great potential. Also, students ranging over the University include those who come here for a "liberal education" in the old sense. Today, this must include a more rigorous quantitative knowledge of technology and its relations to society. While others may teach the philosophical aspects of this subject, it should be the role of the engineering educator to teach the quantitative and technologically accurate aspects.

Considering Again the Professional Engineer. He is faced with problems of increasing sophistication as well as breadth of content, and the possibility of acquiring suitable preparation for such activity in four years is diminishing. For this reason, to reach the primary goal of providing an adequate education for the professional engineer will require continuing growth of the graduate program and in the number of students. The graduate program should not be viewed as solely for an elite group of scholars. It is a major source of education for practitioners.

Chemical Engineering. Delaware is known as the Chemical Capital of the World—not that its output of chemicals is so large, but Du Pont, Hercules, Imperial Chemicals Industries (ICI-Atlas) and Thiokol are heavily represented. The first three have their main offices and research laboratories here. Du Pont also does its engineering design here—over half a billion dollars a year of design projects. Chemicals are high technology goods in the U. S. economy and contribute significantly to our balance of payments with the rest of the world. In 1971

Chemical exports	\$3,700 million
Chemical imports	1,660 "
Chemical net export	2,040 "

The State of Delaware has 0.27% of the national population, has a chemicals industry payroll of 1.46% of the national total, a value of chemicals shipped of 1.31% of the national total. We produce about 1.1% of the national total of degrees granted in chemical engineering annually and about 3.2% of the total chemical engineers in the nation reside in Delaware. The cost of the department to the citizens of Delaware amounts to about 0.1% of the value of chemicals shipped annually. Thus the overall picture is approximately the following: Delaware is very much a center of the chemical industry nationally but only a very small fraction of the revenues of the chemical industry are in fact used to support the chemical engineering department here. By and large, the State imports a large fraction of the technology it needs in the chemical industry. The Chemical Engineering Department's utility is a fairly broad one and not restricted to the State of Delaware only. In other words, there is perhaps some reimbursement in this way for the "free" contributions which other states make towards supporting the chemical industry here in Delaware. Some states which are making such contributions rather heavily are Wisconsin, Minnesota, California, Illinois, Michigan and Indiana.

Water Resources Center. This center has operated for 7 years under the provisions of the Water Resources Research Act of 1964 under the U. S. Department of the Interior. Research efforts are combined with the primary educational responsibilities of the University.

Delaware's water problems result from the conflict caused by the demands made by an expanding population for quality water and the concommitant necessity of improving the quality of the environment. This conflict is intensified by inadequate sewer and waste treatment facilities, water demands which exceed dependable, potable supply, and increased storm-water drainage resulting from urbanization. Based on the strength of future needs, the problems facing Delaware at present will worsen in the forseeable future.

Perhaps the most significant problem in Delaware and elsewhere is communication. The problems have been recognized and alternative solutions have been offered. However, progress toward approval and implementation of promising techniques has been slow. The public and the decision makers must become knowledgeable of the situation. The water resources research community must become active, aggressive public educators. This is the principal challenge.

This Center is committed to the concept that reclaimed water is vital to the solution of Delaware's specific water problems and is evidenced by the emphasis of the research program. Examples of projects are wastewater treatment in tubular contactors, transport properties of dilute polymer solutions, anaerobic treatment of starch wastewaters and catalytic oxidation of organics in wastewater.

College of Marine Studies. The College of Marine Studies, established in June 1970, brought together under one unit a number of marine activities begun in 1950. The College integrates and coordinates marine activities in six broad areas: marine biology, chemical oceanography, marine geology, physical oceanography, ocean engineering and marine affairs. The College of Marine Studies is largely a graduate program.

A grant from the National Oceanic and Atmospheric Agency of the Department of Commerce of $450,000, a grant of $240,000 from the Geography Branch of the Office of Naval Research, and a grant of $180,000 from the State of Maryland for a three-year study of the effects of widening and deepening the Chesapeake and Delaware Canal are evidence that the College of Marine Studies is already highly regarded for its research capabilities. In recognition of these developments the University of Delaware was designated a Sea Grant Institution by the U. S. Department of Commerce, joining twelve other institutions with strong marine programs.

Institute of Energy Conversion. The Institute houses the research activities in the field of energy conversion which are not conducted within the different departments of the College of Engineering or within other departments of the University.

Initially emphasis will be given to research in the domain of solar energy conversion and related fields. This covers the different concepts of solar energy harvesting, energy converters, energy storage, system analysis, statistical and economical aspects. Other areas of energy conversion are expected to develop in line with participating faculty interest.

The Institute will also coordinate interdepartmental programs in which it is chiefly involved, especially in the domain of solar energy conversion. It will offer its assistance to other departments to stimulate research in energy conversion and to obtain research contracts from outside sources. It will cooperate with and contribute to other programs in this or related fields at the University. Accompanying the technical research will be projects in other University departments in Environment, Market/Social Acceptance, Economics and Political Interface.

Delaware Institute of Medical Education and Research. With no medical school in the State cooperative arrangements have been developed between the University, the Wilmington Medical Center and the Jefferson Medical University in Philadelphia. The College of Engineering is engaged in Biomedical engineering research supported largely by federal funds. The State has established the Delaware Institute of Medical Education and Research with a separate state budget but its program centered in these three institutions. Examples of research in engineering include mathematical models of the mechanical strength of bones, polsatile flow (blood), transport and chemical reaction mechanisms as related to cancer and electronic instrumentation applied to medicine.

Relations with Delaware Technical and Community College & Delaware State College. These two institutions and the University work in cooperation to provide maximum opportunities for students. Delaware Tech contracts with the University for a two year College Parallel Program. The combination of that program and Tech's associate degree programs in engineering technology offer a path to transfer to the College of Engineering which has beneficial aspects for certain students, especially for blacks and minority groups. A program has recently received initial funding from the Du Pont Company with the objective of involving these three schools as well as the Wilmington high schools. The program is interested in students from high school through graduate programs in engineering. Attention is called to the Conference on Engineering and Science Programs for Minority Group Students held at Drexel University last October.

The University's City Science Center. This center with offices in Philadelphia is a non-profit organization wholly owned by its 25 participating educational institutions and hospitals in the Delaware Valley region. It performs a function, where beneficial, of coordinating research activities on projects in which several institutions participate.

The Governor's Council on Science and Technology. This Council was established by the Governor to advise him on the Science and Technology aspects of problems in the State. Examples include:

(1) A complete recycle system of solid waste disposal now being pilot planted by the Hercules Company;

(2) The impact of industrialization and proposed port facilities on the coastal region of the State;

(3) The management, support and development of sewage systems in the State and their relation to water use.

Engineering and Society. The College offers a 5 year program leading to two degrees—Bachelor of Engineering (in each curriculum) and Bachelor of Arts and also a Bachelor of Engineering Administration in cooperation with the College of Business and Economics. An experimental Master's degree is under consideration to provide time and incentive for the engineering student to relate with students and faculty in the social sciences. This program proves two-way participation. The engineer learns about and uses the tools and methods of the social scientist. He also uses the tools of the engineer to quantify the social sub-systems of society.

University of Delaware Development Plan, December 1971. This plan, prepared by an off-campus consultant with the benefit of communications from campus individuals and groups, is directed chiefly to University land use and architectual planning. In consideration of the effects of the growing size of the University, both on its internal activities and on the City of Newark, a limit of 15,000 undergraduates and 3,500 graduate students is proposed for the main campus. Growth beyond that would require other campuses or the greater use of other Delaware institutions.

The Design Commission on Academic Structure and Organization, December 1971. This report recommends some redistribution of responsibility among the vice presidents and within the structure of certain colleges. It addresses the problem of interdisciplinary programs and intercollege communication by proposing the establishment of Councils.

The Economic Impact of Educational Institutions

A Discussion of
Estimating the Impact of a College or University on the Local Economy
Part 1

John Caffrey

We knew when we put together the study, *Estimating the Impact of a College or University on the Local Economy*[1], that knowledgeable people were going to take a good look at it and say, "No, no, no, that's not the way to do it at all." But we also knew that it would start them thinking about the problem and about some items that ought to be on the agenda. If you hunt for the small print in the book you will find somewhere a disclaimer to the effect that we know that this is simpleminded and that experienced economists will have nightmares over some of the things we have ignored. In fact, it has been useful to a lot of people if only as something to tear up and use as fodder for their own thinking.

There were two things I was anxious to do when I conceived the study. First was to draw attention to the negative effects, because all of the economic impact studies that had fallen into my hands at the American Council on Education revealed the amount of money that a university and its staff and visitors spent locally, but no one mentioned what the community lost. I felt that actually it puts one in a

John Caffrey is Executive Vice President of Rockland Community College, Suffern, New York.

better position to concede all of that and *still* come out in the black. Second, no two studies that we collected had used anything like the same methodology. We thought that if people used the same methodology, studies might be comparable from one place to another.

It turns out that the first objective has been realized. More studies do now look at the negative side of the picture, but the picture is still as chaotic as ever. People just take the Caffrey-Isaacs recipes and tear them all up and rewrite them to their own specifications. I want to cite three examples that I have in hand, recent illustrations of results of attempts to use the ACE book as a base.

One was done by the New England Board of Higher Education, a college impact study from Hampton County and Amherst, Massachusetts. It is available, I believe, from NEBHE. Another was a study in which I had some role, because I was part of the group of consultants that used the ACE book as a basis for a study of the University of Pittsburgh and its impact on Allegheny County. The study report is in its third draft; and the fact that this is the third draft and that there will be a fourth illustrates something we dimly suspected but had not realized would be so important—namely, that every fact and every allegation of a fact in here can be seized upon by some local political entity for its own purposes. They can say, "Aha, you see!" and pick this or that out of context and perhaps misuse it. Hence the University and all of its friends are being very careful, not only in what they say, but in exactly how they say it. The study report will be available from the University in Pittsburgh, which I think plans to distribute it to its alumni and to people in the community.

The third study, the best known of several recent ones, was done at the University of Wisconsin on that area. This study I like particularly, because the method of presentation is so clear and because, although they tend to follow the same general approach that is in our book, they also sensibly depart from it when necessary.

There are a few lessons that we have learned from this experience to date that have led to some further conclusions. For one thing, many people have said that the recipes appear too complicated, but that when you actually try to follow them they are too simple-minded. A second edition that is somewhere between appearing too complicated and being too simple-minded is obviously necessary.

However, one conclusion that everybody has come to so far (and I am referring not only to the published reports) is that *it ought to be a policy of the administration of every institution to look at its economic impact.* "It is our policy to recognize that we have an impact on this community, outside of the field of education and culture, a simple-minded, cash flow economic impact, and furthermore, some of the factors in that impact are within our control." If, as they discovered at Madison, they were buying an enormous amount of their material outside of the area, it is quite possible that they might turn around and say, "Perhaps we should buy more locally or perhaps some things we are buying at a distance we should encourage local traders to produce for us." Hence, there may be an effect of the observer on the thing observed.

Another corollary is that such studies ought to be repeated periodically, not perhaps annually (too much headache to ask of anybody every year), but certainly every two to three years. Such a study can be repeated by simply filling in the blanks with some of the new data already gathered from local sources. I think this policy is a matter of social and political consciousness and conscience, and should be the policy of any institution, whether public or private, that recognizes that it has this narrower kind of impact.

Many issues emerged from this study, and I want to touch on them briefly. One of the rules that emerged from Pittsburgh—more or less a hypothesis that I would like to test—is what I call Mowbray's rule, named after my colleague, the economist George Mowbray. He proposes that if you count all the capital that has been spent on the institution since it was founded—all capital development, not counting land—the sum total gives a good estimate of the annual operating budget. This turns out to be remarkably true in institution after institution. The University of Pittsburgh has spent about 125 million dollars on its capital development over the last half century, and its annual budget is about 125 million dollars.

There also emerged from this the remarkable observation that there is not nearly the administrative control of expenditures we think there is. It is quite possible in a large institution to find out that there is no one person or no one place in the institution where one can find out about all the expenditures of the institution! It has been a shock to several administrators to discover that many departments and divisions in institutes and centers, and other wards and outgrowths in the institution have control of discretionary funds that the university knows nothing of. Therefore, it cannot easily sum up its own estimate of what its impact is! In fact, I discovered one large institution near the national capital that had 80 faculty on the payroll that were unknown to any dean or department or division head.

Another real problem is estimating *human capital*, which has been touched on today both lightly and heavily and is a bottomless pit. It is difficult to know where to stop analyzing human capital with its secondary and other effects; and so much time and energy can be wasted on arguing about what ought to be included that you never get the study finished, particularly if there are economists on the advisory committee, because they are going to argue rather than gather information.

Another tough problem that emerges in practice is determining what the regional limits are. It is easy in Yellow Springs, Ohio, to decide where the impact of an institution starts and stops, but very hard in mid-town Manhattan. We fell into a trap in Pittsburgh, where we got conned into looking at three concentric rings of impact: the City of Pittsburgh, Allegheny County, and the rest of Pennsylvania. There was really a fourth one, namely everything outside of Pennsylvania.

A further controversial issue, which I have not been able to solve, is how to account for university-owned businesses, such as presses, bookstores, concerts, and similar activities. Is it fair to regard them as competition with local business? One can get into a terrible hassle about that, because it can be argued that if the university owns the bookstore and if the university makes a little profit on that, may 5% a year, at most the community loss is some of that 5% profit. Otherwise, books are bought as any bookstore would buy them, and the personnel are paid in the same way as in a privately owned business.

Another conclusion is that there has to be some kind of an advisory committee that can turn off the bickering now and then and say, "We won't talk about it any more. We are arbitrarily going to say this is over here and that effect is going to be counted over there." The minute you get more than two or three professors together, you know what is going to happen. We also discovered that there is much to be said for both in-house and out-house conduct of the study. Doing the study with an internal committee takes advantage of local resources and expertise, etc., but using an outside consulting agency or group to come in and do it for you takes every one off the hook a little bit. Then you can say, "We hired some experts, they came, and this is what they found." For political purposes, this has certain advantages. The trouble with it is that it costs out-of-pocket money, whereas by putting the resident slaves to work extra hours, the institution can get the study almost for nothing.

We found from our experience that it is important to decide who will be the audience for the report of the study. This decision has to be made fairly early. Are we going to address the citizens and taxpayers, the city council, the alumni, the people we want to give money, the institution, the state legislature? It makes a tremendous difference in the way in which conclusions will be stated. We concluded in Pittsburgh that we had two or three audiences and needed two or three different reports, depending on who was going to pay attention to it.

In the Amherst study, the people who used our book have been kind enough to tell us what they thought of it and what needs to be done to improve it. They recommended that a standardized computer program, perhaps in four panels, be written so that the users can just plug in the data from a local study. I am not optimistic about that, but they made the astonishing comment that there are 47 mathematical models in the system requiring only 27 variable inputs. Now, that makes it *seem* very simple, but it takes from four to eight man months of work to get those 72 inputs; at least a third of them we finally had to abandon because they were not obtainable from university records or from the local community! (There are some things you think you would be able to find out in a university but cannot.) Twenty-five of those values may be obtained from questionnaires.

This raises another controversial issue, however: whether to take the risk of drawing up a faculty questionnaire, asking how they spend their money and finding resistance to it from the average faculty member, or using estimates and guessing what they probably spend and what visitors probably spend. I think the inexactness is about equal both ways. We concluded from a spot check that those who returned the questionnaires tended to be the higher income people.

At Amherst it was also recommended that this study be done every year; that was their most extreme recommendation. They felt the situation would change as a result of the study, but I think that will remain to be seen.

In the Amherst study, there were some very interesting local findings. I will quote a couple of these to whet your appetite. One that I found rather negatively amusing was that when they used our model and came up with the estimated local business volume of the College, the estimate exceeded by almost one million dollars the actual total business activity of the community! Think about that—there is a lot of optimism there. They also discovered that the tax loss to the community was actually negative; that is, the College was paying more in taxes than comparable business and residential areas would have yielded to the community if all the things that were yielded in time were counted. The personal income of local individuals just in Amherst was $47,556,000, attributable to all the five institutions in the Amherst area.

One of the most consistent findings from all these studies, almost an absolute truth, is that, however you do it, however different models you use, and however many you throw out and decide not to use, the results are overwhelmingly the same: a college or university is a tremendous asset to a community. There seems to be no question about this. You may disagree as to whether the asset is great or small; however, there is no question but that, when you sum it all up, the university or college has nothing to fear from opening its heart and soul and its books to look at this problem. The public relations advantages are obvious.

If we had been able to use factor analysis (and we were not), we could have reduced the number of variables to estimate the impact from a very small number of factors. This would of course influence the cost of the study. Based on our experience, the cost of a study, whether one counts in-house or external costs, seems to run between $20,000-$25,000 for a large institution, considering computer runs, personnel time, interview time, questionnaires, key punching, etc. If the study is to be done well, it is not something that can be done by a graduate student working weekends.

REFERENCE

1. Caffrey, J. and H. H. Isaacs, *Estimating the Impact of a College or University on the Local Economy*, Washington, D.C.: American Council on Education, 1971.

The Economic Impact of Educational Institutions

A Discussion of
Estimating the Impact of a College or University on the Local Economy
Part 2

Morton J. Marcus

The college corporation today is necessarily engaged in acts of public relations and image development that are different from those of the past. Scholastic and athletic efforts do not convince the college's neighbors of its essential virtue. A Nobel Prize or an NCAA championship does not justify its existenct to the Mayor and his revenue-hungry Council. The modern ivy-covered tower is capital not Camelot. As such, the productive value of the asset must be estimated and the distributional consequences of the activities based within must be divined.

It was my pleasure to be associated with the effort by the American Council on Education to assist college officials and members of the general public in the continuing examination of a college's economic impact on its community.[1]

Let me outline the approach we took to this important topic. The ACE study looks at the realities and relationships of colleges today. There are three pressing considerations:

Morton J. Marcus is a Research Economist with the Division of Research, School of Business, Indiana University, Bloomington, Indiana.

1. Colleges occupy land and buildings that generate considerable activity and are often exempt from local property taxation.

2. Many urban governments, large and small, perceive increases in input prices and increases in the demands for local government services without concomitant increases in locally-generated revenues; in other words, a fiscal crisis.

3. Colleges and universities have scarce resources, fiscal and administrative, with which to meet the challenges of adjusting to changing enrollment patterns after an era of rapid growth, and to answer the questions and demands, "What are you doing to (for) us?"

The ACE team sought to generate a handbook for college officials similar to that excellent report on economic base analysis developed for the CED by the late Professor Charles Tiebout.[2] It was necessary to develop some very elementary models, or perhaps more appropriately, statements of relationships that could be used by various college officials without too much difficulty. The report also contains numbers from different sources that can be used in these models.

We began with the fundamental realization that no single number represents the impact of the college on the community. The issue is multidimensional and requires a plurality of models that constitute a consistent entity. Our second basic concept was objectivity or balance. A decent impact study should satisfy more than the partisan public relations interests of college administrators. Long-range interests are best served, we felt, by credibility. Thus, the techniques of the impact study should serve anyone studying the relationships of the college and the community. The ACE study suggests considering impacts that may be negative as well as those sure to be viewed as positive by college and community leaders. The economist might call these the costs and benefits of the college's presence in the community, but I would warn against using such terms here since all of the usual technical specifications for cost-benefits studies are not met.

Some college activities merely transfer transactions from one set of resource organizers (firms or institutions) to another. For example, colleges do compete, and often most effectively, with local business firms. This item is often overlooked or thrown out in the typical college report. Colleges run hotels, eating places, bookstores, sporting events (if not sporting houses), and entertainments. Some may look on these as benefits for the community, cultural enrichment, expanded services or opportunities, but to the local movie theatre operator and his business community counterparts, these college-sponsored efforts are competition.

The ACE study does not attempt to encounter all the intriguing issues of a college's potential economic impact on the community in which it is located. Long-term impacts on human resources and the fundamental structure of the community are not discussed. Primary focus is on the short-run, although some attention is necessarily given to the physical capital implications of current activities. The short-run impact of a college on its community varies with both the nature of the institution as well as the structure and scale of the community. Los Angeles is less subject to variations in the activity levels at UCLA than is Bloomington to the activities of Indiana University. The relative magnitudes of the interacting forces differ considerably in these two cases. It is when a school has great *relative* size that the question of impact most often arises.

Typically college administrators, with their public relations offices, develop reports on the economic impact of their institutions. Our reading of these reports led to the belief that these could be improved by being more complete conceptually, less prone to pro-college bias, and thereby more credible.

Such reports would still not encompass sufficient sophistication to satisfy professional economists. To generate a study adequate for the professional literature requires some application of basic input-output techniques detailing the direct, the indirect, and the income-induced effects of transactions within the community and between the community and the rest of the world. Studies of this nature have been produced at Western Washington State and at the University of Colorado. The natural complement for such scholarly virtue is a high level of direct and indirect expense. Most college presidents are not willing, without extra-mural funding, to see their resources consumed in this manner. In addition these endeavors require long periods of time, while pressures on the college are often immediate. There are demands to be met and impressions to be made tomorrow, with the result that the information must be available yesterday.

Not only are the ACE relationships short-run, ignoring amenity and locational factors, but they specify average, not marginal values. So little is known of the specific interrelationships between the college and the local economy that it would be bad guesswork to rely too heavily on average relationships. If enrollment increases by two or three percent, can one really assume that the city's police costs, for example, will increase according to the existing average relationship between such expenditures and the student population? We have no evidence specifically in such cases, but we may often suppose that, for large numbers of functions, marginal costs will be less than average costs where the latter include capital costs as well as operating items.

At the other extreme, the question is often raised, "What would this town be like if the college were not here?" There is no sensible answer to this question. Most college communities have evolved with their colleges; the answer to that question is speculation dependent on the imagination and biases of the respondent. I recommend that a college president pick out the most unattractive, depressed and stifling nearby town and offer that as the alternative result of a different course of historical growth. On the other hand, the Mayor might observe the lack of any reference to a university in the biblical reports of Paradise.

Thus, the ACE report suggests an overall approach that is "rough and dirty" and weighted with limitations, but one which is feasible for those charged with the responsibility of preparing a college-impact study. Let us look at the components of this approach.

Normally the college will report its payroll, the amount spent directly by the administration on supplies, and the capital budget as the sum of the economic impact. Supplementary statements may be made about the fire engine donated to the city a few years back and the free concerts by faculty members available to all citizens. Figure 1 suggests the simplest dollar-flow model that should be considered. There are four major groups of college spending agents: the college itself, its faculty and staff, its students, and its visitors. Part of the expenditures made by each group is received as income by local business and government units.

To these effects must be added consideration of the indirect purchases and income-induced purchases in the community stimulated by the initial transactions. For example, the student buys a hamburger (or some marijuana) at the local drive-in restaurant. His purchase results in wages to the vendor and purchases of supplies to replenish the seller's inventory. The wages of the cook (or pusher) result in further local purchases of goods or services. There is some leakage because the basic ingredients are imported from another community. In small towns these leakages may be considerable portions of the total initial amounts involved in the transaction, particularly if the cook (pusher) lives in a nearby town.

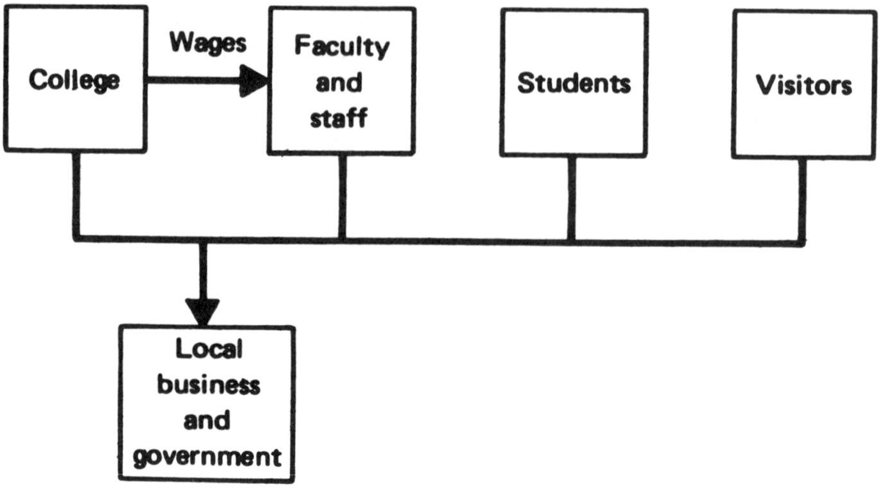

Fig. 1 A simple expenditure model*

*John Caffrey and Herbert H. Isaacs, *Estimating the Impact of a College or University on the Local Economy*, Washington, D.C.: American Council on Education, 1971, p. 5.

An extended cash-flow model, as shown in Fig. 2, depicts more of the initial flows that need to be considered in the complex economic activity of modern college organizations. Funds are received by the college in many forms for services provided to residents and non-residents. Admissions to concerts or basketball games are examples. The college receives grants from foundations and from the federal or state governments. It derives income from earnings on its own assets. To the extent that some of these income-earning assets are in the community, the college has some degree of impact on the local economy by its investment and operating decisions. On the other side, the college is sending dollars into the household sector through stipends, grants, loans, and wages. It is also making purchases from the local business community. Students, faculty, and staff are buying in local shops. These in turn hire employees from the community. There are flows in terms of taxes and fees from all of the above to the local government, which again in turn generates purchases and wages. If all the transactions outside the local community were included in the diagram, we would require a large fold-out page and several printing colors.

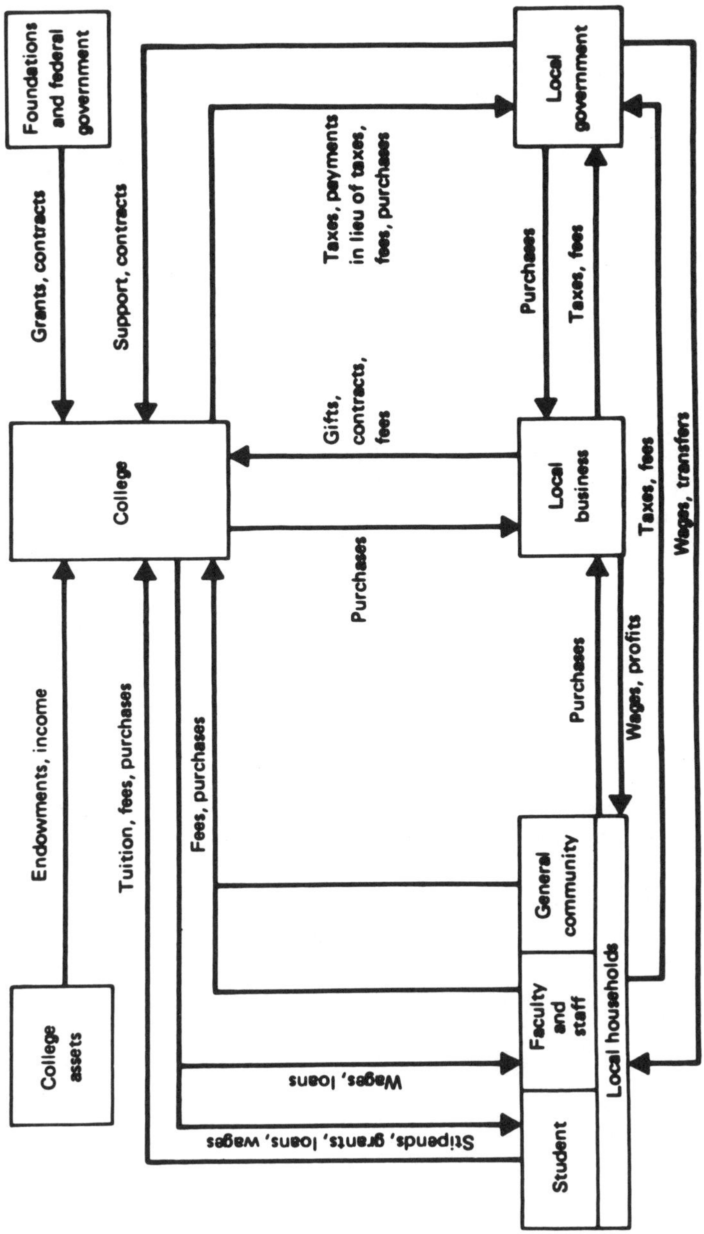

Fig. 2 An extended cash-flow model*

*John Caffrey and Herbert H. Isaacs, *Estimating the Impact of a College or University on the Local Economy*, Washington, D.C.: American Council on Education, 1971, p. 7.

Where possible, the college can employ its own records to start estimating the magnitudes implied by these diagrams. Our experience indicated, however, that often college records are not easily interpreted in these terms. The ACE study gives some guides and several national bench marks to guide the investigator.

The heart of the ACE study is a set of easily computed equations that consider not only the cash flows discussed above but also the employment impact of the college and the capital assets associated with college-generated economic activity. The emphasis is as much on the government sector as it is on the local business community; the interrelationships between local business and government are given full attention. Figure 3 gives some indication of the range of topics included in the full presentation of the ACE study. Some of the items noted in Fig. 3 are standard. For example, no Mayor would fail to note in his analysis of the college's impact the taxes foregone through the tax-exempt status of the college. But these must be counter-balanced by the value of municipal-type services that the college provides for itself and its personnel. If the city had to provide such services could it do so for as little as the college spends? Or phrased another way, would the foregone tax revenue compensate the city for rendering such services?

It is difficult to ascribe a given number of jobs in the community to the presence of the college (other than direct college employment), but with the guidance of several input-output studies we are able to offer some guidelines on income and employment multipliers for the investigator. Such topics are subject not only to much speculative debate, but also to solid research.

It was our hope that the ACE study would stimulate college officials and local community leaders to discuss their interrelationships more realistically. The study should also demonstrate to all the value of supporting research than can contribute to the fundamental knowledge participants in such discussions urgently require.

REFERENCES

1. Caffrey, J. and H. H. Isaacs, *Estimating the Impact of a College or University on the Local Economy*, Washington, D.C.: American Council on Education, 1971.
2 Tiebout, C. M., *The Community Economic Base Study*, New York: Committee for Economic Development, 1962.

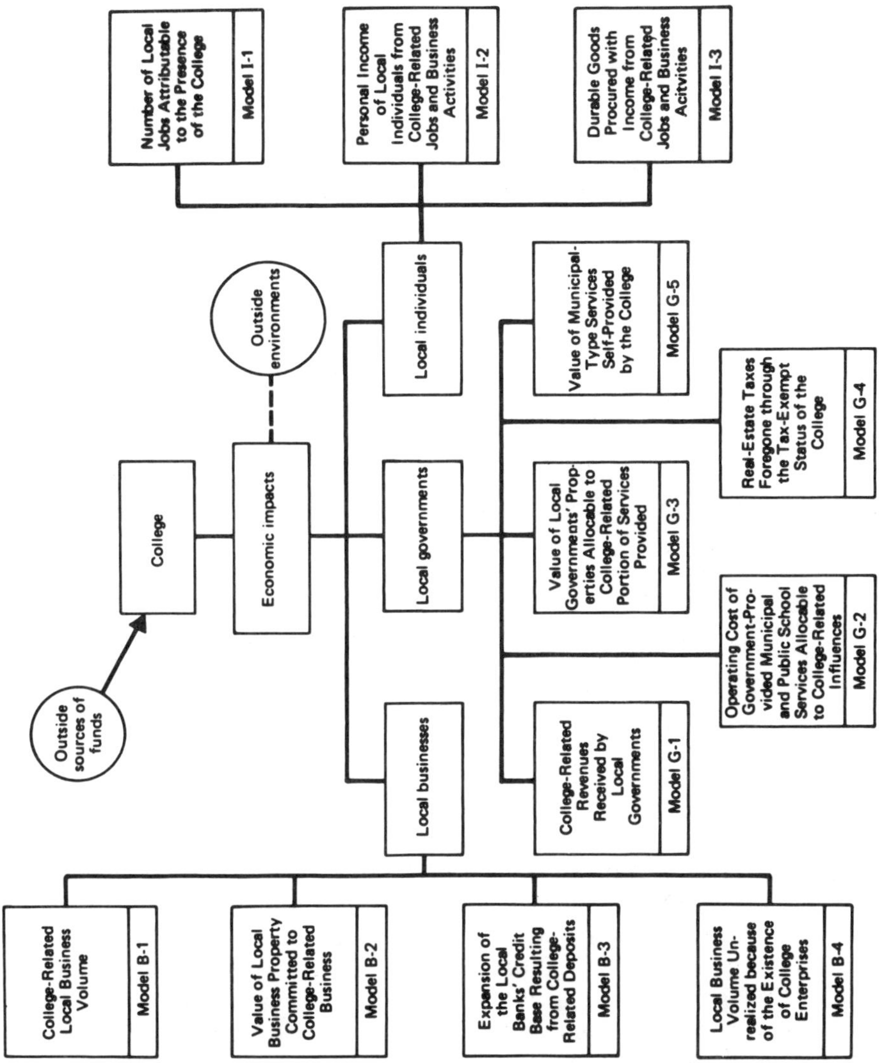

Fig. 3 Economic impacts on local businesses, governments, and individuals*

*John Caffrey and Herbert H. Isaacs, *Estimating the Impact of a College or University on the Local Economy*, Washington, D.C.: American Council on Education, 1971, p. 10.

The Regional Role of Engineering Colleges Conference

Report on Workshop I

Methodology for Stimulating the Regional Role of an Engineering College:

Industrial-Economic-Academic Mechanisms

Lawrence N. Canjar and Lucius Gregg, Co-Chairmen

Participants

John A. Duffie	James H. Leonard
Frederick H. Jackson	Morton J. Marcus
Thomas W. Knowles	Eric Whitman

Dr. Canjar: We first gave our attention to the real reasons faculty do the things they do. The stated goals of a university have to do with national and local interests, but the real goal of most universities is to attain status. In the 1950's, this involved recruiting productive scholars—people who published and impressed fellow scholars at other universities. Money was needed to support their activities and the federal government provided the great bulk of the funds.

The implication here is that universities are not always altruistic. When national goals (promotion of fundamental research) and selfish interests (development of prestige through research) are congruent, the behavior of faculty is easily established and predictable.

With this backdrop we asked ourselves, How do you change the direction of a university?

We recognized the fact that in the past faculty were preoccupied with national recognition as a means to enhance their professional status and potential mobility. At the present time, with engineering colleges stabilizing or contracting, there is decreased mobility and faculty may become more concerned about their contributions and importance to the local scene.

At this point in the discussion, members of this workshop fell into two camps, depending on their backgrounds. Some come from very well established schools with strong faculties and well established research programs. One has an experiment station that is very successful. Obviously this kind of school encounters a great deal of inertia when it considers changing directions. Rather, there is a tendency to talk about small or large programs which may be additions to what is being done already. Some other workshop members come from what might be called emerging colleges—those that have not established themselves as yet and are still seeking a direction for a principal thrust. These two types of schools approach the problem of response to local and national needs in different ways.

We learned about how the University of Wisconsin takes care of regional needs. Wisconsin has a long tradition of being public service oriented. We noted that this is probably so because the University is situated in the State capital and faculty-legislator interaction takes place constantly.

There is an entirely different situation at a school like the University of Cincinnati which is separated from the State capital. At Cincinnati, we heard that they have something called the Technical and Business Services, which is an agency made up of some staff people, professors and personnel from the Chamber of Commerce, such as the chief engineers in local companies. This is essentially a link or a liaison between the University and the surrounding community. It is only an idea. They have not had enough activity for proper evaluation at this time.

The University of Detroit's College of Engineering is an emerging professional school. Through its Doctor of Engineering interns, it has intimate involvement with local industry and recently has extended this involvement to the City of Detroit. The students of the College, graduate and undergraduate working in concert, have designed the City's emergency ambulance service and have proposed solutions for alleviating congestion and difficult health care delivery at the Detroit General Hospital.

Dean Duffie suggested that we develop a list of premises upon which future discussions might be based and provided the group with ideas which were accepted by all. They may be summarized as follows:

1. Research, teaching and public service are inter-related.
2. Engineering needs to be problem oriented and based on situations within a region.
3. Engineering cannot operate in public service in isolation from the rest of the university.
4. Each institution must design its own program or relations with the region, depending on its traditions, abilities and the nature of the region it serves.
5. Effective public service should be the responsibility of the faculty and be recognized and rewarded. (Faculty are the only ones who can provide this service and they have to think it is a good thing. If you have not sold them, you have not accomplished what you wanted to accomplish.)
6. The development of the college must be planned with public service to the region in mind. (Those colleges which are already developed will have to rethink their goals.)
7. Engineering curriculum can influence and be influenced by effective public service.

The last item is most important. It is not enough for the faculty to continue to teach its traditional courses in its traditional way and try to function as public servants. If they are really involved in the region's or community's problems, somehow, this will be reflected in changes in their courses and curricula on campus.

We developed a list of things a university or college of engineering might do. It can perform applied research, provide continuing education, operate a referral service (patterned after the Technical and Business Services of the University of Cincinnati), and make available university facilities, such as library and information services.

The major problem is faculty education and recognition of public service as a vehicle or the equivalent of academic research.

As for mechanisms, the group felt that some sort of an agency within the college is necessary to facilitate public service and that it is important not to isolate it from the rest of the university. It has to be able to draw on other disciplines in the university. Engineering schools are recognized as having the talent to attack large scale problems, but they do not have all the talent necessary. They have to draw on political science, sociology and other departments outside the college.

There is a danger in developing a permanent agency which would tend to perpetuate itself and become a large overhead burden. We preferred agencies that operated on an ad hoc basis with a minimum staff that would identify a problem or an area, get the college or university involved and then disband. The kind of staff for this agency would have to consist of personnel who were well accepted by the faculty (faculty members) and who had good contacts in the community (outside men). These characteristics do not exist in one person and the group debated the advisability of converting a faculty member into a development man or bringing a man with good contacts into a faculty organization.

The question was raised, Should engineers be involved at all? Dr. Lucius Gregg gave what the group felt was the most convincing argument. Urban problems have to be attacked by some one. Should it be the social and behavioral scientist or the engineer? Should an effort be made to develop broader quantitative skills and methodology in the social sciences or should engineering with an already sophisticated quantitative orientation be encouraged to move in on social problems? Dr. Gregg argued that if there is anything all social problems have in common it is that they have many variables, and engineering schools are the only ones that are producing students, consistently and in high volume, that are not afraid of five or more variables. In fact, their interest is proportional to the number of variables involved in a problem.

The whole question boiled down to where can one get the kind of manpower output that has the greater likelihood of looking at the number of variables that are involved in many of the social problems of today. In fact, the number of variables may indeed be the reason why the problems evolved in the first place. City and regional governments lack the competence to handle these complex matters in any way except for one or two variables at a time.

The Regional Role of Engineering Colleges Conference

Report on Workshop II

Methodology for Stimulating the Regional Role of an Engineering College:

Political Mechanisms

Samuel K. Gove and Robert Granda, Co-Chairmen

Participants

Samuel Baxter	Judy Millaney
Lynton K. Caldwell	Herman Sievering
William Geekie	

Mr. Granda: We defined the major problem areas in the methodology of political mechanisms for stimulating the regional role of an engineering college, without arriving at a suggested solution. I will go over the problem areas and say a few words about each.

The first one was that, in order for interaction between the engineering colleges and government to come about, some one needs to start it. The next point is that there seems to be a mutual distrust and lack of knowledge between the people in government and in the University. Also, the reward system seems inappropriate to the problem in the schools. There seems to be no premium given for public service. These points emerge from the samplings of our discussion, reported below.

Comment: It is really not the system but the access to the system. I assume that people want provisional rewards—the promotions, the salary raises. The issue is who is going to get them.

Comment: There is not an appropriate reward system in government either.

Comment: We were concentrating our discussion on the reward of the faculty. The faculty member who specializes in special services does not get the reward, although we go through the lip service of saying that he does. There is one other thing that our workshop discussed briefly and which may be a good point to come out of this conference: I think we have to turn around our professional organizations and think of ways of rewarding people who contribute through public service. Right now, the full focus is on those who do research and publish an article. Somehow these organizations have to figure out a way that the professor who really accomplishes something in a neighborhood where some of the worst problems are will be congratulated by his peers in his professional organization.

Comment: The problem of publications is a serious one. If you begin as an engineer involved in problems that have a strong social interface, you have difficulty knowing where to publish. If you publish in a sociology or in a political science medium, you are going to be attacked because you are neither a sociologist nor a political scientist; you will be judged in terms of the standards of those fields. There is really no way for the people who edit those journals to make allowances for the fact that some one is trying to cut across fields and, while he may be weak in that area, he brings other knowledge and skill to the topic. So even when using the traditional medium of publication for reward, there are problems which have a bearing on public service.

Comment: I think, too, that a reward for public officials is very pertinent to this discussion. We spent much time considering the lack of communication between the university and government. While it may be inappropriate to mention, I notice that the number of public officials attending this conference is very limited, even though quite a few were invited. It has been suggested that we need to change the universities to promote greater communication with government; but from the discussion in our workshop, which included a representative of the City of Chicago, we concluded that about twenty-five percent of the blame for the lack of communication could be placed on the universities and about seventy-five percent of the blame on the government.

Comment: Do you feel that it is helpful to bring together engineering, political science and other social sciences? From your experience do any of you feel that the interdisciplinary approach is effective?

Professor Gove: Certainly. There are several examples of this kind of approach and effective output here at this conference.

Comment: We spoke previously about the publications of professional associations and the difficulties of publishing papers that involve an interdisciplinary approach. There are associations and publications that are specifically aimed at such an approach. They have a tendency to be dominated by one or two professions, largely because of lack of knowledge about their existence or their purpose on the part of others who possibly could be involved. An example is the Regional Science Association which has considered some of the topics discussed at this session. I have never seen anyone from a college of engineering at one of those meetings. Yet there are Regional Science Association meetings held in this country and throughout the world to deal very specifically with sub-national problems. The emphasis of this organization is one of economics and geography because these two groups have collaborated on many campuses. Now a small number of sociologists and political scientists are entering into the Association. It is such professional associations and professional journals that have a different structure that should be investigated.

Comment: I certainly did not know about this Association, and I suspect that very few people in engineering groups about the country know of its existence.

Comment: I think that is irrelevant, though. What you really want is peer group respect on campus. Take the case of the electrical engineer who is publishing in varied publications. If he does not publish in the IEEE journals, the electrical engineers will not be reading what he is writing and he will not make it in his electrical engineering department. We have to face up to this kind of problem.

It really is a double problem. If you are an engineer writing for a sociology journal, it is not just a matter of having your paper turned down because you do not know the vocabulary. We all know an engineer, a Nobel Prize winner, who is very concerned about genetics and who has earned for himself and others a deserved or undeserved degree of criticism. This unquestionably deters a great many engineers from looking at other social problems because to do so is dangerous. Every good university has mavericks who operate outside their fields. The reaction of the departments to these people typically is that they are dangerous or risky because the departments cannot judge the adequacy of performance outside the field. These people may get into the newspapers in a way that the universities do not like. Therefore the universities fear them. If a maverick is a young man, he does not stay around long enough to become an old one. This has been the traditional experience that has caused men continually to refocus on their specialty. We have not the ability to judge cross-disciplinary people. It is the hardest job in the world for an engineer to look at an artist and say he is a good artist or for an artist to look at an engineer and say he is a good engineer.

Comment: Under present conditions, one must remain current in his field if he is to advance.

Comment: Are you not condemning engineering school administrations since, after all, it is their policies which cause this?

Comment: It is not the administrations so far as I can tell. It is a matter of the peer group.

Comment: I disagree strongly. I think there is an enormous amount that an administration can do to focus attention on problems like this to get results.

Mr. Granda: You touched on a few of the other points I had in my report. One of them concerned the reward system. Another was that the state may not be as interested in interaction as it should be, as evidenced by the lack of representatives here and by the discussion about the response from government to potential contacts with the universities. Further, there was general agreement that there is a need for some kind of linking mechanism and for the interdisciplinary approach.

Comment: Have you given any thought to the nature of this linking mechanism?

Mr. Granda: I have, and I am afraid I am prejudiced. I think it should be either a legislative council or some central receiving place to ensure the possibility of getting help from a variety of disciplines. We need an organization and people who can be mediators and who can "translate" the languages of the various participants. I think that would be one of the most important and valuable contributions. I have seen legislators whose eyes become veiled when they encounter someone who speaks a jargon different from their own; they just stop listening. Of course it is not only legislators—there are many people who do that. Therefore it is essential that there be somebody to "translate." In addition, someone needs to initiate this kind of contact.

It seems that everyone agrees that there is a need for this interaction, but no one has really tried to start it. Finally, I think there needs to be some kind of public relations work done to show that we can perform; to show that, in fact, we are all human beings interested in solving the same problems.

Comment: Most of what we have said today has dealt with a problem of communication. We say that we need to know each other, that we need to talk the same language, etc. But is there not something more direct and more massive at stake here? The state has to decide how much it should invest to support its own technological resources. Do we have any criteria to aid in this decision? Do we have any criteria whereby a state could go to the Federal government and say, "We represent this kind of resource nationally and we need such and such help in order to develop the requisite technology"? Communication is one thing, but I think probably the thing that will make the states develop in suitable technological areas will be adoption of a good state policy and, second, a massive infusion of support.

Professor Gove: We gave some thought to the notion of having a state NSF. This would show that the state is really committed to this idea. So far, such a development seems far off and not in the mainstream, but I have been told that Pennsylvania has done something in this direction. Such a step could get things going.

Comment: I would like to amend that statement. The proposal is not for an NSF, rather it is for a state applied research foundation. There would have to be an immediate problem orientation for the research activity. Incidentally, NSF is now sponsoring about six Governor's R&D Priority Conferences around the country and, while this is a preliminary step, it does serve to marshall the thoughts of the scientific community in those states concerning state R&D needs.

Comment: I was wondering whether anything could be said concerning the efforts of Governor Kerner's State Science Advisory Council?

Comment: The focus at the time was an attempt to bring together academic, industrial and political factions within the State of Illinois to do this type of forecasting and to make available money for risk capital. Their thought was to develop regional types of activities to benefit the entire region, ergo the country. In fact, for Illinois we established one priority, biomedical engineering. It seemed a more promising field for this region than anything else that we considered in 1968. We recommended a ten year period of development, after which the effort would be reviewed to see whether significant progress had been made. The intent was not to proceed initially into very basic research, but to begin with areas that had already been developed, such as prototype development or similar areas which would bring funding based on national priorities into the program. I do not know what happened to the plans since I have not been in contact with any of the parties involved with the project.

Comment: I would like to add two comments. First of all, I think that engineering has a lot to learn from agriculture which is the discipline that has been most successful in coordinating research, public service and education with regional needs and regional resources. The other point I would like to make is that this State, unlike other states with which I have been familiar, apportions a great deal of State money—millions of dollars—to the universities for research. I am not talking about only agricultural research. I am talking about State dollars for research in other disciplines. In a sense, the resources are there. Resources are at the disposal of the institutional officers. These resources are not allocated programmatically on the basis of any sort of Statewide or regional scientific research priorities. They are allocated to institutions and disbursed as they will. So if, for example, those resources were used by the institutions to supplement Federal monies in attempting to meet Federal research programs or used by institutional officers to aggrandize their institutions on a basis of national priority or national prestige, the State would really not have anything to say about it. But this situation could be turned around so that more and more of these resources could be directed toward the regional and State research programs to fill government needs.

Comment: May I just add that the institutions would not do this themselves. In every state where something has been accomplished toward this end, a new institution was created to work with the existing institutions and give the programs focus.

Mr. Granda: Let me summarize. There are six points:

- Someone needs to start.
- There is a lack of understanding between the people in government and the university.
- The reward system in the university may be an inhibiting factor.
- The state may not have the interest that it should have in cooperating.
- There needs to be some kind of interactive mechanism.
- The mechanism should be of an interdisciplinary nature.

The Regional Role of Engineering Colleges Conference

Report on Workshop III
Alternative Goals for Interaction of an Engineering College and Its Environment
John L. Cain and Barclay Hudson, Co-Chairmen

Participants

Egon Brenner	David H. Mintzer
Edward W. Comings	Charles M. Overby
Lucius Gregg	Albert H. Rubenstein
Glenn H. Keitel	George A. Russell
Robert Large	Frederick E. Terman

Professor Cain: Our report will be in two parts. I will make an opening statement along with our general recommendations, and then Barclay Hudson will provide a still broader set of alternatives.

Specifically, we focused on the theme: Interaction of an engineering college and its environment. Initially we had limited the environment to the area or the region about the college and then discussed both some short range goals and some long range goals. I think the general feeling of the group was that we should be somewhat pragmatic in answering the questions, "What can an engineering college really do?" and "How do you decide what you are going to do so as to be creative and not just talk about things that we should do or that we are concerned about?" In discussing interactions it is necessary to think of both on-campus interaction, including things such as interdisciplinary programs, as well as off-campus action.

At this point differences among institutions are important. Certainly the University of Detroit is much different from Auburn University or the University of Illinois. Our settings are completely different. The goals of the institutions, the traditions, and so forth, within which we must operate and function, differ. In this light we discussed what may be said in general about what an engineering college should really do.

A further problem is that we Deans and Department Heads do not command faculty. Crucial to altering present policy is the role of the reward system and the recognition of faculty. Where do the faculty get their kicks? We know where the kicks have been in terms of promotion, tenure, rank and money. A dean does not necessarily legislate these things.

We made a list of some things, short orders and long orders, that we felt might be given some attention in terms of engineering colleges undergoing change: how is change effected, what are the goals of the college itself, what are the goals of the faculty, how do you develop policies, how do you implement change in terms of programs? I was reminded of the statement that I heard a while ago which seems particularly a propos—"Behold the turtle. He maketh progress only when he sticketh his neck out."—and I think there have to be some people who are willing to stick their necks out. We may be conservative in this regard. Maybe we need more radicals. I think we need to be aware of what influences we exert within our institutions, on our fellow faculty members, and on our friends. Our needs are pressing. Some have been outlined in a report of the Organization for Economic Cooperation and Development by Dean Harvey Brooks. Another, sponsored by the Club of Rome, investigates the limits on the growth of our civilization. These reports are not encouraging. Most certainly we need to be concerned about these issues because of their impact on engineering education and on education in general.

There were several areas of and avenues to interaction that we identified as having large potential returns. I will go through these briefly and then ask Professor Hudson to make some additional comments:

1. Increased interaction with industry.
2. Becoming aware of the problems of the industry and community

within the region that we serve with graduates or other programs.
3. Reaching an understanding of these problems.
4. Viewing these problems through an awareness of the technological.
5. Increased interaction with the local government, either community, urban, county, state, etc.
6. Developing person to person relationships with industry and community groups.

In general, it was felt there is a need to develop understanding and articulation and improved communication, both on campus with other departments and with elements in the region. How is this to be done? Possible ways suggested were through communication with the public and the high schools, guidance efforts and continued attraction of talented youth to engineering careers. This lead to discussion of the area of manpower development—needs, demands, and market place. How do we relate to the problems of education for minority students in our community? Again, we considered the matter of manpower development—productivity, etc.—and serving the needs or desires of our constituency. We discussed continuing education. Mr. Terman suggested the need for a national system in this area for part-time advance degrees, non-credit. We asked what could be done in that area. Possible relationships with other engineering colleges in the vicinity, community, state, etc., were explored as a means of attacking some of these problems.

In summary these were some of the factors that we identified and that engineering colleges might consider in establishing new goals, realistic goals, action goals for their interaction with their particular environment.

Professor Hudson: Mine is something of a minority report, not to disagree with other members of the workshop, but to emphasize some additional points.

Our task was to identify "alternative goals" and, specifically, alternatives to the traditional role of an engineering school. The major point that I wish to make is that the alternative goals are already there. They are clear in the speeches of politicians and evident every time you open a newspaper. These are the goals of finding solutions to *social* problems, ranging from transportation, pollution and housing for the poor, to issues like the design of feasible small-scale industries that can absorb locally unemployed persons and provide needed local services for impacted areas.

The "new" role of the engineering school is implicit in these goals; it consists of no longer waiting for clients to knock on the door, but of *creating* clients. This requires two things: first, supporting these new social priorities, and secondly, designing the means to make these objectives feasible. One of the most important points brought out at this conference was the one made by Betty Vetter: there is no problem today about the *need* for engineers; the problem is translating these needs into an effective market *demand* for them. This is largely a technological problem, a problem for engineers, because it consists of showing that there are feasible ways to respond to these social needs in terms of specific packages of investments. This does not mean engineers have to produce miracles or final solutions, just some concrete alternatives to doing nothing.

I have the strong impression, however, that many engineers consider such a role highly unrealistic. In particular, I have heard four different kinds of objections raised by people here at this conference. I will try to respond to these in turn.

First is the matter of tight money. The argument is that it is useless to advocate new social priorities because nobody will be able to pay for them anyway. But this may be a pretty shortsighted premise. We are looking at things now from the perspective of the current retrenchment in the aerospace industry and projecting this situation into the future of our engineering students. Twenty or thirty years from now, when today's freshman entering engineering schools should be reaching peak responsibility in their careers, the American economy will have an additional trillion dollars to spend, give or take a couple of hundred billion. The problem then will not be where the money will come from, but how to spend it, or at least how to spend it wisely. A lot will depend simply on the know-how of applying technology to evident needs, and that will depend to some extent on what we are teaching our engineering students in the educational pipeline today.

Secondly, the argument is made that there is no power in ideas to change social priorities. We have all seen—it is true—how vulnerable these "new" social priorities are to the impractical utopianism of philosophers, the cynicism of politicians, and the urgency of "defending our national interests" against the foe in Indochina. Yet academics tend to underestimate themselves. For example, much of the social legislation in this country rests on the academic theory of the legitimate role of unbalanced budgets, first used during the New Deal. The power of an idea can also be seen in the interstate highway program, a concept amended into our "defense" strategy with incalculable effects on our national life. Another example might be the idea of the land grant colleges and their effect on giving American higher education a whole new character, distinct from the earlier European prototype.

Thirdly, I have heard people tell me, "Why should engineers tell society what to do? What right do *they* have, or what special providence allows them to set social goals?" I would say in response that engineers are going to be held accountable along with everybody else for the things our society as a whole is leaving to the next generation. In other words, engineers, too, are citizens with an "existential" responsibility for things over which no one can exercise total control. But engineers at least have the advantage of a close association with universities. Remembering the literal and original meaning of the term "university", it can be said that engineers have a far better opportunity than most citizens to discuss and develop ideas that affect society at large.

More than this, however, engineers have had a far more direct hand than most citizens in helping to create the "technological" society and all its consequences. There is little distinction between the view I have heard here, to the effect that "we have no right to determine social processes," and the view that "since we have not consciously determined these processes, we can not be held responsible for them if they have happened."*

Eventually, however, society will probably make it a part of the engineer's job to account for the social spillovers associated with his designs. Just as pharmaceutical houses are required by law to specify the deleterious side effects of their products, so engineers will become increasingly involved in the role of specifying the full range of outcomes associated with their own "medicine" in terms of such things as pollution, unemployment, income distribution effects, safety, esthetics,

*One person who recognized this problem was Albert Speer, ex-Minister of Armaments under Hitler. Speer noted that during the war, "The more technical the world imposed on us . . . the more dangerous was this indifference of the technician to the direct consequences of his anonymous activities . . . Because of what seems to be the moral neutrality of technology, these people (the technicians) were without any scruples about their activities." (Albert Speer, INSIDE THE THIRD REICH, New York: Avon Books, 1970, p. 283).

and resource depletion. With the Environmental Protection Agency, this accountability has already begun. So it is not a matter of whether engineers have a "right" to get concerned about social objectives; it is a question of how they have managed—like all the rest of us—to avoid it so long.

The fourth and final objection that I have heard to the "retooling of engineers" is that the new priorities do not have any institutions to lobby for them, while the old priorities have vested interests like Boeing or the Department of Defense. The answer to this seems to me quite simple. There is an enormous amount of money to be earned in doing good. Preserving the environment costs money. Therefore someone is going to have a chance to earn it; so there will be jobs and profits to be had for cleaning up pollution, just as there are for the old priorities. One reason why engineers have been employed in building missiles and smarter bombs is that they have gotten pretty good over the years at the technology of killing people. But when it comes to the technical problems of building a feasible steam powered auto or even a better carburetor, nobody seems to know *how*. At least that is what advertising paid by American industry would lead us to believe about engineers.

To the extent engineers do not actively work to refute this conservative image of their potential role, the American public will remain technologically illiterate to future possibilities of finding real solutions to social problems. In other words, if engineers remain pessimistic about their potential usefulness, the public will take their word for it. Consequently, there will be only limited public support for new social programs which need engineering skills, and any pessimism about the limited emergence of new lobbies in these fields will be a self-fulfilling prophecy.

To go back to my original point, in the gap that exists between evident needs and the yet-to-emerge demand for new social priorities, there is a necessary stepping stone that consists of *designing products* that can show how money can be spent fruitfully. This is the very basis of our consumer society: designing products for people to spend money on and thus creating new kinds of demands. Designing products to solve the social problems is largely a technological problem and one that engineers should be prepared to deal with.

Now that we have gone to the moon, we know that anything is possible. But we are now beginning to see also that an older backlog of social problems have not been solved either by getting to the moon or by creating a mass consumption society while we were going there. Everybody else is just as vague about the solutions to these old problems as engineers but perhaps not quite as professionally capable of generating some new options.

As suggested earlier, whether engineering schools get the money to even begin with such tasks probably depends a good deal on the image that engineers can create about themselves in the public mind and also among their own colleagues. Let me close by listing a few possible ways that this image might be enhanced.

First of all, I think that you can get out into the social environment itself, both where the problems lie and where the financial and political support for solutions is going to come from. This means tapping resources in legislative offices, academic departments outside your own, political groups, government agencies, and major foundations—places where people are evidently thinking about getting these problems solved.

Secondly, you can do something about establishing a popular journal of engineering, along the lines of SCIENTIFIC AMERICAN, or TRANSACTION in the social sciences field. Such a journal can provide an important forum for exchange between professional and the public on critical areas of public policy, focusing on the feasibility of alternative solutions.

Thirdly, greater emphasis can be put on the social effects of engineering practices. What I mean by this is being explicit in the engineering curriculum about the social side effects of technology and including some imaginative thinking about the future role of engineering practices as they affect—whether intentionally or not—the nature of society and social policy. In effect, this refutes the view that there are no politics of garbage collection. Where our solid wastes go, how it goes, whose goes, how much it costs, and what alternatives can be designed are political questions in which engineers will be finding their future professional markets. It is important to anticipate that engineering schools are going to have to hire other kinds of professionals to help deal with such questions. If not, I would predict that other academic departments with a greater endowment of optismism will begin to hire away your most imaginative engineers and attempt to do the job themselves. I know this is already happening at UCLA.

Fourth, you can encourage and support students who are concerned with these kinds of social issues. At UCLA and also at Stanford, students have taken a major initiative in developing and inviting professors to teach special courses that deal with specific problems involved in pending state and Federal legislation. In some cases, these have involved direct contracts with legislative agencies. This just illustrates that students have changed since ten years ago. It is not just that they are less patient with the older generation and more concerned about being relevant, but that they are also more serious, more mature, more imaginative, and more capable of accepting responsibility for programs that seek real solutions. If engineering schools do not recognize these changes in their own students, it may reflect an unwillingness to recruit them actively or to encourage them with curricula materials "relevant" to new problem areas.

Finally, engineering schools might explore the design of new institutions to support these emerging priorities. The concept of "earth-bound NASA's" is one possibility, except that we are not dealing with building a vehicle for a handful of specially trained astronauts, but systems for more popular consumption. A better analogy might be the old idea of land grant colleges, which had a unique concern for combining research with its practical applications to the everyday lives of common people. Significantly, feedback went both ways, not only from theorists to practitioners, but from practice to theory as well. That was a hundred years ago, and it is perhaps time higher education had something like that again. But since those days we have been changing from a rural to a technological society, and it may be that this time, the engineers get their turn.

The Regional Role of Engineering Colleges Conference

Report on Workshop IV

Technological Manpower Retraining: The Role of an Engineering College in Retraining Engineers as Technological Needs and Employment Patterns Shift

Hugh Folk and Betty M. Vetter, Co-Chairmen

Participants

Michael Babcock	John E. Pearson
William Haga	George W. Reid
W. Ryland Hill	Thomas Smith
Thomas B. Jefferson	Richard Swalin
Bradley Parlin	

Professor Folk: Ours is a report on manpower training and retraining rather than specifically retraining. Our view of the manpower supply and demand situation was for a temporary short term surplus and longer term—5 to 10 year—shortages. Special field demands we viewed as uncertain. There is a need for conscious policy decisions to build flexibility into the engineering education system and the engineering labor market. We see very little that engineering colleges specifically can do on the engineering labor market but certainly the ideal of flexibility in engineering education has to be recognized.

Second, we foresee recurrent repetitions of the shortage, over-expansion, surplus cycle which affected aerospace engineering. Today the shortages in environmental areas, water and sewage design and so forth, may be reflected in a decade or so in overproduction of specialists in this area—this is the acceleration effect. Then we may be faced with retraining of these people rather than the current aerospace displacees.

Third, we see no general need for engineering colleges in Illinois to become engaged in large scale retraining efforts at the moment. This is in view of the regional role aspect. Were we in California, Washington or Massachusetts, our recommendation would very probably be different. We generally felt that on-the-job training was the best job training and the university has at best a supplementary and supporting role in this. Engineering college faculty however may play important roles in stimulating organizational interest in using some existing programs for these purposes. It is delightful to find one social problem in which the binding constraint is not an insufficiency of funds. Opportunity for some work study activity involved with on-the-job training could be made available on an individualized, subsidized basis so that any particular background course could be taken on campus in conjunction with work.

Fourth, we see a continuing need for better knowledge of what skills a specific engineering sub-specialty involves. An aerospace engineer, for example, may have reasonably good knowledge of the skills involved in the sub-specialties in aerospace engineering but may not know how his particular sub-specialty relates to other specialties in engineering. Of course, with the need to change specialties, the aerospace engineer must move into the new labor market which does not know and which may not be particularly hospitable to an interloper. We notice the sociological fact that as objective differences become increasingly small subjective distinctions are drawn increasingly sharply. Thus, Baptists may feel a greater need for social distance from Methodists than they do from Buddhists, and sociologists may need to differentiate themselves from political scientists more than from electrical engineers.

Fifth, preventing fluctuations between shortage and surplus of jobs has four aspects: undergraduate education, continuing education for graduates, graduate education, and mid-career change. First, an undergraduate engineering education must be designed to recognize the fact that a full work life in a single sub-specialty is not in the outlook for many, perhaps most, graduating engineers. The uncertainty of the course of future demand dictates that this will be the case. Undergraduate education must stress fundamentals, for never again will the engineer have the opportunity to learn the fundamentals. At the same time, market realities require that the undergraduate engineer have a specialized background and training to get a job. General engineering and the other attempts at broad engineering are not generally very successful in the market. Premature specialization should be avoided in so far as possible, if only to make the system more responsive to the demands without changing the quantitive content of the speciality in the undergraduate engineering degree. The undergraduate engineer must be led to understand the cost and benefits of specialization and generalizing in his career.

In particular, we see the growth of the importance of environmental and social problems in the engineering careers. Biodynamics and social-dynamics will become at least as important as thermo-dynamics. This need for the understanding of environmental and social aspects of engineering problems is not solved by the requirement that engineering students be subjected to the mercies of social science courses. We believe engineering colleges should not and cannot avoid direct responsibility for assuming the leadership in the development of teaching research programs in social technology and the social implications of technology. We were impressed by the Illinois School of Technology experimental program which abandons traditional course structure and emphasizes project groups. No one knows if the approach will effectively produce good professional engineers and we do not prejudge this. We assume the quality assessment problem will be attacked. We see the possibility that these programs will be useful in attracting the kinds of engineers who used to enter engineering because they were motivated by the need to do good technological things.

Continuing education of engineering graduates is a clear, regional educational function of an engineering college. This involves not only short courses and conferences, a need which some engineering colleges meet markedly better than others, but also the development of long term degree programs, consisting of correspondence courses, short courses and conferences, aimed at maintaining the broad general competence of engineers, at least at the level of currently graduating engineers. This program should be aimed at maintaining currency, increasing breadth and reducing obsolescence. It may be that such continuing education degree programs should try to award the master's degree.

We were particularly concerned with the problem of the mid-career engineer who, because of displacement, reduction of social interest in his sub-specialty, or other reasons, finds himself at loose ends. The importance of mid-career programs, such as the Sloan Fellow program at MIT, can be great both in helping promising and successful men, and giving others a new start. The problem of older and retired engineers was of some importance to us. There are several possible activities in this regard for the colleges. One could be the establishment of expert panels which would provide part-time employment or, perhaps in some instances, full-time employment in partially subsidized consulting service to local governments. Another could be the coordination of joint employers of a single engineer or the referral and authentication of expertise, although we recognize the problem of competition with established commercial consulting organizations.

Sixth is an area of social needs. Considering only results and not the process by which these results occurred, engineering colleges must be considered the most successful racist and sexist organizations in American society. Regional needs of women and minorities for occupational upgrading is great. The demands of business and other employers for women and minority engineers is also great. Salaries and recruiting activity reflect these needs. We recognize the problems of recruiting women, but we believe that no engineering college except Montana State has made a serious effort to recruit women engineers—that is, women students in engineering. We would like to see every college make an effort, hopefully with some assessment, to develop ways of interesting more young women in engineering careers. One possibility is the formation of groups of women and men, industry engineers and professors to take an active part in recruiting.

Increasing the number of Blacks, Spanish, Indians and other minority groups in engineering poses more severe problems because of the general educational inadequacies which many in these groups have. Many universities have successful programs for minority students. Georgia Tech, for example, starts with high school students to identify aptitudes for engineering and to support developing interest in engineering careers. Financial support, counseling and encouragement to assist educationally disadvantaged students are probably necessary to gain adequate completion rates. The over-all white survival rates of 40-50% of the freshmen entering engineering education may not be acceptable for the minority. As in the case of women, it is necessary to utilize and make known successful role models in joint industry-college teams for recruitment.

Mrs. Vetter: We also talked about the necessity for not only continuing education, but for refresher education. This is particularly needed for the veteran who in many cases entered military service immediately upon completion of his college work and did not do work during his military service that was in any way connected with engineering, so that when he left the service he really was an obsolete man who had never practiced his profession. He is just entering civilian life now. What he needs principally is a refresher course to pick up and review the things he knew two years ago. It was pointed out by two or three people that courses already exist on most campuses—not for this purpose, but to get engineers ready to take the professional exam—and that perhaps this should be pointed out to returning veterans. It is a good way of catching up on those things that have been forgotten during two years of disuse.

We gave considerable attention to the need for more women in engineering. The greatest obstacle to recruiting women has been their fathers. They are tied to the mores of our society which say a daughter is not a success unless she is feminine; otherwise she will not attract a husband but will become an old maid. The recruiting effort to attract women into engineering at the University of Montana at Boseman is worth studying, since it builds its appeal effectively on the existing mores. They point out to girls, in particular to high school seniors who have interests and high aptitudes in math, etc., that if they want to meet men, where are the male/female ratios better than in an engineering school? Also, if they want a good salary, here is the place to prepare. The salaries offered to women engineers are far and away higher than those offered to women graduates in any other field. For an equivalent educational level, women can do no better. In addition, there are opportunities for both active, travel type engineering jobs as well as for stay-at-home type engineering jobs. Most women really do not understand much about what an engineer does. There are engineering jobs that allow girls or men to get their hands dirty. There are also other kinds of engineering jobs which do not. The opportunities in engineering should certainly be explained to women.

A final point is that an engineering background has a general usefulness, a point which we stress to men all the time. There are an awful lot of women who, faced with the necessity of running a household, wish they knew more about why the lights do not work. If changing a fuse will not get them working again, most women have to wring their hands. The opportunities to use an engineering background even in a domestic area ought to be emphasized for women because this advantage has really never been sufficiently stressed. You engineers can do this much better than I, because you know what an engineer can do.

Question: Do you know what percentage of women they have at Montana State?

Mrs. Vetter: It is about 6-7% now, which is much higher than most schools have.

Comment: Northwestern brings in 10% a year.

Mrs. Vetter: The other great thing about women, at least the last statistic I saw put out, I think, by the Society of Women Engineers, is that the loss ratio of men who start engineering programs is over 50%. This is true across the country. Some colleges lose more than that. Of the freshmen starts, over half never attain the bachelor's degree. The retention ratio for the women who graduated in engineering in 1971 was somewhere above 90%. At the moment engineering is attracting only those women who are so determined to be engineers that they will do it in spite of the discouragements, rather than because of the encouragements. However, even if the figure is brought down a bit, there is still a marvelous background of potential from which to work.

Conference Schedule and Attendees

Conference Schedule
Monday, February 14, 1972

CHICAGO CIRCLE CENTER—ROOMS 509-510

8:30-9:00 A.M.	*Introduction*
	George Bugliarello, Dean College of Engineering University of Illinois at Chicago Circle
	M. Frank Hersman, Director Office of Intergovernmental Science Programs National Science Foundation
9:00-12:30 P.M.	*Study Reports*
	Clarke Schneider, University of Illinois at Chicago Circle Joseph Schofer and Amado San Mateo, Northwestern University Thomas Knowles, Illinois Institute of Technology Nancy Nihan and William O'Neill, University of Illinois at Chicago Circle William Haga, University of Illinois at Urbana

	Betty Vetter, Scientific Manpower Commission Hugh Folk, University of Illinois at Urbana
12:30-1:45 P.M.	*Lunch*
	Warren B. Cheston, Chancellor University of Illinois at Chicago Circle
2:00-5:00 P.M.	*Study Reports* (Continued)
	Judy Millaney, City of Chicago Department of Public Works Robert Granda, Illinois Legislative Council Barclay Hudson, University of California at Los Angeles Selma Mushkin, Georgetown University John Caffrey, Educational Systems Research Group and Morton Marcus, Indiana University Eric Whitman, Allied Chemical Corporation
5:30-7:00 P.M.	*Dinner*
	Representative John H. Conolly, Chairman Illinois Legislative Council
7:15-8:15 P.M.	*Study Reports* (Continued)
	Edward Comings, University of Delaware John Duffie, University of Wisconsin at Madison
8:15-9:15 P.M.	*Workshops*

Tuesday, February 15, 1972

8:30-12:30 P.M.	*Workshops*
12:30-1:45 P.M.	*Lunch*

Steven B. Sample, Deputy Director for Programs
State of Illinois Board of Higher Education

2:00-3:00 P.M. *Workshop Reports*

3:00-4:00 P.M. *An Overview of the Conference*

G. Bugliarello and F. Hersman

4:00 P.M. *Conference Adjournment*

WORKSHOPS

I. Methodology for Stimulating the Regional Role of an Engineering College:
Industrial-Economic-Academic Mechanisms
Lawrence Canjar and Lucius Gregg, Chairmen

II. Methodology for Stimulating the Regional Role of an Engineering College:
Political Mechanisms
Samuel Gove and Robert Granda, Chairmen

III. Alternative Goals for Interaction of an Engineering College and Its Environment:
Policy Development, Technological Assessment, Technological Forecasting,
Information Repository Development, Early Identification of Critical Problems
John Cain and Barclay Hudson, Chairmen

IV. Technological Manpower Retraining: The Role of an Engineering College in
Retraining Engineers as Technological Needs and Employment Patterns Shift
Hugh Folk and Betty Vetter, Chairmen

1* Arnold Addison, Professor
Industrial Engineering Department
The Pennsylvania State University

4 Michael Babcock, Research Assistant
Center for Advanced Computation
University of Illinois, Urbana

2 Vinton Bacon, Professor
Department of Civil Engineering
University of Wisconsin-Milwaukee

2 Samuel S. Baxter, Commissioner
Department of Water
City of Philadelphia

3 Egon Brenner, Dean
College of Engineering
City College of New York

George Bugliarello, Dean
College of Engineering
University of Illinois, Chicago Circle

1 John Caffrey, President
Educational Systems Research Group
Washington, D. C.

3 J. L. Cain, Director
Engineering Program Development
School of Engineering
Auburn University

2 Lynton K. Caldwell, Professor
Political Science Department
Indiana University

1 Lawrence N. Canjar, Dean
College of Engineering
University of Detroit

Vivian G. Cardwell
Assistant to the Dean
College of Engineering
University of Illinois, Chicago Circle

4 Pauline V. Chapman, Placement Direc
College of Engineering
University of Illinois, Urbana

3 E. W. Comings, Dean
College of Engineering
University of Delaware

1 John J. Desmond, Associate Director
Engineering Experiment Station
University of Illinois, Urbana

1 J. Duffie, Associate Dean
Engineering Experiment Station
University of Wisconsin-Madison

*Number before name refers to suggested Workshop assignment. (See Conference Schedule)

4 Hugh Folk, Professor
Department of Economics
and Institute of Labor
and Industrial Relations
University of Illinois, Urbana

2 William Geekie
Department of Children and
Family Services

Christopher Gibbons
Systems Engineering
Department
University of Illinois,
Chicago Circle

2 Samuel K. Gove, Professor
Director, Institute of
Government
and Public Affairs
University of Illinois, Urbana

2 Robert Granda
Illinois Legislative Council
Springfield, Illinois

4 Robert S. Green, Associate Dean
College of Engineering
The Ohio State University

3 Lucius Gregg
Alfred Sloan Foundation

4 William Haga, Research Assistant
Department of Sociology
University of Illinois, Urbana

2 M. Frank Hersman, Head
Office of Intergovernmental
Programs
National Science Foundation

4 W. Ryland Hill, Associate Dean
College of Engineering
University of Washington

3 Barclay Hudson, Assistant Professor
School of Architecture and
Urban Planning
University of California at Los
Angeles

1 Frederick H. Jackson, Director
Committee on Institutional
Cooperation
Evanston, Illinois

3 John E. Jacobs, Professor
Biomedical Engineering Center
Northwestern University

4 Thomas B. Jefferson, Dean
School of Technology
Southern Illinois University

2 Boyd Keenan, Professor
Institute of Government and
Public Affairs
University of Illinois,
Chicago Circle

3 Glenn H. Keitel, Dean
College of Engineering
Bucknell University

Dennis Kerber, Research Assistant
Illinois Institute of Technology

1 Thomas Knowles, Assistant Professor
Department of Industrial Engineering
Illinois Institute of Technology

3 Robert Large
Technological Institute
Northwestern University

1 James H. Leonard, Head
Department of Chemical and Nuclear Engineering
University of Cincinnati

1 Morton J. Marcus, Lecturer
School of Business
Indiana University

4 Elizabeth J. McLean, Assistant Commissioner
Department of Public Works
City of Chicago

2 Harold L. Michael, Associate Director
Joint Highway Research Project
College of Engineering
Purdue University

2 Judy Millaney, Research Assistant
Department of Public Works
City of Chicago
Chicago, Illinois

3 David H. Mintzer, Acting Dean
The Technological Institute
Northwestern University

2 Nancy Nihan, Professor
Systems Engineering Department
University of Illinois, Chicago Circle

4 William D. O'Neill, Professor
Systems Engineering Department
University of Illinois, Chicago Circle

3 Charles M. Overby, Professor
Acting Chairman, Department of Industrial and Systems Engineering
College of Engineering and Technology
The Ohio State University

3 Walter S. Owen
Vice President for Science and Research
Northwestern University

2 Vernon Ozarow, Associate Direct
New York State Education Department
Office of Science and Technology

4 Bradley Parlin
Institute of Labor and Industri Relations
University of Illinois, Urbana

4 John E. Pearson, Professor
Assistant Director, Environmental Studies Program
University of Illinois, Urbana

1 Milton Pikarsky, Commissioner
Department of Public Works
City of Chicago

4 George W. Reid, Director
School of Engineering and Environmental Science
University of Oklahoma

3 A. H. Rubenstein, Professor
Department of Industrial Engineering
The Technological Institute
Northwestern University

3 George A. Russell, Associate Dean
Graduate College
University of Illinois, Urbana

2 Steven B. Sample
Deputy Director for Programs
State of Illinois Board of Higher Education

Amado San Mateo
Department of Civil Engineering
Northwestern University

Clarke Schneider, Project Coordinator
Center for Urban Studies
University of Illinois, Chicago Circle

Joseph L. Schofer, Professor
Department of Civil Engineering
The Technological Institute
Northwestern University

2 Herman Sievering, Professor
College of Environmental and Applied Sciences
Governors State University

3 Henry B. Smith, Associate Dean
Research and Graduate Studies
School of Engineering
North Carolina State University

4 Spencer Smith, Professor
Department of Industrial Engineering
Illinois Institute of Technology

4 Thomas Smith, Professor
Information Engineering Department
University of Illinois, Chicago Circle

Herbert J. Stein, Associate Dean
College of Engineering
University of Illinois, Chicago Circle

4 Richard Swalin, Dean
Institute of Technology
University of Minnesota

3 Frederick E. Terman,
Vice President and Provost Emeritus
Stanford University

4 Betty M. Vetter, Executive Director
Scientific Manpower Commission

2 Lynn Weaver, Assistant to the President
Argonne Universities Association

1 Eric Whitman
Allied Chemical Corporation

4 Robin S. Wilson, Associate Directo
Committee on Institutional Cooperation

APPENDIX III
ANCILLARY REPORTS AND COMMENTS

*State Science and Technology Interaction: A Cybernetic View**

George Bugliarello

It is perhaps not inappropriate in a gathering such as this, that deals with key functions of government and with their effect on the relation of man to the environment, to introduce the word cybernetics. The word was coined by Norbert Wiener[1] to mean the discipline of governance of a complex system—and there can be little doubt that federal and state governments, science, and technology, form a most complex and interacting system. It is also perhaps not inappropriate to point out that Norbert Wiener's programmatic concern in establishing the bases of cybernetics was, in his words, "the human use of human beings,"[2] and that this has been, in broad terms, the essential concern of this conference also.

A cybernetics viewpoint can be helpful as a framework for dealing with the complexity of the problems which have been posed throughout this conference. But it is not panacea. It does not possess implicitly miraculous qualities. Simply, it is the best model that we prob-

George Bugliarello is Dean of the College of Engineering at the University of Illinois at Chicago Circle.

*From *Proceedings*, Midwestern Conference on Science, Technology and State Government, Arlington Park Towers, Illinois, November 1970.

ably have today for linking science, technology, the political process, economics, ecology, and man, and for avoiding polarizations in the process.

The essence of a cybernetic model is too well known to warrant a detailed description here. In the simplest terms we may consider a system that has certain goals, and certain inputs and outputs, and ask ourselves how can we modify its operation so that the results of its activity will agree with our goals. One of the processes whereby the operation of the system can be modified is feedback—the return to the system of some signal that enables comparisons to be made between the actual outputs from the system and what our goals, our intended outputs originally were. Feedback is a fundamental concept in all systems, but the nature of feedback signals varies from system to system. A political man or a political system is supremely sensitive to votes as feedback while a businessman or a business system is more sensitive to profits. Man, in general, is sensitive to recognition, love, and pleasure. He is, of course, also motivated by money, but money is only one element of the feedback signals that control his activity. Were money the only stimulus which controls people, we would have neither public servants nor teachers or researchers and our hopes of improving our environment would be much dimmer.

In addition to straight forward feedback, there are other, more complex and effective processes for controlling a system. We shall come to them later.

Schematic as it may be, this view of a system—of any system—suggests to us two tasks. In the first place, we can think of analyzing the system and determining what its responses will be to particular goals or inputs. Secondly, we can attempt to use our understanding of the system to modify it in order to suit our needs.

Analysis of the system means determination of its extension, composition, and characteristics. That is, we are seeking answers to questions such as: What are the parts and subparts of the system? What are the other systems with which it interacts? What are the characteristic times with which each component of the system responds to stimuli or performs its function?

This conference has been extremely successful in pointing out that if we intend to look at the relation between science, technology, state and federal government in the context of the ecological-environmental problem, we must consider a series of interacting systems: What can be called the science and technology system, what can be called the political system, the ecological-environmental system, and so on, all the way to the human system—man himself, with his own goals, characteristics, and feedback processes. These systems are interconnected through their feedback loops, their outputs, and their goals. Thus we cannot discourse fruitfully about our science system without considering our political system, as we cannot discourse about our environment without considering our technology. Indeed, we have learned that many of our contemporary problems stem from our failure to consider these interconnections—and it is legitimate to suspect that, through this failure, also the full potential of our society and our humanity remain unfulfilled.

However, our pivotal concern must be man. All the other systems must be viewed and operated in relation to their effects on man. Ralph Nader's opening talk has emphasized the tangible, physical, biological needs of man while Herbert Hollomon has emphasized the intangible ones. It is clear that our man-made systems, our science and our technology and our governments, must take into account and be driven by both needs, tangible and intangible. The "whole man," as it has been expressed repeatedly throughout this conference, must be both our point of departure and our objective.

The key question then becomes: What are the characteristics of our science and technology system that have exerted, and continue to exert, such a portentous effect on man and his environment? Specifically, what are the elements of this complex system? What drives it? Above all, what should its characteristics and goals be? Clearly, here we can only pose these questions and suggest some of their implications for a science and technology system for the states.

Don Kash has pointed out that perhaps the greatest achievement of our federal government in the domain of science and technology has been successful integration of science with technology and with the other institutions which utilize knowledge.

There has been no corresponding integration on a regional and state basis. No effective mechanisms exist at this moment for bringing science and technology to bear on regional problems as it has been brought to bear on some of our national needs.

If the science and technology in a region is to be integrated with the needs of the region, it becomes evident that the states have a supremely important function to perform. This is a function that the federal government cannot assume for them but only assist in performing. It is also a function that cannot occur at the exclusion of the national view but must be developed as a complement to it.

Many of our problems with the environment stem from the fact that most states have not had sufficient scientific and technological "clout" to react effectively and constructively to scientific and technological trends and to develop a well articulated science policy of their own. There have been exceptions, of course, examplified by the control standards for automobile pollution in California or by the Delaware River Basin Compact, but they have been limited both in number and scope. The very fact that we think of these same examples so frequently underscores their nature as exceptions.

Establishment of a state science policy will make it possible to develop joint federal-state policies for science and technology. The need for such policies is evident. In their absence, for instance, anti-pollution measures taken by a state with a highly consumer-oriented electorate and a capability for monitoring and enforcement are likely to encourage exodus of industry to other states where both these elements are weaker.

THE ELEMENTS OF THE SYSTEM

How can we develop a state policy for science and technology to complement the federal one? In the first place, we need to describe and define the science and technology system of a state—or of a region. We need to know, for example, the contribution that engineering schools or scientific institutions make or can make to the economy of the state or to the development of the state's social goals. We need to know what instruments the state can use for receiving unbiased advice in the area of science and technology. What is the relation between the roles that technological and scientific institutions play in their urban, state or regional community and the role that they play in the national community? Is there an effective coordination at the state level of the activities of federal agencies?

Information of this kind is the point of departure for evolving a rational state science policy for the states. More generally, we need to develop what the systems analyst call production functions. These are the functions that describe the relation between the inputs and the outputs of each element of the state science and technology system and of the related systems with which it interacts.

THE CHARACTERISTIC TIMES OF THE SYSTEM

Each of the interacting systems that we need to consider, and each component in each system, has its own characteristic time of reaction to stimuli (feedback time). Thus, the reaction time of a political system is typically the time between elections. Mobilization and development of the science and technology system in any given direction requires its own time—at times considerably greater than the time that activates the political system. The reaction times of a bureaucratic system are still different, frequently geared to the tenure periods of the members of bureaucracy—substantial change being possible only with the retirement of a particular individual. The ecological system too has its characteristic times, different for each component, at times so long as to encompass several cycles of technological development and to prevent perception of the results of current technological developments, and at times so short as to lead to irreparable environmental damage before a technology can be altered. The human system has other response times yet.

Consideration of these times leads to two important conclusions. In the first place, while there is relatively little that we can do about biological and ecological time cycles, we can act on the scientific-technological and on the political cycles. Therefore, the scientific-technological and the political cycles must be adjusted to the human and ecological cycles and not the other way around. We cannot move nature by legislative fiat or legal process. This is the first time in history that a society faces the moment of truth concerning the full implications of this point.

Mechanisms must be evolved to insulate the development of our scientific and technological approaches to the problems of the states from the instabilities arising from the different characteristic times of the different interacting systems. Achievement of this objective has been somewhat easier at the federal level because of the existence of a federal bureaucracy for science and technology that has provided some degree of programmatic continuity. There have been, however, examples of severe inadequacies and malfunctions—the present unemployment of 65,000 to 100,000 engineers being a case in point.

At the state level the current practice makes the achievement of stability extremely difficult. Some governors are in office for only two years; the governors of neighboring states that ought to establish common environmental policies frequently have different and unsynchronized terms of office; rare is the science advisor who has survived more than one state administration. A clear consequence of these factors is the great length of time necessary to reach interstate compacts as pointed out by Dr. Zimmerman's talk. Periods of eight to ten years are more the norm than the exception. We need a stabilizing mechanism with adequate staff and with control over its destiny for a period of time longer than the ephemeral one characteristic of the political cycles of a state.

BETTER FEEDBACK CHANNELS

Our third need, beyond understanding the kind of systems we are dealing with and beyond considering the implications of the different characteristic times of these systems, is to develop more effective feedback channels. Channels must be established that can more rapidly and directly feed back information as to what are the consequences of the operation of the system. Ralph Nader's proposal at the beginning of this conference for an ethics of "whistle blowing" that would give every individual the opportunity to communicate directly his reactions is addressed precisely to this. The creation of an interstate "free discourse agency" advocated by the student group that participated in this conference is another proposal for more effective feedback.

On the other hand, a feedback channel cannot operate without filters. The more we increase the flow of information, the more difficult it becomes to give proper weight to the various channels through which such flows occur, to distinguish between considerate reactions and unconstructive emotions, to separate the opinion of the expert from the sensationalism of the press or the influence of the pressure groups. One of the most dangerous characteristics of a large flow of feedback signals is the executive fatigue that it generates in the administrator or manager of a system. In the attempt to listen to so many voices, the manager becomes exhausted, with little energy left for action. Furthermore, if we were to design complex governance systems in which every scrap of information is acted upon, we would find the systems to be uncontrollable. A delicate balance must be struck between information and action. Excessive information can paralyze action and endanger the system just as much as action based on too little information.

Thus the design of suitable filters for the feedback process becomes a key problem in the design of the system. The problem cannot be solved mechanistically. We need to determine the goals of the system, that is the values to which the operation of the system is to be responsive.

GOALS AND VALUE JUDGMENTS

While it may be fairly clear what our goals for the scientific-technological system ought to be in terms of the tangible, physico-biological aspects of our environment and of our own bodies, it is far less clear what they ought to be in terms of the intangibles and in terms of the structure of the decision process in our social system. Probably most of us can agree that, at least ideally, our science and technology ought to provide us with abundant food and shelter, with good health and decent health care. There is, however, far less agreement as to what, in this conference, has been called the "ethics of living." Perhaps the problem lies in the fact that in the development of such ethics, it is not the absolute extremes of the problem of living that count and can provide an effective guide for political action—it is not Camus' existentialist concern with the problem of suicide—but it is the enormous middle zone between extremes. For instance, in the design of feedback filters for our social-political systems, should we give more weight to biology or aesthetics? To prosperity or ecology? And, if so, in what degree? How? What is the base for our value judgments?

Our major difficulty in dealing with the goals of the systems stems from the fact that, in reality, these goals are not wholly imposed from outside but, to a very large extent, are generated from within the system. In the parlance of the cyberneticist, a socio-political system or the science-technology system are self-adaptive systems with a multitude of hierarchical levels of goals and goals setting mechanisms, and with the ability to modify these goals—within a certain range—according to the environment of the system. Each component of the system and each hierarchical level within such a component has its own goals. As we well know from experience, it is impossible to impose goals totally from above.

Indeed, total imposition of goals from above is not only impossible but also dangerous. Thus, there is a need to continually study and develop mechanisms whereby the perceptions of the grass root levels of the system can be transmitted, with proper filters, all the way up to the highest level of the system and vice versa. If we succeed in doing so, we may succeed in conciliating the two diametrically opposite recommendations that Ralph Nader and Albert Cox made at the beginning of this conference. We may succeed in balancing the means of avoiding environmental violence against society, with the requirements to use science and technology to produce growth.

TOWARD PREDICTIVE SYSTEMS

If up to this point we have discoursed on cybernetics in terms of a feedback system, we have done so because this is the simplest and most immediate cybernetic system and because the concept of feedback has been stressed time and again, implicitly or explicitly, throughout this conference.

In reality there are several ways in which one can govern a system. Feedback is only one, and not necessarily the most effective. It has the advantage of simplicity, but it is an after-the-fact reaction; if the system has strayed too much from our intended goals, it may become a postmortem reaction, too late to do us or the system any good. To avoid this shortcoming, we could attempt to predict in advance the reactions of the system by the use of a regulator mechanism acting in parallel to the system—what can be called a feedforward mechanism. Thus, a politician, rather than waiting for the votes at the next election to tell him whether he has acted effectively (or successfully) during his tenure of office, can attempt to forecast the reaction of the voters to particular positions he may be taking during his tenure in office and alter his positions accordingly. In fact, every consummate politician does so. Through this kind of feedforward scheme, if the predictive capacity of the regulator is good, it is possible to achieve a better regulation of the system. On the other hand, if the predictive capacity is poor, the regulation can be much poorer than with a simpler feedback scheme.

To go from feedback to feedforward is to go in the direction of evolution, to go from reaction to prediction. We have a clear example of this in the biological domain. Our brain, which is the most complex regulator ever devised by nature or by man, has feedforward capacities while an oyster has only feedback capacities.

Transition from feedback to feedforward has important policy implications for the systems we have been considering in this conference, and for that matter for the government of any system. In the first place, if we desire to avoid unnecessary oscillations in the output of our systems, we should move toward predictive regulation. There are some beginnings of a predictive regulating mechanism for science and technology at the federal level—the technological assessment of Mr. Daddario being an important element in point (the other element being the more complicated business of how to act on the results of the assessment). On the other hand, there is at this moment absolutely nothing at the state level to permit predictive regulation of the state science and technology systems.

A second policy implication stems from the fact that the regulator, in order to operate, must have the ability to predict the behavior of the system in response to a given input. Filters are no longer sufficient for this kind of regulation. It is no longer sufficient to sit passively and to weigh the information we are receiving (which is, in effect, all that a single, lonely, part-time science advisor to a governor can do). We now need, instead, to possess a good model of the system. Such a model implies a regulator possessing essentially the same degree of complexity that is present in the system that we intend to regulate.[3] This can be achieved only through a highly developed and diversified staff with skills that enable it to model and forecast both the operation and the evolution of the system—in the case of a state government, a highly sophisticated and well articulated staff capable of predicting the development of the system of science and technology in the state.

Therefore, it is pertinent to suggest at this point that before we go too far, both at the national and the state level, into the development of purely reactive regulator systems—of a "whistle blowing" feedback system, for instance—we should give a great deal of attention to the possible design of predictive regulator mechanisms. Such mechanisms, of course, should incorporate "whistle blowing," but should be able also to predict all of its consequences, the unfavorable as well as the favorable ones. Should we become convinced that a predictive regulation is better than a reactive one and that the time has now come to develop such a kind of regulation, we must also be prepared to accept the consequences of our convictions. We must accept—and plan for—the development of the appropriate mechanisms such as, at the state level, a substantial bureaucracy involved in the regulatory prediction of science and technology.

PERTURBATIONS AND THE RAISON D'ETRE OF THE CYBERNETIC VIEWPOINT

In ultimate analysis, the most important *raison d'être* of the cybernetic viewpoint is that it enables us to deal with the perturbations of a system. These perturbations can appear in every component of the system: there can be perturbations in the goals, in the structure of the system upon which the goals act, in the feedback pattern, and so on. The system should be designed so as to be able, within limits, to accept these perturbations and to react to them. For instance, the first objective of our combined federal and state science and technology system could be the stability of the system and of the other social systems with which it interacts. Not a static stability, but a dynamic one, which allows for change without destroying the essence of the system—the ability of its components to work together. The problem becomes expressed by these two questions: How far can we allow the system to be perturbed without being destroyed? How can we reduce unacceptable perturbations?

Perturbations can be positive as much as negative. We have to fear recessions and depressions that slow down the economic development of our system—the fear expressed by Dr. Cox. We have equally to fear growth so rapid that it will give the surrounding systems no time to adjust ecologically and socially—the fear expressed by Ralph Nader.

CONCLUSIONS

If we hope to preserve the integrity of the cybernetic system of individual man—which, as we said at the beginning, must be our ultimate concern—an essential element is the development of an intelligent and well-articulated science and policy mechanism at all levels of governance. This must be a policy mechanism that would complement and counter similar mechanism at the federal level. It must also be a mechanism with well-defined articulation of goals, effective feedback loops, filters and anticipatory regulation components and with well-articulated interactions with other governance and social techno-economic systems, all the way down to the system of individual man.

This conference has brought with great efficacy to the fore many of the considerations and criteria that must govern the design of such a mechanism. It is now up to all of us—in the federal government, in the state governments, in industry, in the universities, in the professions and as concerned citizens—to build the system and make it work.

REFERENCES

1 Wiener, N., *Cybernetics: or Control and Communication in the Animal and the Machine*, 2nd. ed., Cambridge: MIT Press, 1961.

2 Wiener, N., *Human Use of Human Beings*, Boston: Houghton Mifflin, 1954.

3 Conant, R. C., "Information Transfer Required in Regulatory Processes," IEEE Transactions on Systems Science and Cybernetics, Vol. SSC-5, No. 4, October 1969, pp. 334-338.

*Summary Report of National Evaluation Committee on Summer Faculty-Student Program in "Application of Interdisciplinary University Resources to Local and State Government Problems"**

Merritt A. Williamson

The evaluation committee consisted of Dr. Elmer C. Easton, Dean, College of Engineering, Rutgers University; Mr. Porter Homer, President, Public Technology, Inc.; Dr. Merritt A. Williamson, O. H. Ingram Distinguished Professor of Engineering Management, School of Engineering, Vanderbilt University. The committee was appointed by

Merritt A. Williamson is Director of Engineering Management Programs at Vanderbilt University in Nashville. Dr. Williamson served as a consultant to the present study.

*Draft report from a communication by Dr. Williamson, May 1973.

Mr. John L. Cain, Director, Engineering Program Development, School of Engineering, Auburn University, in his position as Chairman of the Government Relations Committee of the American Society of Engineering Education.

We were asked to visit Auburn University, Oklahoma State University, and Rhode Island State University, where the National Science Foundation had placed three grants. These grants were awarded based on proposals submitted by these institutions. A separate report has been prepared on each grant, setting forth our findings and our evaluation of how well the proposed tasks were being carried out.

This report presents our generalized observations and contains our thoughts on the value of improving the linkages between engineering schools and state and local government.

THE ENVIRONMENT OF GOVERNMENT-UNIVERSITY RELATIONSHIPS

As a guide to understanding and interpreting our remarks, it was felt desirable to set forth some of the environmental differences between government and university organizations. These differences are important and we believe that an understanding of them is necessary for a full appreciation of the difficulties of achieving effective rapport between the two.

The university environment might be characterized as one of primary emphasis on knowledge, contemplation, scholarly concern and endeavor; a pre-occupation with generating new knowledge and transmitting new and old knowledge to students who attend. Government, in contrast, appears to be characterized by a need to solve pressing problems, and to search for knowledge, processes and procedures which may be applied to satisfying urgent needs now. There is little time for emphasis on training and educating.

Government deals with real problems on a real time scale involving real and very vocal publics. There is pressure to get a job done. Universities, on the other hand, have maintained that scholarly work requires time for discussion. The collegial process resists pressures to meet deadlines except those associated with the academic year teaching activities. In universities, the pace is leisurely compared with some of the departments of government.

Problems in government may be defined and limited more easily than those in universities where desired knowledge is hard to specify in advance. Government emphasizes pragmatism—what works is the key criterion. Universities, on the other hand, can be and should be more idealistic and should explore untried and theoretical possibilities.

Real world problems rarely involve only one academic discipline. The real world is not a respector of academic departments. Problems arising within universities, on the contrary, usually originate in a particular department and are so defined that their solution will involve knowledge available in a rather narrow field.

In government, the environment is one of advocacy. Personnel involved must be attempting to influence certain courses of action, whereas in universities the emphasis is on neutrality, getting the facts and letting them speak for themselves.

The continuity of personnel and policies may be quite different. Top persons in government may be removed due to the dictates of the people expressed through the ballot box. In universities, department heads and deans remain in the system. Although their titles and responsibilities may change, their influence may continue. The rate of change is much slower.

By its nature, government is highly hierarchical in structure, whereas in universities, power is alleged to come from the faculty and they consent to be led by fellow faculty members, who assume hierarchical positions. Important policies and directions are arrived at by the democratic process within the universities, whereas in government they are imposed by the public and transmitted through the chain of command. Civil service personnel have a certain guaranteed tenure, but job security for the majority, particularly in state and local governments, is not as great as it is within the faculty of universities.

Individual responsibility is very great in government. The operations are public and the press has ready access to all actions and can and does criticize freely. University faculty actions are more sheltered and protected, and individual decisions have much more limited effect on persons' lives. Fewer people are affected by poor academic decisions. At a university, learning may take place even though an experiment fails to produce the desired result. In government failure may result in a political disaster with attendant loss of jobs.

Emotion and "gut feelings" play a greater part in government than in the academic world where logic and rationality reign supreme. A leadership role in the university is usually occupied by a specialist who generally has little interest in becoming a generalist. Leadership in government is more than likely carried out by a person who does not function as a specialist, but rather has to appreciate many points of view in his decision making.

One associates greater conformity with government personnel in contrast to the individualism, which is found on campus. This is understandable since governmental policy must be complied with, whereas there is no similar source of conformity in an academic discipline.

A government agency is a servant of the people and must be ever conscious of being service oriented. An academic department, on the other hand, rarely has an overriding consciousness of direct public service. Although universities do derive their support from society, mechanisms for direct influence do not exist as they do in government organizations.

The above broad generalizations are presented because we believe that they do help to describe and illustrate the difference in the two worlds of politics and academe. There are some similarities, but these are not likely to produce difficulties in working together. For example, a similarity that plagues both worlds is that of departmentalization. It would appear that the human drive toward empire building, specialization, and compartmentalization are common to both.

The Professional School and Chicago Circle

A Position Paper, May 7, 1972

George Bugliarello

THE PROFESSIONAL SCHOOL AND THE UNIVERSITY

Although the independent medical school of Salerno antedated Oxford, Paris and Bologna, and although in the United States professional schools such as Rensselaer or Polytechnic were established before many other institutions of higher education, it must be recognized that the professional school has been an uneasy addition to the university. It is only with the land grant movement that it acquired full fledged citizenship in state universities. In private universities, the role of the professional school is still far from assured; medical schools, engineering schools or business schools are not universally present in universities; recently several universities, mostly private, have eliminated engineering; state boards of higher education use increasingly as yardsticks for the approval of professional schools the market demand, however perceived.

These are all phenomena underscoring the widespread if unstated viewpoint, namely that an environment containing a professional school is not essential to the attainment of a university education. Undoubtedly there are justifications for such a viewpoint. A strong theory of the role of a professional school in a university has been lacking, and the case for the intellectual contribution of such schools to the mainstream of liberal education has seldom been made, let alone been made convincingly. The faculties of the professional schools themselves have been strangely mute on this issue.

Thus, the presence of the professional schools is more often the result of pragmatic decisions connected with training needs, rather than of the attempt to place the schools in the perspective of the university as an organically articulated and intellectually coherent institution. Yet, the ever stronger pressures on the university for involvement in a multitude of often conflicting tasks, the growing constraints of a financial nature, and above all, the need to redefine the meaning and content of a liberal education make it imperative to attempt to consider the professional school in more than pragmatic terms. The school's thrusts and activities must be considered in terms of their relation to the rest of the institution, as a vehicle not only for catering to the needs of a particular profession, but also as essential contributors to the mainstream of the university tasks and operations. This poses a series of questions that acquire a particular significance in the case of Chicago Circle, as a developing institution and as a major public university in a large urban area.

OPERATIONAL AND NON-OPERATIONAL DISCIPLINES

To place in perspective the role of the professional school, it may be convenient to think in terms of disciplines or thrusts and to consider them as falling in two categories: operational disciplines and non-operational disciplines (Table 1). The operational disciplines will be defined here as those primarily concerned with problem solving, with the how to, with methodology. They range from engineering and architecture (how to build), to medicine (how to heal). The non-operational disciplines will be defined here as those concerned with knowledge per se, with analysis and critique, with providing principles and values, with the "why?" or "you must!". There are of course many disciplines that have both aspects. For instance, political science, law or painting are in part concerned with the how to, and in part concerned with principles.

Table 1 Operational and Non-Operational Disciplines.

Operational Disciplines	*Non-operational Disciplines*
Business	Archeology
Architecture	Music
Medicine	English
Nursing	Philosophy
Engineering	Languages
Social work	Painting
	Physics
	History

DESIRABLE MIX OF OPERATIONAL AND NON-OPERATIONAL DISCIPLINES

How much emphasis should be given to operational disciplines, relative to the non-operational ones?

Clearly, an institution encompassing almost exclusively non-operational disciplines can have impact only in the area of analysis and it will educate its students in a limiting environment. It is an open question whether those institutions that have recently tended to de-emphasize professional education will remain viable, in the long run, as first rate educational institutions. Equally limiting is the environment in the institutions at the other extreme—the institutions encompassing almost exclusively operational disciplines.

Neither kind of institution can be said to be, in the true sense, a university. An institution such as MIT, originally devoted almost exclusively to operational thrusts, has found it necessary in its evolution to develop strong non-operational departments, from philosophy to linguistics to physics to biology, and has been very successful at doing so. A major strength of Stanford lies in its broad spectrum of strong professional and non-operational disciplines, which has made it among the private institutions the one of probably greatest overall strength and capacity for innovation.

The question of the appropriate mix between operational and non-operational disciplines needs to be answered at three levels. The first is the societal level: What emphasis in general should our society place on operational disciplines as compared to non-operational ones?

The second level is regional: What should the emphasis be for a particular state, region or city? And finally, the question needs to be answered for a particular institution, such as Chicago Circle.

While the answers at each one of these levels are clearly interrelated, the answer at the national level is probably the most difficult, as it strikes at the core of our culture. If attempted on a philosophical basis, it demands value judgments that we may not be prepared to make, and possibly a prescription for the development of our culture that, in the present status of our society, may have to be more devisive than salutary—no matter how intellectually honest the views guiding it.

The same factors operate at the local and at the institutional levels, but the decisions at these levels allow for more variability in the answers, since no state or no institution alone can be expected to resolve the problem by itself and to provide the whole answer. The real place where the philosophical debate as to the desirable directions of cultural development of our society becomes critical, and must be waged, is at the national level.

The question of the balance between operational and non-operational thrusts needs also to be examined pragmatically, in terms of the expected thrusts of the development of our society, of the thrusts that are inexorably pressing upon us, whether we philosophically agree with them or not. Thus, the trend toward an ever more professionalized and specialized society, and the shift of the center of gravity of our economy from production toward services, spell the need for vast cohorts of citizens possessing increasingly diversified and sophisticated operational skills.

The thrust of the public demand at Chicago Circle is a clear indication of this trend—the large enrollments acquired in a very short period of time by the professional schools: social work, architecture, engineering, business administration and education. That these enrollments have been attained in general not at the expense of the enrollment at other institutions in the state is an indicator of the existence of a demand that was not being fulfilled.

Without doubt, as new operational or professional activities will emerge, new demands will be placed on the university to provide appropriate training. Inevitably if the sizes and budget of the university are not to expand indefinitely, such demands will come in conflict with the demands of the non-operational disciplines. An overall institutional quota—as is the case at Chicago Circle and, increasingly, throughout the country—means that under the present system gains in the operational disciplines can only occur at the expense of the non-operational ones. But should the distinction between operational and non-operational disciplines, between professional and non-professional disciplines, continue to be regarded as an inevitable dichotomy? Cannot an engineer be trained to issues of values and a philosopher to the methodology of the planning of transportation? Are there no mixes, no curricula, no departmental arrangements that can be brought into being? Have not the critics of higher education —the public at large, the students themselves—by focusing on the issue of "relevance" of higher education, diverted the attention from the central question of how to provide non-operational disciplines with operational literacy and vice versa?

Is the essence of the problem an organizational one, or does it lie in the deeply ingrained intellectual views of the roles of the operational and non-operational disciplines? Can the sociological structure of each group of disciplines—the attitudes and composition of the peer groups, the rewards, the career objectives—be altered to facilitate the compenetration of operational and non-operational concerns?

Specifically, what could be done at Chicago Circle to approach this problem?

What should be appropriate roles and content of the operational thrusts?

How can the university be organized to achieve an optimum interplay of operational and non-operational thrusts?

ROLES OF THE OPERATIONAL THRUSTS

What should the roles be for the thrusts of the operational disciplines?

The roles of an operational thrust at a university can be many-faceted—some overt, others of a more subtle meta-nature (Fig. 1). They include:

(1) *Methodology for dealing with specific operational problems.* At this moment the methodology is fairly well understood in traditional areas. It is far less understood in new areas such as urban problems or environmental control or educational technology or health care delivery.

(2) *Research,* to acquire the knowledge necessary to solve the problems.

(3) *Training* the personnel to deal with the problems. In general this occurs today as a single pass affair. Once the students have been trained and leave the university, the maintenance of the knowledge thus imparted, the bringing up to date methodologies, information, and attitudes, is not done systematically—extension activities notwithstanding—and is not perceived as a primary task of the university. Only a fraction of those trained maintain contact in a learning mode with the university. Furthermore, training frequently sins of excessive specificity, impeding later moves to new areas of concerns, prompted by new societal or personal needs. The aeronautical engineer that designs airframes has difficulties in designing buildings —let alone in applying his skills to the delivery of health care or other societal services. Similarly the teacher of high school mathematics does not know how to apply his mathematical knowledge to the solution of traffic problems—and so on.

(4) *Policy inputs*, to provide information and guidance for the formulation of policies at the national, state, or local level. At this moment the process is a haphazard one, with certain schools in certain universities—the MITs and the Harvards—exerting a disproportionate influence in providing advice, particularly at the national level, while other schools and concerns do not make themselves heard. At the regional level, policy inputs by universities have been fewer and far less effective. Even rarer have been the inputs at the local level, the level of the municipalities and the countries—the truly forgotten level.

(5) *Involvement in the solution of critical problems*. The skills that the university possesses are at times so critical that the university can make a unique contribution to the solution of an actual operational problem. The nature of the university as a hopefully neutral ground for many antagonistic forces also adds to the unique function of the university. It is thus inevitable that great and growing pressures are exerted on the university—from without and from within—that it become involved in the solution of operational problems. The ability to respond favorably to such pressures becomes an important element in the mission of the university and in the goals of its operational thrusts. While caution must be exerted, lest the ability of the university to respond to the other components of its mission be weakened, it is clear that at this moment responses have been weak and haphazard and the vast potential of the university remains tantalizingly untapped.

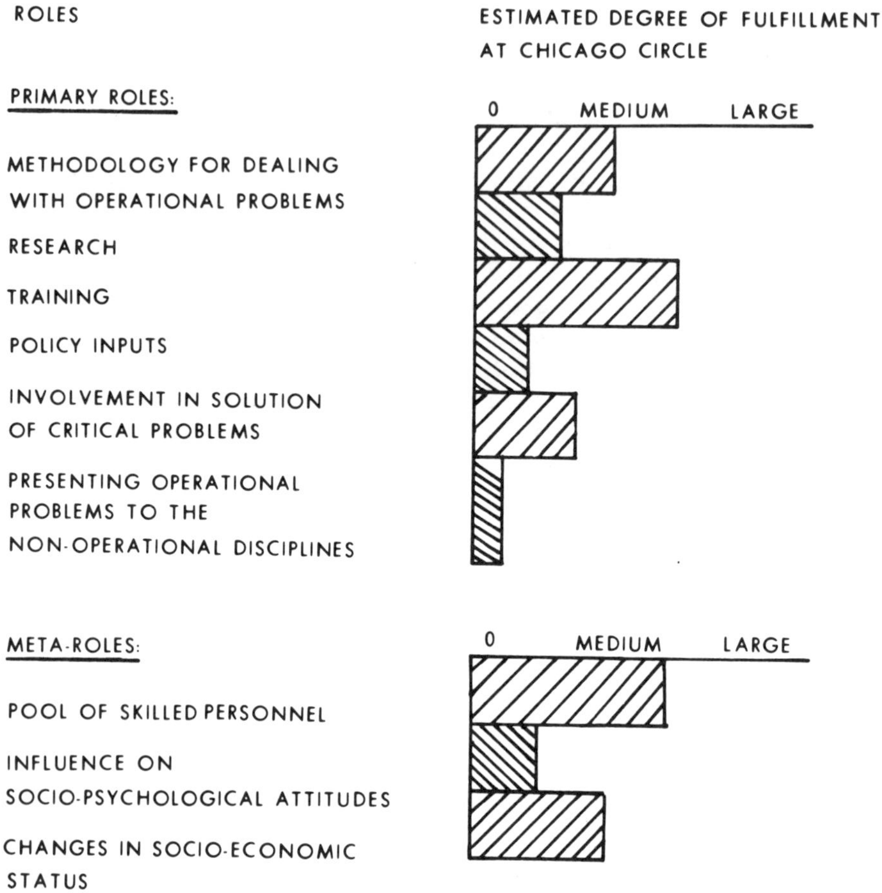

Fig. 1 Estimated Degree of Role Fulfillment of Professional Disciplines at Chicago Circle.

Particularly at Chicago Circle, a great opportunity exists for developing the ability to respond, both because, hopefully, the organizational arteries of the institution have not hardened, and because of the enormous opportunities for meaningful action offered by the metropolitan area in which the campus is embedded. But the experience at Chicago Circle and elsewhere seems to indicate that this activity cannot be carried out as a simple appendage to the existing program—as one more thing to do. It must be recognized as a major thrust and commitment of the university, it must be given means and visibility in the organizational structure of the university, and it should not be set up as separate from the other academic concerns.

(6) *Presenting operational problems to the non-operational disciplines*. Lest the current insularities in the university be propagated—and echoed by the structure and attitudes of our entire society—the operational thrusts need to engage in two essential tasks. They must make the case for their relevance in the intellectual fabric of the university, and they must continuously pose to the non-operational thrusts the problems of values which are likely to arise from operational advances—from the ethical problems in architecture, to the social problems posed by computers. At this moment, it must be recognized that such a role of the operational thrusts is seldom exerted in most universities, Chicago Circle included.

In addition to these overt roles, an operational thrust can fulfill a number of *meta roles*. These include:

(1) *Injection* in the community *of a pool of skilled personnel*—the faculty of the operational discipline—that can provide advice and other public service functions.

(2) *Influence on the socio-psychological attitudes* of the community, which are affected, e.g., by the presence of a medical school or a college of agriculture.

(3) *Changes in the socio-economic status* of the community. Although the university as a whole influences such a status, e.g., through its purchasing policies, the operational thrust may exert a major influence. For instance, traditionally in the United States engineering schools have been vehicles for upward social mobility. Can they continue to do so for the urban minorities?

Some Specific Questions for Chicago Circle

The questions then that confront Chicago Circle and its professional disciplines are of a multiple nature:

What should be the major emphasis, in the spectrum of roles, of the operational thrusts in the university?

Should the greatest emphasis be given to training? To the development of new professional knowledge through the study of methodology and through research?

What role should be given to providing inputs for the formulation of policy, at the national, regional and local level?

To what extent should the disciplines in professional schools respond to the continuous pressures from the outside for involvement in the solution of operational problems?

What should the role of the professional schools be vis-à-vis the system of community colleges? Vis-à-vis the professional schools in private universities?

Should deliberate steps be taken to exert a particular influence on the socio-psychological attitudes of the community?

Should specific policies be adopted to effect in certain directions the socio-economic status of the community? For instance, should certain purchasing policies be emphasized to encourage the economic activities in the immediate neighborhood of the university? Can the university do this?

The answers to some of these questions are easily reached by considering the location and other characteristics of Chicago Circle, and other factors on which there is fairly general agreement. Thus, without question training is a major goal of the operational disciplines at Chicago Circle, and so is the development of methodology and research, with particular focus on methodology and research for solving operational problems in the urban context.

Training poses the problem as to how to more precisely define the client population—the socio-economic characteristics and the geographic distribution of the students to be served by the campus. For instance, there are indications that the economic levels of the families of students in private universities are increasingly less represented in the *middle* income range. What are the implications for the professional schools at Chicago Circle?

Although the involvement of Chicago Circle in policy formulation, either at the national or local level has been very small, there is probably little question that such an involvement is both important and desirable. There is a need and an opportunity to fill a gap at the state and local levels in the area of policies for urban problems, industrial research, and the environment, that no other university with professional schools is filling at this moment in Chicago. There is also a similar need at the national level, to provide policy inputs for engineering, social work, business and education resulting from our experiences with operational problems at the local level. For instance, already we have been able to exert some influence on the structuring of Federal training programs for urban mass transportation.

Again, there is wide spread consensus that Chicago Circle must become involved in the solution of critical practical problems of the community—the only question being one of magnitude of effort and of how capillarily such an effort should permeate the university.

There are other questions on which there may less readily be agreement. For instance, to what extent, if at all, should the operational thrusts at Chicago Circle be planned to influence the socio-psychological attitudes and conditions of the community? Or, to what extent should the professional schools at Chicago Circle establish collaborative and even symbiotic programs with Urbana—or with other institutions?

THE CONTENT OF THE OPERATIONAL THRUSTS

Setting goals for the operational thrusts of the university involves not only the assessment of the roles and meta roles of such thrusts, but also decisions as to the contents of the thrusts. What operational disciplines should be present at Chicago Circle? Should there be law, or nuclear engineering, or printing management or library science?

The desirable thrusts clearly vary from institution to institution, and from time to time in the trajectory of the institution. At a certain moment in time particular thrusts may lose their importance or urgency, and thus confront the university with tough decisions: Should they be continued? Should they be abandoned totally? Should they be kept alive in a limited fashion, because they may become useful again at a later time, or because it may be desirable to keep alive the "culture" of the thrust?

This is a peculiarly important and yet neglected problem of the operational disciplines. Once we have learned something in an operational context which seems to be no longer useful, should we abandon it altogether? Railroad engineering, for example, was abandoned perhaps too rashly. Sanitary engineering became for a while unpopular, while aerospace attracted most of the attention of technological schools, to the extent that certain universities phased it out completely. The teaching of practical technological skills, also abandoned for many years, is now returning in great strength and to many universities.

ORGANIZATION OF THE UNIVERSITY

The university can be viewed as an institution with three cardinal concerns (Fig. 2): Values (generated through knowledge, and the quintessence of knowledge), methodologies, and effectors (the processes, such as teaching or public service, whereby values and methodologies are propagated and brought to bear on the problems of the world outside). Values have been traditionally the realm of concern of the non-operational disciplines, while methodology has been that of the operational disciplines. Effectors are the concern both of operational and non-operational disciplines but, save for teaching, still represent a haphazard and uncoordinated process.

Until now, the operational disciplines have contributed little to values. Engineering groups, for instance, have contributed little to consumerism, or to environmental alert. Similarly, the non-operational disciplines have contributed little to operational methodology —although more than non-operational disciplines have contributed to values. Physics and chemistry have contributed to the methodology of engineering and medicine more than the latter have explicitly contributed to new ontologies, to ethics or sociology. (Not that engineering and medicine have not affected Weltanschauungens, values and gestalts; simply they have not contributed actively and systematically at the intellectual level to their formulation.)

How can the operational and non-operational thrusts be organized so as to enable the university to achieve its greatest potential as a generator of values and methodologies, and as an effector?

1	2	3
KNOWLEDGE—▶VALUES	METHODOLOGY (of operational and non-operational thrusts)	EFFECTORS (propagation and utilization of 1 and 2)

Fig. 2 The Three Cardinal Processes of the University

Traditionally, most operational thrusts are in professional schools and most non-operational ones in schools of liberal arts and sciences. There are, of course, some exceptions, but rare is the business school that has a philosopher on its faculty or the liberal arts college that has an M.D. Thus, one major component of the problem is how to achieve an optimum interplay between the professional and non-professional schools. The question needs to be examined from several viewpoints: That of the organization of professional schools as a group; the viewpoint of the academic coordination of operational and non-operational disciplines; that of the structure of departments and schools (or colleges); and the viewpoint of the organization of effector mechanisms.

The Professional Schools as a Group

The first and probably the key issue in achieving an interplay between operational and non-operational thrusts is how to make the case for the relevance of the operational—the professional—thrust in the intellectual fabric of the university. Can such a case be facilitated by providing an organically and organizationally coherent view of the combined role of all the professional schools? An equally coherent view is necessary of the non-operational thrusts, but organizationally these tend already to be grouped mostly in one college—the liberal arts and sciences college. There is no counterpart of such an arrangement for the professional schools. Each one stands alone and isolated.

Thus, frequently a professional school may relate more to a liberal arts and science college than it does to another professional school, even though the two schools may have many methodological problems in common. With few exceptions, in the past the links among professional schools have been very tenuous at best, at Chicago Circle as well as elsewhere (although some links are currently being actively pursued on our campus, such as business and engineering, social work and engineering). New disciplines can emerge through such links—through the interplay of business and social work, medicine and physical education, architecture and business, education and engineering—that can exert a profound effect on our theories of man, society and living systems, and hence on the operational disciplines. What new combinations of operational thrusts should be encouraged at Chicago Circle?

Secondly, it is certainly a fallacy to assume a priori that a professional college contains or should contain exclusively operational thrusts or that a college of liberal arts and sciences contains or should contain exclusively non-operational ones. In mathematics, physics, philosophy, political science, in sociology and in biology, operational involvements may be quite strong—or may need to be so, both for the health of the discipline and for that of the soul of the faculty member. Yet, in the quest for an operational thrust by the non-operational disciplines, a major difficulty is often represented by the fear of treading upon unknown territory, or by the fear of lack of intellectual respectability of the effort, by the difficulties in homing on significant operational problems, or by the inability to have access to a community or to effector mechanisms that can translate the intellectual inquiry into a prescription for action. Furthermore, to an increasing degree today major operational problems, from housing to ecology to traffic control, have acquired a degree of complexity that cannot be approached within the context of a single profession, let alone a single discipline.

Can these fears and difficulties be reduced by a greater cooperation among professional colleges that would provide common policies, focal points for information, and a broader and stimulating common ground with the operational thrusts?

In the third place, the professional units have many problems in common: How to approach the community, how to benefit from commonalities in methodologies, how to structure their graduate programs, how to evaluate the professional activities of the faculty, how to share facilities, or, above all, how to devise new effector mechanisms that would enable them to bring to bear more successfully their operational expertise on the solution of practical problems. For instance, a professional school able to reach a community, an industry, a local government only in a fragmental fashion when acting individually, may be far more effective when acting jointly with other professional schools. To an increasing degree, the major problems of a municipality, a manufacturing plant, or of a community cannot be categorized as being exclusively of an engineering, or management, or health care nature.

Should not the university—above all Chicago Circle—respond accordingly with a coordinated approach?

Would it be advantageous for Chicago Circle to administratively and academically coordinate the professional units, so as to enable them more effectively to:

i) solve their common administrative, academic, research and public service problems?

ii) achieve common and compatible operational thrusts?

iii) communicate with the other academic units on campus?

iv) provide a more symmetric arrangement of resources and enrollments (Fig. 3)?

If so, what administrative arrangements should be desirable? A coordinating committee? A high-level administrative officer responsible for the professional schools?

Organization for the Interplay of Operational And Non-operational Disciplines

In general—and Chicago Circle is no exception—few universities, if any, have attempted to provide systematically for a greater interplay between the operational and non-operational thrusts, so as to join them in the development of values, methodologies and effector mechanisms and policies.

What kind of mechanisms can facilitate the interplay of the operational and non-operational thrusts?

A System of Common Thrusts

A first step would be to search for and organize intellectual focal points that can provide for a broad interplay among these disciplines. For instance, information as a basic concept and as a process, has economic, sociological, engineering, biological and political aspects and implications. It can thus become a fertile meeting ground—a "commons" or common thrust—for a number of operational and non-operational disciplines. Similarly, common thrusts can be envisioned that would be concerned with the process of change, the process of value formation, the establishment of societal goals, esthetics (e.g., science and esthetics, philosophy and esthetics), etc.

One of the important challenges to the university is how to organize itself so that these common thrusts can grow in depth and in numbers, pervading all the disciplines of the university.

The common thrusts need to be organized as flexible structures, with tasks and responsibilities in all the three key roles: in the formulation of values, in the creation of methodologies and in the effector mechanisms. Thus, they need to be engaged in research, in teaching, in public service and in whatever else can serve to propagate values and methodologies.

The common thrusts should be held accountable for their productivity in each of these areas and the unproductive ones disassembled to give way to new thrusts. Thus, they represent a mechanism whereby the university's operational and non-operational thrusts can be integrated, and the output of the university evaluated in terms of what really matters. And what matters are not only the conventional parameters—the number of students taught, or of papers published, or the research dollars—but, and above all, the contribution to ideas,

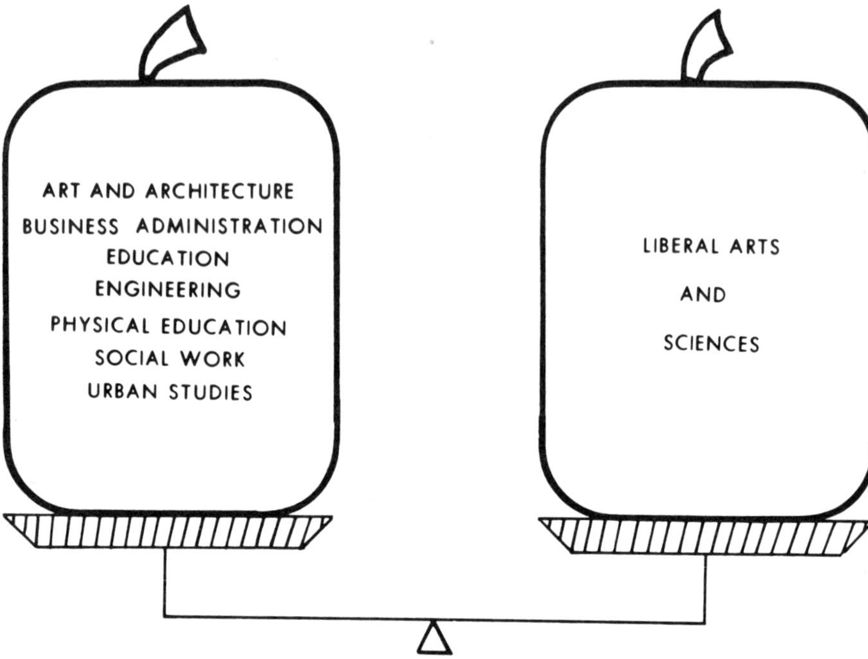

Fig. 3 Equilibrium of Resources and Enrollments of Professional Colleges as a Group and the College of Liberal Arts and Sciences.

values, know-how and to the intelligent solution of societal problems—solutions in which pragmatism does not drive out values.

To achieve these objectives, the common thrusts would need to be given importance in the structure and operation of the university (Fig. 4). The pit-falls of interdepartmental and intercollege activities have been, traditionally, inappropriate leadership, inadequate budgetary support, or insufficient representation or advocacy in the decision-making process of the university. Thus, the common thrusts need to be led by faculty members with vision and strength. They need to be given the budgetary means to carry out their mission and they need to have strong inputs in the promotion and reward systems and in educational policy. They must be placed in the position of being able to constantly and effectively reiterate to students, faculty and administration that values, methodology and propagation are the raison d'être of the university, and that everyone who is part of the university is expected to contribute to each of these processes.

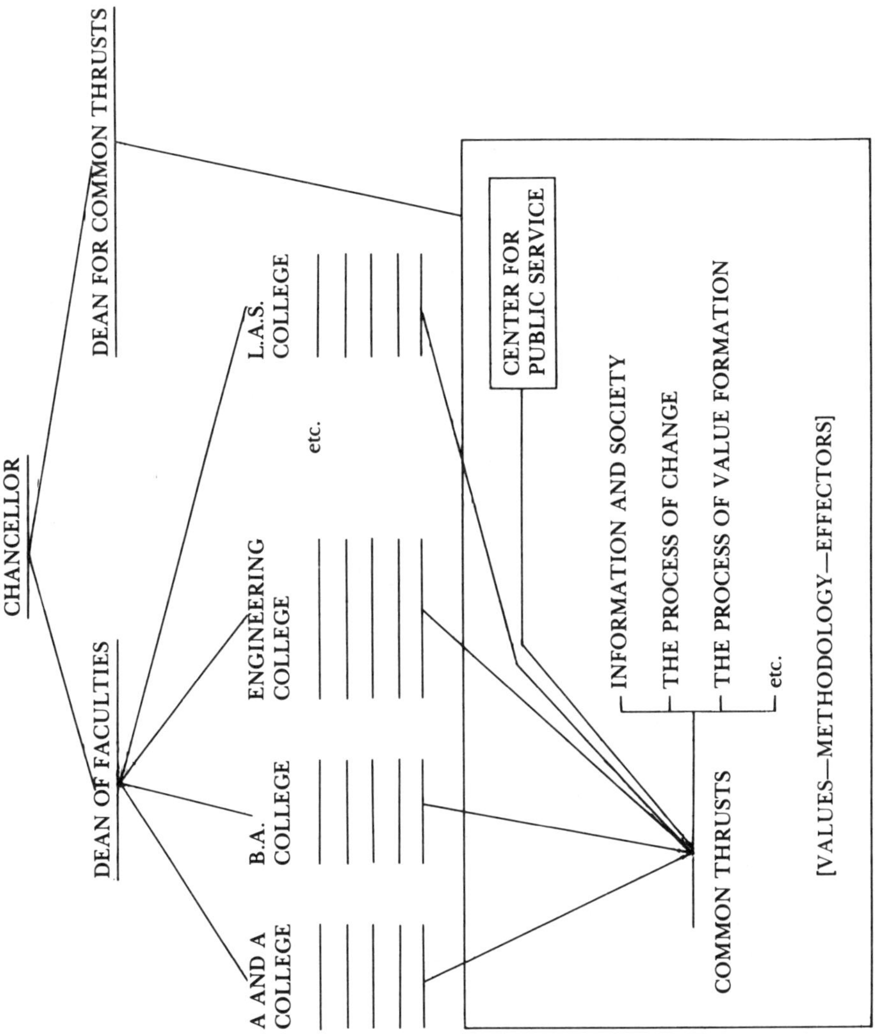

Fig. 4 Organization for Common Thrusts

Above all, the common thrusts would need to be represented at high level in the administrative hierarchy of the university—e.g., by a dean reporting directly to the Chancellor, and appointed from among the most charismatic and distinguished faculty members for a term not longer than two years.

Departments, Schools and Colleges

The organization of departments, schools and colleges needs to be continuously scrutinized, to determine whether new alignments of disciplines and groups of disciplines can facilitate the tasks of the university. For instance:

Is the quest for basic knowledge and values best served by housing all the non-operational disciplines in different administrative units from the operational ones? Should the professional colleges include departments encompassing non-operational disciplines? Should for instance, the College of Engineering encompass physics, in the same way that the College of Business encompasses economics, and that of Architecture and Art encompasses painting and art history?

Even within the existing college structure, can different departmental groupings of disciplines facilitate communication with the rest of the university? While little change can be expected in the departments housing basic disciplines—English is English is English—the re-alignment of other disciplines in certain groups may be profitable. For instance it is difficult for fluid mechanics to communicate directly with philosophy or esthetics, but the grouping of fluid mechanics in an energy engineering department can facilitate such a communication as part of an overall energy problem, which has indeed obvious philosophic and esthetic implications.

Effector Mechanisms

Public pressure will continuously and relentlessly be exerted to make the university become involved more decisively and broadly than in the past in the problem of the society outside its walls—to propagate to the outside its skills, to apply its values and its methodologies. Traditionally, and rightly, teaching has been the major mechanism, the primary effector, on which the university relies.

Other important effector mechanisms have been the experiment stations or the non-profit research institutes. It must be recognized, however, that their success in involving the entire fabric of the university in their mission-oriented projects has been, on the balance, rather limited. Their programs have tended to unfold autonomously rather than in close interplay with the main academic programs of the university, at best with exchanges, at worst with reciprocal indifference and lack of communication. Thus, rapidly these mechanisms have tended to acquire a life of their own, largely independent of the rest of the university, with their own staffs and facilities.

One reason for this fatal tendency toward autonomy can be found in the rather close disciplinary identification of many of these endeavors—typically with engineering or agriculture or with the sciences—that has de facto excluded vast groups of other disciplines in the university. Another reason lies in the failure of the traditional academic organization to provide sufficient relevance, leadership and support to the effector mechanisms. The appointment of ad hoc administrators (frequently of limited academic charisma, and therefore unable to exert leadership on the faculty) or the creation of separate divisions of the university have proved to be in general ineffectual.

The problems to whose solution the university is being called to contribute today are far more complex than those for which the mechanisms devised in the past were created—the experiment station, the extension division, the center or research institute. There are at least three major conditions that the university must meet, if it is to be effective:

(1) Close cooperation of a number of operational and non-operational disciplines—a cooperation such as can be achieved only through a restatement of the purpose of the university and the establishment of a rich network of common thrusts.

(2) Establishment of supportive mechanisms and services—from repositories of information, to liason personnel with state and local government, to satellite centers in the community, to communication networks (e.g., microwave links emanating from the university), to a series of external advisory committees, to special laboratories.

(3) Creation of ad hoc legal and academic mechanisms that would enable the university to act more effectively as a catalyst for a variety of outside forces and institutions. For instance, a central challenge to our society is how to make the know-how of the private production sector available to the public sector. Can the university become the neutral meeting ground where private enterprise and public services can be brought together to provide the first with a better understand-

ing of the needs of the latter, and public service with an opportunity to avail itself of the know-how of the first? Can new imaginative programs be proposed for the Chicago area? There are indications that the Federal government is becoming interested in this concept and is indeed beginning to provide some support for its implementation.

Should a Center for Public Service be established at Chicago Circle, with the task of satisfying these three major needs, and closely integrated in the fabric of the university, by reporting, e.g., to the dean in charge of the common thrusts?

CONCLUSIONS

This paper has endeavored to touch on what appear to be some of the key questions concerning the role, content and operation of the professional schools at Chicago Circle, by placing them in the perspective of a more general theory of the university.

The principal thesis of this paper, to which most of the questions address themselves, is that the complexity of the problems to which the professional schools at Chicago Circle need to respond requires much closer links between operational and non-operational disciplines. To achieve such closer links, two questions must be asked in earnest. One is that of a closer administrative coordination of the professional schools. The other question is the organization of the concerns that are the raison d'être of the university—knowledge and values, methodology, and effectors—and their representations at the highest decision-making levels in the university.

There are many additional questions concerning the professional schools that have not been considered in this paper. Although they are generally of a corollary nature, they also need to be discussed and resolved in the formulation of a long-range plan for the university. They include the level of budgetary support appropriate to each professional endeavor, the role of the dean of the professional college (vis-à-vis that of a liberal arts and sciences college, described extensively in the literature), and the desirable personnel policies to maintain the expertise of professional faculty members and to facilitate their involvement in new effector mechanisms.

Regional Role: A Simple But Human View

Herman Sievering

1. Introduction

Confronted by change, most universities have neither remained loyal to their traditional goals nor claimed a new role as ministers of change. They have rather drifted between past and present and been altered out of recognition by the forces they should have been involved in shaping. The result is increasing anxiety on the part of administrators unable or unwilling to determine the implications of change; increased need to predict future changes in the role of colleges and universities; development of simulation schemes to forecast changing roles. Unfortunately, computer models constructed to predict change ordinarily monitor the effects of changes rather than predicting the causes for change. Whether college role models can be made to function as predictive tools is questionable (noting particularly that too little data is normally available to allow independence from the set of data used to formulate the model).

The task of predicting change in the regional role of an engineering college will require radically modifying the assumption that a university's function is simply to produce knowledge or pursue truth, and that the public applications of knowledge can safely be left to others. Above all else, we need knowledge with a conscience.

It is generally recognized that universities perform functions

Herman Sievering is University Professor of Applied Science, College of Environmental and Applied Sciences of Governors' State University, Park Forest South, Illinois.

categorized by (1) research, (2) teaching, and (3) service. Every university plays a role in service by its sheer existence. However, rarely is it recognized that the service function may be the best predictor of a university's changing regional role. Service in most engineering colleges is not so much an alien concept as it is empty rhetoric. By paying more than lip service to the concept and role of "service" a functional and practical model for engineering college regional roles may result.

II. Rationale for the Service Model

A systems view of science education makes it clear that the feedback mechanism to citizens has not functioned properly. In the area of technical problems this can be perceived in a number of ways:

A. Those with immediate problems have difficulty communicating them to those in the university who could respond effectively to the problems. More importantly, it is not just the public, but many of those dealing with the public as well, who do not know the answers to technical questions or how to find the answers.

B. Similarly, those charged with operating local programs often lack the expertise to use sophisticated techniques in problem solving. Application of scientific and technical innovations to existing programs usually in itself requires specialized knowledge in order to adapt the innovations to the specific situation.

C. Those with local concerns often lack long-term perspective; the structure of most bureaucracies does little to counteract this tendency.

D. Unfortunately, those at the universities with general concerns frequently lack an appreciation of local needs, local problems, and the requirements for implementation of scientific and technical innovations. It is not sufficient to solve waste disposal problems "in principle"; the garbage must be removed from the street. Under present conditions this creates a dilemma, for the most effective laboratory personnel are likely to be those who specialize in their laboratory activities; without more contact with "real world" problems, it is dubious whether they can be more effective in responding to social and regional needs.

E. Given the social discontinuities that create these problems, there is a further and perhaps more important consequence: because communities and other discrete segments of our society perceive innovations by other segments (particularly the universities) as threatening, it often is difficult to gain widespread support for the allocation of resources to support university-based solutions to technical problems. Thus, innovative solutions cannot become part of the normal organizational process unless widely accepted at the local level.

These broad generalizations lead to recommendations falling into three categories:

(1) Improve regional communication channels in order to move information across organizational boundaries more effectively.
(2) Develop ways of periodically changing the role structures of individuals as a means of breaking down organizational barriers.
(3) Involve those with the problems in the problem statement/problem solution process.

Based on these recommendations, a service model is generated.

III. Regional Service Model

The most basic need, as described above, is for a more authoritative information transfer network with sufficient feedback to allow complete recycling of ideas, problems and alternative solutions. The most adequate solution appears to lie in putting necessary information into the hands of the concerned public. Thus, the regional service center with concomitant service council suggests itself.

Some of the functions of the service center would be:

A. to act as free consulting and information dissemination service to community groups, local government and the public at large.
B. to make more effective use of state capabilities for problem solving and technological extension.
C. to assist in community or local planning.
D. to provide reference service.

Teaching in the formal sense would be avoided. Staff would consist of a technical service agent and university faculty called in to deal with specific problems or on leave with the temporary label of community professor. A community professor on returning to his home campus would bring back to the university the problems of the community, thereby making the university less of an "ivory tower" and, hopefully, more "relevant". Organizationally, the chart on the following page describes the administrative functioning of service centers and their relationship to the cooperating university.

The purpose of the council is to represent the interest of the people in the community—community being here defined as that body of people which can bring together resources sufficient to physically house the service center. No member of a community council would be paid a salary for services rendered to the council. Members of the council would be appointed by a regional service board. The principal function of the council would be to cooperate with the environmental service agent for that community in planning programs and facilities utilization.

A regional service board, established to act as direct liaison with the state university serving that region, would consist of women and men who are members of the community councils, plus a minority from the cooperating university. The regional board, after reviewing requests for funds (as related to programs) by the community councils, would submit budget requests as separate service itmes to the appropriate governing board. It is at this level that regional as well as long-range planning would—hopefully—take effect. Placing review of budgets and programs in the hands of this board would remove redundancy and bring integration and perspective to the overall service program. It is at this level, also, that a major departure from the agricultural extension arises, for the agricultural extension has no such regional and long range planning mechanism.

Each cooperating university would have a director of service to draw upon and to help service agents draw upon the human and physical resources available at the cooperating university. The fact that financial resources are placed in the hands of the community councils would be reason enough to believe that universities would indeed turn their attention toward local community problems. Actual administration of funds would be carried out by the technical service agent—reporting both to the community council and director of service. No portion of the above finances would be used, directly or indirectly, to purchase or repair any building or land. Rather, funds

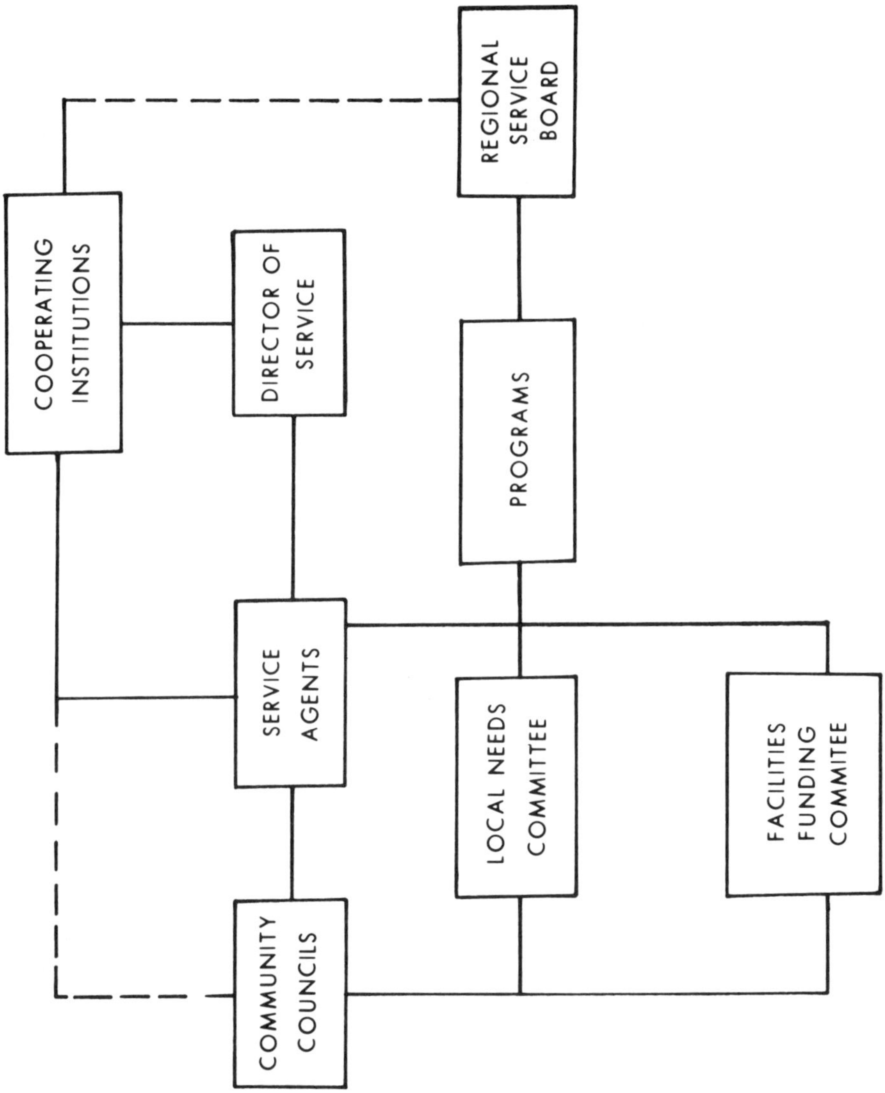

ORGANIZATION CHART

would cover the salaries of the service director and technical agents, service office and supplies costs, and for imparting information on programs through demonstration and publications, including necessary printing and distribution expenses. Thus, this mechanism would allow local community people to define the problems and the university to present alternative solutions with both its normal long-range perspective, as well as the regional perspective, monitored jointly by the director of service and regional service board.

Among the programmatic functions of service centers for both the private and public sectors of neighborhoods, communities and the region would be:

1. Delivery techniques for comprehensive planning, community organization, self-help and private sector motivation.
2. Methods for monitoring and surveillance, on a common grid, of data for epidemiological analysis of combinations of environmental conditions and forces germane to physical health and safety and well-being: human, social, physical and biological.
3. Develop and continually update a systematic subject matter categorization of research and information available at the university.

Note the similarity to the agricultural extension service. This is intentional for over the years the agricultural extension has performed so well in its service. In analogy to the agricultural extension service a first small step in the development of an engineering college extension service is forthcoming.

IV. A First Small Step

Talk of the environment as more than clean air and protected wildlife, and the use of phrases such as "comprehensive planning for environmental health services" have come into fashion recently. However, few of these phrases have been translated into actions that actually improve people's surroundings and health. The University of Illinois Engineering College at Chicago Circle resides in the midst of a slum (as do many other engineering colleges). The rhetoric of environmental improvement could be made reality by an extension service which would train unemployed men from the community surrounding the Circle campus to paint walls, repair plumbing, supervise buildings, fix boilers, and organize tenants into building associations. Such activities may appear at first sight as less then dynamic solutions to environmental/technical problems of the engineering college's region but preventive maintenance of housing could do more for the physical health of the region than can thousands of traditionally-trained engineers and doctors. The program would aim at cooperation with the region's landlords by offering them a variety of reasonably priced services intended to turn the dilapidated and dangerous tenements into safe and decent places to live.

The multifaceted aspect of the extension service is seen most clearly in the training program for what might be called urban extension agents. Each man would receive 2 to 4 months of in-school vocational training. Training would emphasize boiler maintenance, building repair, plumbing, electricity, and other maintenance skills. Important to the training process is to teach the significance of the work in terms of improving the health of area residents.

Although this proposed first step may seem an insignificant one, it is the combination of many such small steps that will define a regional role for engineering colleges in *human terms*. Programs of this nature are being proposed not only by frustrated young engineers at new universities but also by distinguished scientists from around the world. Julian Huxley, C. H. Waddington and 31 other British scientists have proposed a "Blueprint for Survival" in which the key concept is "the formation of communities small enough to be reasonably self-regulating and self-supporting and education toward and for such communities".

*An Alternate Interface Mechanism**

Merritt A. Williamson

". . . I would conceive of an office established by the state with a director selected by a board to whom he would report. The board would consist of persons selected from deans of engineering, the directors of government agencies, the state chambers of commerce, state association of manufacturers, and perhaps retired industrial vice presidents of engineering or research. Such a knowledgeable group should be able to select a non-political director. This director would have a small staff of permanent personnel but would be supplemented by persons from colleges of engineering on loan completely for a quarter, a semester, or a year or two years. Thus, while working on projects funded through this agency, each person could devote his attention exclusively to the work on a full time basis. The university would receive an amount of money equal to his salary, plus, say 20%. He would receive his salary plus 10%, and the university would receive 10%, thus providing an incentive for both to cooperate. Generally, he would work out of his regular office and would have access to the laboratory and equipment of his institution. If not on an assigned time basis, any individual could undertake small one or two day consulting jobs. These would be arranged with him as an individual. I believe that this would be practical since I believe that most

Merritt A. Williamson is Director of Engineering Management Programs at Vanderbilt University in Nashville. Dr. Williamson served as a consultant to the present study.

*Excerpted from a personal communication by Dr. Williamson, May 1973.

deans of engineering would welcome added income to pay for their faculty in these days of declining enrollments.

The key to this operation would lie in the quality, the vision, and the managerial skill of the director and his office. He should be staffed with persons who would have the respect of the academic world and yet be able to communicate with the government, legislators, or anyone in the public who would seek assistance. It would be in this office that the real problems would be understood, broken down into component parts which could be farmed out for work; and it would be here that the various pieces would be put together for a coherent and useful answer if extensive research were involved. If it were a case of consultation, this office would be able to arrange for a knowledgeable person to bring in. Copies of reports would be classified and filed in the director's office.

I visualize that this office would be funded by state money for the essential continuing service functions, but would be set up to receive Federal monies, donations and grants from industry, etc. In the event of a job, all persons qualified would be asked to indicate their interest. Quotas would be set up as guidelines for funds to be spent at the different colleges so that all money would not be funneled into one institution. Decisions about assignments would be made by a small committee of central office personnel. There should be an ample budget for travel because first-hand, face-to-face contact between the customer and the provider of the service will be essential.

Many of the same arguments present in the Report could be used to support this structure and need not be repeated here. This does have the disadvantage of removing the control from the engineering colleges, but it does allow a working arrangement with which most institutions are already familiar. It would not preclude anyone from working with any established centers, but it could serve as a coordinating agency between centers. The plan would, however, centralize for the state the information, monitoring and innovating activities, and in turn take care of "farming out" this work to other locations as local needs require. It would also serve as the central location for financing and for contact with other states.

This central office would not attempt a teaching role, or a research role, but it should facilitate the more efficient allocation of available intellectual resources for the benefit of government, industry and the people as a whole. The colleges of engineering should benefit. It must be remembered, however, that assistance in technology or technical matters is only one area of assistance that is needed by government and others. The pattern proposed could be expanded to include the identifying and locating of expertise in areas of sociology, political science, etc., and could be of help to the various arrangements taking place with individual institutions or with individual faculty members. It is a service which should function economically and hence be used.

One advantage to this organization is that the personnel employed as a small competent nucleus can be public relations conscious without embarrassment or without losing face among their colleagues, which would not be true of university based operations.

As I stated above, I have not planned this structure in any detail but am merely presenting it for discussion . . ."

*Training and Environment For New Technical Ventures**

Lamont Eltinge

Our society, to progress, needs new technical ventures. America's history has reflected an independent creative ethic exploiting and capitalizing upon technology and creating new ventures and progress. Individuals have recognized needs and coupled them with new technological solutions. The result has been benefits to the consumer of the product, to the area in terms of jobs and tax base, and to the innovator/venturer as well.

Trends in technology, organization, and society seem to have tended to increasingly inhibit new technical ventures. There is a concern about that trend and an interest, as evidenced by President Nixon's 1972 State of the Union message, in finding ways to stimulate technical development and its commercialization.

An intellectual/academic community could and should be able to perceive the situation, the causes of the trends, and ways to move in the desired direction. Through its contribution of trained students, understanding of the phenomenon and ways to influence it, and provision of an environment encouraging to new technical ventures (by both new organizations or existing ones), it can discharge some of its responsibility to society.

A center for new technical ventures must be built on the raw mate-

Lamont Eltinge is Vice President for Research, Cummins Engine Co., Columbus, Indiana.

*A suggestion to the College of Engineering, University of Illinois at Chicago Circle.

rial of a community of varied intellectual and entrepreneurial interest *concentrated* in selected, identified, similar areas. It is most effective if augmented by a hospitable environment for venturesome individuals and new ventures.

A dynamic model would contribute substantially to training of individuals, both experienced and inexperienced, and help existing organizations create and evaluate plans to grow and serve over the next decade or two by meeting the needs of those periods with new technical products ventures. In addition, a "new technical venture park" would lower the barrier to new ventures and provide "laboratory" training for a much larger group of people who would subsequently use the skill developed in existing or new organizations to improve the effectiveness of those organizations in new technical venturing.

Consideration of some of the outstanding examples of new technical ventures and organizations should provide a data base and valuable case studies. One can consider McCormick and the reaper, the Caterpillar Tractor Company and large earth-moving equipment, John Deere and the plow, Clessie Cummins and the truck diesel, and Hybrid corn. It is also worthwhile to consider the success of the Minnesota Mining and Manufacturing Company, Dupont, General Electric, Route 128 around Boston, Bill Lear, Motorola Radio, Allis-Chalmers, and others. Dr. James B. Hlavacek previously of the University of Illinois and now at the State University of New York at Albany and others are apparently studying interesting areas. Such studies are an important and necessary base, but it is necessary to go much further to a simulation; a "laboratory" and the real world. That's in the engineering tradition of applying the findings of science to help mankind reach its objectives. Development of information is necessary, but it's necessary to go on from that to the training of people by lecture, games, simulations, "laboratory," and real world experience. In addition, increasing difficulties may make worthwhile, in addition to training, some contribution to the environment of existing small, medium and perhaps even large organizations that would help them perceive future opportunities in time to respond efficiently and productively.

A computerized model of a few selected industries would be a central training and planning tool. While no one company or effort might justify it, broad usage throughout an area could. The model should include 1) evaluated forecasts of general demand and requirements for equipment *in terms of performance*, 2) forecasts of the rate at which technology might be developed with application of engineering research effort, 3) projected resources, both natural and human, 4) interactions between this industrial element and others, 5) the environment provided by the more global business community in terms of business cycles and growth and, 6) input of the chance factors that face every venture. It would be a submodel of a more global one and conversely could be augmented by and used to interpret smaller-segment models that were more detailed.

Clearly, the crucial factor would be clear perception of the industries involved and accurate forecasts. The model would be developed around a structure but would be evolutionary in nature, hopefully, reflecting inputs from a variety of creative and reflective individuals, possibly as a result of their use of the model and experience with its correct results and failures.

The model could be used as a teaching tool. Students, either individually or in small groups, could be treated as the management of companies and asked, on an accelerated time scale, to make decisions and observe results. I think this is already done at Harvard Business School, but on a much shorter time scale and without allowance for decades-time scale technological change and forecast. Students could be asked to play this "game" for specific companies currently in the region or state. That would certainly improve the reality of the game. Review of the results with the corporate managements could be mutually beneficial both in terms of getting realism and detail to the students and indicating an outside view and different creativity to the managements. The managements could use the model to explore their options before making decisions. The government, industry or scholars might use the model—and individual experience with it—to determine how changes in our organizations or society (reflected in model parameters) affect the creation and function of new technical ventures and evaluate the benefit of "feed forward" communication efforts.

Much of the rationale for a center for new technical ventures is nonspecific as to geographical area, state, or city, and the underlying concepts are universally applicable "laws." But it is necessary to move from a general case to specifics to make the training and modelling "real." This requires selection of and restriction to specific areas of practice and technology. In that selection, *regional* emphasis comes to the fore. For Illinois, and Chicago, the emphasis could be on agriculture of the feed and livestock variety, farm, and industrial machinery of the kind that has evolved from the agriculture requirements, and big city needs epitomized by Chicago.

The technical venture park would carry the student from the practice stage to the real stage and hopefully within itself create and foster some new venture. In the "park," typical ventures would be smaller and shorter in time scale, than those modelled, because of resource and time constraints. But opportunities for application and testing of venture concepts would exist, and receive emphasis. The park would operate as a confederacy providing capabilities of a number of kinds on a relatively easily accessible (but cost defraying) basis. These need to include marketing and financial advice, technical and design consultation (probably from other students) manufacturing and machining and processing capability, personnel supply information and sales advice. To some extent this is like an expanded Junior Achievement. It might evolve into a continuing entrepreneurial confederacy.

Support for the model and park could come from the State, university, venture capitalists, federal government (National Bureau Standards-Experimental Program), foundations (Ford, Lily, Cummins or Shell), the U.S. or Illinois Chambers of Commerce, and/or companies or trade associations in the industries modelled.

*Project Real World**

(A Mechanism for Involving a College of Engineering in the Solution of Small Technical Problems Encountered by its Alumni)

Objective:	To provide a meaningful link between alumni, faculty and students, through the joint solution of engineering problems which arise from alumni work experience.
Need:	Students enrolling in special independent research courses or structured courses requiring special term projects are often unable to develop meaningful subjects on relevant problems to use as the subject of their research or special term project. Engineering alumni, in an everyday work environment, occasionally encounter engineering problems requiring detailed analysis or further academic study but are unable to allocate the research time for such projects. Students have the time and academic freedom to pursue some of the problems alumni would like to investigate further.
Mechanics:	Establish a joint alumni-faculty committee composed of one alumnus and one faculty member from each department to meet

*A project undertaken by the University of Illinois at Chicago Circle Engineering Alumni Association in May 1973.

twice each quarter to review alumni project requests, refer feasible projects to appropriate course instructors, and monitor progress of program.

1. Alumni submit standard one page description of project.
2. "Real World" committee reviews project and if project has potential, refers project to faculty in area to be explored.
3. Faculty members can refer to students as term project topics, special extra credit project, or suitable for independent research.
   a. If project finds an interested student and a willing instructor, student is referred to "Real World" committee for name and phone number of alumnus with project. Project is assigned to student. Final report issued to alumni.
   b. If project does not find an interested student, or willing instructor, within two quarters, project is returned to alumni.

Benefits:

1. Exposure of engineering students to "Real World" engineering problems. Graduate more mature engineers.
2. Continued exposure of alumni to "academic solutions" to "real world problems."
3. Improve student, faculty, alumni relations through joint problem solving.

*Additional Comments on the Problems of City-University Interaction**

Milton Pikarsky

". . . In reading through [the Regional Role of Engineering Colleges Report], I was struck by two things: first, how far universities have come in directing their activities towards the needs of the students and therefore their communities. If we look at universities as they were over 20 years ago, and compare this with what is described in this report, we see how truly sweeping these changes have been. Evidently, many of these changes have come from such small increments that many of us have scarcely noticed.

At the same time our needs, especially in the Urban Areas, have risen exponentially. Consequently, our demands on the educational institutions have also increased sharply so that no matter how much is done, it never seems enough. Our crisis philosophy is good in that it creates a sense of urgency which promotes change, but it also blinds us to some of the very real accomplishments of our time. One of the salient features of your report, and one which I especially enjoyed, was the way in which these ideas surface to the reader.

Your point in your cover letter that an additional study is required to develop truly innovative and progressive mechanisms relating university activities to urban problems is well taken.

Milton Pikarsky is Chairman of the Chicago Transit Authority.

*Excerpted from personal communications in September 1973.

Elizabeth J. McLean

"... Identifying mechanisms whereby cooperation and interaction can be enhanced between universities and local governments is the most significant portion of the document [the Regional Role of Engineering Colleges Report]. Information, Monitoring, and Innovation centers, as suggested by the text, may prove a desirable mechanism for university/government interface ...

The problem of bringing applied science to government is indeed a difficult one. While it is true effective links must be established between engineering colleges and government, it is also true many barriers prevent communication.

Notably, interaction with the public sector, particularly at regional and local levels, is very limited. Lack of an information exchange mechanism appears to be the major obstacle to effective interchange. Who should initiate cooperative action, universities or local government, should be explored. For too long, both sides have waited for the other to act.

We agree with the Report when it notes universities at times fail to understand the outlook of an outside institution, and assume an educational motivation. Often, outside institutions (cities) are vitally interested in problem-solving rather than the traditional educational role of the university. This parallels the fact that too often research proceeds at a leisurely pace while pressing problems of the community remain unsolved. It is true, as stated in the document, considerable restructuring may be necessary if new efforts are to be effective. We agree the primary role of the university is to educate but problem-solving should not be interpreted as cessation of the traditional educational function.

For this reason, faculty incentive should be reconsidered. The Report states some faculty are eager to devote time and energy to public service. Unfortunately, service of this nature is generally not included in the university's evaluation of performance for salary and promotion. The role of faculty in the public sector should be considered a complementary role to that of educator. Performance in the public sector should be considered a criterion for promotion. We realize the task of revising ideas along this line will be a slow and complex process ...

Elizabeth J. McLean is First Deputy Commissioner of the Chicago Department of Public Works.

INDEX